Graduate Texts in Contemporary Physics

Series Editors:

R. Stephen Berry
Joseph L. Birman
Jeffrey W. Lynn
Mark P. Silverman
H. Eugene Stanley
Mikhail Voloshin

Springer

New York
Berlin
Heidelberg
Barcelona
Hong Kong
London
Milan
Paris
Singapore
Tokyo

Graduate Texts in Contemporary Physics

S.T. Ali, J.P. Antoine, and J.P. Gazeau: **Coherent States, Wavelets and Their Generalizations**

A. Auerbach: **Interacting Electrons and Quantum Magnetism**

N. Boccara: **Modeling Complex Systems**

T.S. Chow: **Mesoscopic Physics of Complex Materials**

B. Felsager: **Geometry, Particles, and Fields**

P. Di Francesco, P. Mathieu, and D. Sénéchal: **Conformal Field Theories**

A. Gonis and W.H. Butler: **Multiple Scattering in Solids**

K. Gottfried and T-M. Yan: **Quantum Mechanics: Fundamentals, 2nd Edition**

K.T. Hecht: **Quantum Mechanics**

J.H. Hinken: **Superconductor Electronics: Fundamentals and Microwave Applications**

J. Hladik: **Spinors in Physics**

Yu.M. Ivanchenko and A.A. Lisyansky: **Physics of Critical Fluctuations**

M. Kaku: **Introduction to Superstrings and M-Theory, 2nd Edition**

M. Kaku: **Strings, Conformal Fields, and M-Theory, 2nd Edition**

H.V. Klapdor (ed.): **Neutrinos**

R.L. Liboff (ed): **Kinetic Theory: Classical, Quantum, and Relativistic Descriptions, 3rd Edition**

J.W. Lynn (ed.): **High-Temperature Superconductivity**

H.J. Metcalf and P. van der Straten: **Laser Cooling and Trapping**

R.N. Mohapatra: **Unification and Supersymmetry: The Frontiers of Quark-Lepton Physics, 3rd Edition**

V.P. Nair: **Quantum Field Theory: A Modern Perspective**

R.G. Newton: **Quantum Physics: A Text for Graduate Students**

H. Oberhummer: **Nuclei in the Cosmos**

G.D.J. Phillies: **Elementary Lectures in Statistical Mechanics**

R.E. Prange and S.M. Girvin (eds.): **The Quantum Hall Effect**

(continued after index)

Harold J. Metcalf
Peter van der Straten

Laser Cooling and Trapping

With 115 Illustrations

 Springer

Harold J. Metcalf
Department of Physics
SUNY at Stony Brook
Stony Brook, NY 11794-3800
USA
hmetcalf@notes.cc.sunysb.edu

Peter van der Straten
Debye Institute
Faculty of Physics and Astronomy
Princetonplein 5
3584 CC Utrecht
The Netherlands
P.vanderStraten@phys.uu.nl

Series Editors

R. Stephen Berry
Department of Chemistry
University of Chicago
Chicago, IL 60637
USA

Joseph L. Birman
Department of Physics
City College of CUNY
New York, NY 10031
USA

Jeffrey W. Lynn
Department of Physics
University of Maryland
College Park, MD 20742
USA

Mark P. Silverman
Department of Physics
Trinity College
Hartford, CT 06106
USA

H. Eugene Stanley
Center for Polymer Studies
Physics Department
Boston University
Boston, MA 02215
USA

Mikhail Voloshin
Theoretical Physics Institute
Tate Laboratory of Physics
University of Minnesota
Minneapolis, MN 55455
USA

Library of Congress Cataloging-in-Publication Data
Metcalf, Harold J.
    Laser cooling and trapping / H.J. Metcalf, P. van der Straten.
        p. cm. — (Graduate texts in contemporary physics)
    Includes bibliographical references and index.
    ISBN 0-387-98747-9 (cloth : alk. paper)
    ISBN 0-387-98728-2 (pbk. : alk. paper)
    1. Laser manipulation.   2. Laser cooling.   I. Van der Straten, P.
II. Title.   III. Series.
    QC689.5.L35 M47   1999
    539.7—dc21   98-55408

Printed on acid-free paper.

Printed in the United States of America.   (AU/MVY)

9 8 7 6 5 4 3

ISBN 0-387-98728-2 (softcover)
ISBN 0-387-98747-9 (hardcover)

Springer-Verlag is a part of *Springer Science+Business Media*

springeronline.com

To the memory of my parents

—

Voor Wilma, Lieke en Anouk

Foreword

When Hal Metcalf and I began to work on laser cooling of neutral atoms in about 1979, we found ourselves in a field that was nearly unoccupied by other researchers, or by any real understanding of what the problems and possibilities were. While the study of laser cooling of trapped ions was well under way, only two other groups had ventured into laser manipulation of neutral atoms, one in Moscow and one at Bell Labs (although the latter had temporarily dropped this line of research). Today, laser cooling and its applications represent one of the major subfields of atomic, molecular and optical physics, with over one hundred active groups around the world. Laser cooling has been the enabling technolgy for a wide range of new endeavors. These range from a new generation of atomic clocks, which are operating or under construction in many of the world's major standards laboratories, to the achievement of Bose-Einstein condensation in atomic alkali vapors, one of the fastest growing fields of basic research at the close of the twentieth century. From the highly practical to the very fundamental, laser cooling has become an important part of many research programs.

With this explosion of interest in laser cooling came the obvious question of writing a book about it. The "right time" to write a book on a new subject is a delicate thing. On the one hand, the subject needs to be well enough understood and developed so one can produce a text that stands the test of time, a text that will not be outdated in a few years. On the other hand, one wants the subject to be exciting and of current interest. Metcalf and Van der Straten are two of the finest scientists in the field of laser cooling and they have written the right book at the right time. Certainly, *Laser Cooling and Trapping* will serve as a valuable reference for researchers working in this field. More importantly it will serve to introduce young people to this exciting field. Now, when someone asks me how to start learning about laser cooling, I'll tell them to read Metcalf and Van der Straten. I fully expect that some of the most exciting developments yet to come will come from researchers who begin their study of laser cooling with this book.

April 1999

William D. Phillips
Gaithersburg, MD

Preface

The purpose of this book is to introduce students to the dramatic developments in electromagnetic control of atomic motions that has emerged since the 1980s. The book evolved from lectures and courses given by each of us at Stony Brook and Utrecht to advanced undergraduates and beginning graduate students. Its three parts have quite different purposes: Part I serves to review, but not teach, those elements of quantum mechanics and atomic physics that are applicable to the material that follows. Its last chapter addresses certain topics in kinetic theory and statistical physics. Part II introduces the experimental tools and techniques that have been used for electromagnetic control of atomic motion. The first such topic is simply slowing down atoms, usually called laser cooling. But trapping them with magnetic or optical fields (or both), focussing and steering them, and other kinds of manipulation are discussed. The theoretical methods developed in Part I are integrated into these descriptions. Part III discusses some of the manifold applications of the spectacular new tools provided to physicists by these technologies. It is divided into two sub-parts: those topics for which the traditional classical description of atoms moving as classical point particles is appropriate, and those topics for which this view must be abandoned and the center-of-mass motion of the atoms must be described quantum mechanically. It is here where some of the most dramatic progress has occurred: atoms in optical lattices, deBroglie wave optics, Bose-Einstein condensation, and the fascinating Schrödinger cat states.

Although there are 50 year-old articles suggesting optical cooling, the topic attracted serious attention with proposals for cooling trapped ions and neutral atoms, as well as trapping neutral atoms, at the end of the 1970s. The experimental aspects really began with trapped ions in the late 1970s and the beam slowing demonstrations in Troitsk and Gaithersburg in the early 1980s. Then in 1985 the first neutral atom traps were demonstrated: trapping in velocity space was done with optical molasses at AT&T Bell Labs, and in configuration space with purely magnetic fields at Gaithersburg. Simultaneous cooling and trapping was demonstrated using the magneto-optical trap at AT&T Bell Labs in 1987. Then there followed an explosion of interest in this field that was culminated by the award of the Nobel Prize in Physics in 1997 to three of its earliest practitioners.

The text is written from our experimentalists' perspective. There are no long, formal derivations, and most of the theoretical material is presented in a conversational rather than formal manner. It is our goal to inspire the readers with some of the beautiful "finger physics" pictures that have evolved in this new field, even though there have been quite elegant formalisms developed by many theorists. Any book intended as a complete, up-to-date, thorough treatment would be obsolete in a few years, and for this lack of completeness we apologize to those whose work may have been slighted or omitted. Instead, we have intended this book to be a guide for students learning the basic elements of the field.

Both of us are indebted to the generations of students and postdocs who have passed through our laboratories, and whom we have encountered in visits to other laboratories, who have taught us so very much. Their fresh approach to the new problems posed by this expanding field have made our research careers fascinating and our time a great pleasure. We also wish to thank Hanneke de Vries and the staff at Springer, including external readers, for all their work on the manuscript.

June 1999   Harold J. Metcalf
                                                    Peter van der Straten

Contents

Foreword   vii

Preface   ix

I   Introduction   1

1   Review of Quantum Mechanics   3
   1.1   Time-Dependent Perturbation Theory 3
   1.2   The Rabi Two-Level Problem 4
       1.2.1   Light Shifts 7
       1.2.2   The Dressed Atom Picture 9
       1.2.3   The Bloch Vector 11
       1.2.4   Adiabatic Rapid Passage 12
   1.3   Excited-State Decay and its Effects 14

2   The Density Matrix   17
   2.1   Basic Concepts 17
   2.2   Spontaneous Emission 20
   2.3   The Optical Bloch Equations 23
   2.4   Power Broadening and Saturation 24

3   Force on Two-Level Atoms   29
   3.1   Laser Light Pressure 29

3.2   A Two-Level Atom at Rest   31
3.3   Atoms in Motion   34
        3.3.1   Traveling Wave   34
        3.3.2   Standing Wave   35

4   Multilevel Atoms   **39**
4.1   Alkali-Metal Atoms   39
4.2   Metastable Noble Gas Atoms   43
4.3   Polarization and Interference   45
4.4   Angular Momentum and Selection Rules   47
4.5   Optical Transitions in Multilevel Atoms   50
        4.5.1   Introduction   50
        4.5.2   Radial Part   51
        4.5.3   Angular Part of the Dipole Matrix Element   52
        4.5.4   Fine and Hyperfine Interactions   53

5   General Properties Concerning Laser Cooling   **57**
5.1   Temperature and Thermodynamics in Laser Cooling   58
5.2   Kinetic Theory and the Maxwell-Boltzmann Distribution   61
5.3   Random Walks   63
5.4   The Fokker-Planck Equation and Cooling Limits   66
5.5   Phase Space and Liouville's Theorem   68

II   Cooling & Trapping   **71**

6   Deceleration of an Atomic Beam   **73**
6.1   Introduction   73
6.2   Techniques of Beam Deceleration   74
        6.2.1   Laser Frequency Sweep   76
        6.2.2   Varying the Atomic Frequency: Magnetic Field Case . . .   77
        6.2.3   Varying the Atomic Frequency: Electric Field Case   77
        6.2.4   Varying the Doppler Shift: Diffuse Light   78
        6.2.5   Broadband Light   79
        6.2.6   Rydberg Atoms   79
6.3   Measurements and Results   80
6.4   Further Considerations   83
        6.4.1   Cooling During Deceleration   83
        6.4.2   Non-Uniformity of Deceleration   84
        6.4.3   Transverse Motion During Deceleration   85
        6.4.4   Optical Pumping During Deceleration   86

7   Optical Molasses   **87**
7.1   Introduction   87
7.2   Low-Intensity Theory for a Two-Level Atom in One Dimension   .   88

7.3   Atomic Beam Collimation 90
    7.3.1   Low-Intensity Case 90
    7.3.2   Experiments in One and Two Dimensions 92
7.4   Experiments in Three-Dimensional Optical Molasses 95

8   Cooling Below the Doppler Limit   **99**
8.1   Introduction 99
8.2   Linear $\perp$ Linear Polarization Gradient Cooling 100
    8.2.1   Light Shifts 101
    8.2.2   Origin of the Damping Force 102
8.3   Magnetically Induced Laser Cooling 104
8.4 σ^{+}-σ^{-} Polarization Gradient Cooling 106
8.5   Theory of Sub-Doppler Laser Cooling 107
8.6   Optical Molasses in Three Dimensions 111
8.7   The Limits of Laser Cooling 113
    8.7.1   The Recoil Limit 113
    8.7.2   Cooling Below the Recoil Limit 114
8.8   Sisyphus Cooling 116
8.9   Cooling in a Strong Magnetic Field 118
8.10   VSR and Polarization Gradients 120

9   The Dipole Force   **123**
9.1   Introduction 123
9.2   Evanescent Waves 124
9.3   Dipole Force in a Standing Wave: Optical Molasses
    at High Intensity 126
9.4   Atomic Motion Controlled by Two Frequencies 128
    9.4.1   Introduction 128
    9.4.2   Rectification of the Dipole Force 129
    9.4.3   The Bichromatic Force 131
    9.4.4   Beam Collimation and Slowing 135

10 Magnetic Trapping of Neutral Atoms   **137**
10.1   Introduction 137
10.2   Magnetic Traps 138
10.3   Classical Motion of Atoms in a Magnetic Quadrupole Trap 140
    10.3.1   Simple Picture of Classical Motion in a Trap 140
    10.3.2   Numerical Calculations of the Orbits 141
    10.3.3   Early Experiments with Classical Motion 143
10.4   Quantum Motion in a Trap 145
    10.4.1   Heuristic Calculations of the Quantum Motion
        of Magnetically Trapped Atoms 146
    10.4.2   Three-Dimensional Quantum Calculations 146
    10.4.3   Experiments in the Quantum Domain 147

11 Optical Traps for Neutral Atoms   **149**
   11.1 Introduction 149
   11.2 Dipole Force Optical Traps 150
       11.2.1 Single-Beam Optical Traps for Two-Level Atoms 150
       11.2.2 Hybrid Dipole Radiative Trap 152
       11.2.3 Blue Detuned Optical Traps 153
       11.2.4 Microscopic Optical Traps 155
   11.3 Radiation Pressure Traps 156
   11.4 Magneto-Optical Traps 156
       11.4.1 Introduction 156
       11.4.2 Cooling and Compressing Atoms in a MOT 158
       11.4.3 Capturing Atoms in a MOT 159
       11.4.4 Variations on the MOT Technique 162

12 Evaporative Cooling   **165**
   12.1 Introduction 165
   12.2 Basic Assumptions 166
   12.3 The Simple Model 167
   12.4 Speed and Limits of Evaporative Cooling 171
       12.4.1 Boltzmann Equation 171
       12.4.2 Speed of Evaporation 171
       12.4.3 Limiting Temperature 174
   12.5 Experimental Results 175

III   Applications   **177**

13 Newtonian Atom Optics and its Applications   **179**
   13.1 Introduction 179
   13.2 Atom Mirrors 180
   13.3 Atom Lenses 181
       13.3.1 Magnetic Lenses 181
       13.3.2 Optical Atom Lenses 184
   13.4 Atomic Fountain 185
   13.5 Application to Atomic Beam Brightening 186
       13.5.1 Introduction 186
       13.5.2 Beam-Brightening Experiments 188
       13.5.3 High-Brightness Metastable Beams 189
   13.6 Application to Nanofabrication 190
   13.7 Applications to Atomic Clocks 192
       13.7.1 Introduction 192
       13.7.2 Atomic Fountain Clocks 193
   13.8 Application to Ion Traps 194
   13.9 Application to Non-Linear Optics 195

14 Ultra-cold Collisions   **199**
   14.1 Introduction 199
   14.2 Potential Scattering 200
   14.3 Ground-state Collisions 204
   14.4 Excited-state Collisions 207
       14.4.1 Trap Loss Collisions 207
       14.4.2 Optical Collisions 209
       14.4.3 Photo-Associative Spectroscopy 213
   14.5 Collisions Involving Rydberg States 218

15 deBroglie Wave Optics   **219**
   15.1 Introduction 219
   15.2 Gratings 220
   15.3 Beam Splitters 223
   15.4 Sources 224
   15.5 Mirrors 225
   15.6 Atom Polarizers 226
   15.7 Application to Atom Interferometry 227

16 Optical Lattices   **231**
   16.1 Introduction 231
   16.2 Laser Arrangements for Optical Lattices 232
   16.3 Quantum States of Motion 235
   16.4 Band Structure in Optical Lattices 238
   16.5 Quantum View of Laser Cooling 239

17 Bose-Einstein Condensation   **241**
   17.1 Introduction 241
   17.2 The Pathway to BEC 243
   17.3 Experiments 244
       17.3.1 Observation of BEC 244
       17.3.2 First-Order Coherence Experiments in BEC 246
       17.3.3 Higher-Order Coherence Effects in BEC 248
       17.3.4 Other Experiments 249

18 Dark States   **251**
   18.1 Introduction 251
   18.2 VSCPT in Two-Level Atoms 252
   18.3 VSCPT in Real Atoms 254
       18.3.1 Circularly Polarized Light 255
       18.3.2 Linearly Polarized Light 257
   18.4 VSCPT at Momenta Higher Than $\pm \hbar k$ 258
   18.5 VSCPT and Bragg Reflection 259
   18.6 Entangled States 261

IV Appendices   263

A Notation and Definitions   265

B Review Articles and Books on Laser Cooling   269

C Characteristic Data   273

D Transition Strengths   279

References   291

Index   317

Part I

Introduction

1
Review of Quantum Mechanics

This chapter presents a brief review of those aspects of quantum mechanics that are important for understanding some of the material to be found elsewhere in this book. Its purpose is not to teach the subject, because that is so very well done in numerous other textbooks [1–9]. Rather, the intent is to bring together certain things that are sometimes scattered throughout such texts, to establish notation and conventions, and to provide a reference point for many important and useful formulas.

1.1 Time-Dependent Perturbation Theory

The time-dependent Schrödinger equation is

$$\mathcal{H}\Psi(\vec{r}, t) = i\hbar\frac{\partial \Psi(\vec{r}, t)}{\partial t}, \tag{1.1}$$

where $\mathcal{H}$ is the total Hamiltonian for an atom in a radiation field and $\vec{r}$ is the coordinate of the electron. The field-free, time-independent atomic Hamiltonian is denoted as $\mathcal{H}_0$, its eigenvalues as $E_n \equiv \hbar\omega_n$, and its eigenfunctions as $\phi_n(\vec{r})$. Then $\mathcal{H}_0\phi_n(\vec{r}) = E_n\phi_n(\vec{r})$. The interaction with the radiation field is described by $\mathcal{H}'(t)$, and thus $\mathcal{H}(t) = \mathcal{H}_0 + \mathcal{H}'(t)$ when the radiation is considered as a classical electromagnetic field, which is appropriate for laser cooling. Since the eigenfunctions $\phi_n(\vec{r})$ form a complete set, the solution $\Psi(\vec{r}, t)$ of Eq. 1.1 is expanded in terms of $\phi_n(\vec{r})$ as

$$\Psi(\vec{r}, t) = \sum_k c_k(t)\phi_k(\vec{r})e^{-i\omega_k t}, \tag{1.2}$$

where the coefficients $c_k(t)$ are generally time-dependent. The Schrödinger equation then becomes

$$\mathcal{H}(t)\Psi(\vec{r}, t) = [\mathcal{H}_0 + \mathcal{H}'(t)] \sum_k c_k(t)\phi_k(\vec{r})e^{-i\omega_k t} \qquad (1.3)$$

$$= (i\hbar)\left(\frac{\partial}{\partial t}\right) \sum_k c_k(t)\phi_k(\vec{r})e^{-i\omega_k t}.$$

Multiplying on the left by $\phi_j^*(\vec{r})$ and integrating over spatial coordinates $\vec{r}$ gives

$$i\hbar\frac{dc_j(t)}{dt} = \sum_k c_k(t)\mathcal{H}'_{jk}(t)e^{i\omega_{jk}t}, \qquad (1.4)$$

where $\mathcal{H}'_{jk}(t) \equiv \langle\phi_j|\mathcal{H}'(t)|\phi_k\rangle$ and $\omega_{jk} \equiv (\omega_j - \omega_k)$.

Equation 1.4 is exactly equivalent to the Schrödinger equation 1.1: no approximations have been made. However, for the case of an atom in a radiation field it is unsolvable, and so approximations are required. One of the most common approaches found in textbooks is the use of perturbation theory. For an atom in the ground state ($k = 1$) at $t = 0$, all $c_k(0) = 0$ except for $c_1(0) = 1$. For the perturbation approximation, one chooses

$$|c_k(t)| \ll 1 \qquad (1.5)$$

for all $k \neq 1$ and does a formal time integration of Eq. 1.4 to calculate these $c_k(t)$ values. The small components $c_k(t)$ of the excited states $\phi_k(\vec{r})$ for $k \neq 1$ that are mixed into $\Psi(\vec{r}, t)$ become the transition amplitudes and their squares are the transition rates.

For transitions to the continuum, such as photoionization, averaging over the density of final states results in the familiar "Fermi Golden Rule" of quantum mechanics. For transitions between discrete states driven by radiation whose spectral width is larger than the natural width of the transition, averaging over the spectral density gives the same golden rule.

1.2   The Rabi Two-Level Problem

Such a textbook approach is not suitable for narrow-band laser excitation of atoms, however, because large excited-state populations are possible, thereby violating Eq. 1.5. Instead, a different approximation is made by truncating the summation of the exact Eq. 1.4 to just two terms, the single ground and excited state connected by the laser frequency, and solving the resulting coupled differential equations directly. Such a calculation for a two-level system was first studied by Rabi [10] in connection with magnetic resonance, and is described very well in textbooks [5,11].

The solution of this Rabi problem begins by absorbing any diagonal elements of $\mathcal{H}'(t)$ into $\mathcal{H}_0$, and then only one nonzero value, $\mathcal{H}'_{ge}(t) = \mathcal{H}'_{eg}{}^*(t)$, remains in the summation (here $1 \rightarrow g$ and $2 \rightarrow e$). Then Eq. 1.4 becomes

$$i\hbar \frac{dc_g(t)}{dt} = c_e(t)\mathcal{H}'_{ge}(t)e^{-i\omega_a t} \qquad (1.6a)$$

and

$$i\hbar \frac{dc_e(t)}{dt} = c_g(t)\mathcal{H}'_{eg}(t)e^{i\omega_a t}, \qquad (1.6b)$$

where $\omega_a \equiv \omega_{eg}$ is the atomic resonance frequency.

Evaluation of $\mathcal{H}'_{ge}(t)$ begins in the most general way by writing

$$\mathcal{H}(t) = p^2/2m + V, \qquad (1.7)$$

where V is the Coulomb potential seen by the electron whose momentum is p, and then by replacing p by the canonical momentum to obtain $\vec{p} - (e/c)\vec{A}(\vec{r}, t)$, where $\vec{A}(\vec{r}, t)$ is the vector potential of the applied field, in this case, the laser light[1]. The Hamiltonian can now be expanded and manipulated, beginning with $\vec{\mathcal{E}} = \dot{\vec{A}}/c$, until the expression

$$\mathcal{H}'(t) = -e\vec{\mathcal{E}}(\vec{r}, t) \cdot \vec{r} \qquad (1.8)$$

emerges (this is the classically expected operator). Here the $\vec{A}^2$ term is not considered because it represents only the energy of the electromagnetic field of the light and is transparent to the atomic eigenstates $\phi_n(\vec{r})$. (The reader is cautioned that deriving Eq. 1.8 is fraught with certain difficulties that have been discussed in the literature over the past 40 years, and will not be considered here [12, 13].)

Using Eq. 1.8 to solve Eqs. 1.6 for the case of only two atomic levels connected by a single, narrow-band excitation requires the use of two very well-known approximations in addition to truncating the sum in Eq. 1.4. The first of these approximations is the rotating wave approximation (RWA), which consists of neglecting terms of order $1/\omega_\ell$ compared with terms of order $1/\delta$, where ω_ℓ is the laser frequency and δ is the laser detuning from the atomic resonance frequency, $\delta \equiv \omega_\ell - \omega_a$ [1, 3, 14]. The second approximation is the electric dipole approximation, which consists of neglecting the spatial variation of $\vec{\mathcal{E}}(\vec{r}, t)$ over the region of the spatial integral of $\mathcal{H}'_{ge}(t)$ because the optical wavelength λ is typically several hundred nm whereas the wavefunctions $\phi_n(\vec{r})$ are almost entirely contained within a sphere of radius typically < 1 nm.

For a plane wave traveling in the positive z direction, the electric field operator is

$$\vec{\mathcal{E}}(\vec{r}, t) = E_0 \hat{\varepsilon} \cos(kz - \omega_\ell t), \qquad (1.9)$$

where $\hat{\varepsilon}$ is the unit polarization vector and E_0 is the amplitude of the light field. The coupling element for this case becomes $\mathcal{H}'_{eg}(t) = \hbar\Omega \cos(kz - \omega_\ell t)$, where Ω is the Rabi frequency defined by

$$\Omega \equiv \frac{-eE_0}{\hbar}\langle e|r|g\rangle \qquad (1.10)$$

---

[1] Although the present discussion refers to a quasi one-electron atom, the formulation is more general and is valid for other atoms as well.

and r is the electron coordinate. Here the electric dipole approximation has been made. For a two-level atom the dipole moment of the atom $e\vec{r}$ is parallel to the polarization $\hat{\varepsilon}$ of the field. This will be reconsidered in more detail in Chapter 4, where "real" atoms with multiple levels are discussed. However, the present approach is perfectly valid for a two-level atom, where the coupling strength between the atom and the field can always be identified with a single Rabi frequency. The reader should note that there are many definitions of Ω in the literature, but this one is chosen because it is the real oscillation frequency of $|c_k(t)|^2$ for $\delta = 0$.

The Rabi frequency is proportional to the matrix element $\langle e|r|g \rangle$, which, in general, is not easily calculated. For the hydrogen atom, the wavefunction of the electron in the bound states is known, and these elements can be calculated accurately, but for all other atoms the situation is more complicated (see Sec. 4.5).

The two Eqs. 1.6 can be uncoupled by differentiating the first one and substituting for c_e to find

$$\frac{d^2 c_g(t)}{dt^2} - i\delta \frac{dc_g(t)}{dt} + \frac{\Omega^2}{4} c_g(t) = 0 \qquad (1.11a)$$

and

$$\frac{d^2 c_e(t)}{dt^2} + i\delta \frac{dc_e(t)}{dt} + \frac{\Omega^2}{4} c_e(t) = 0, \qquad (1.11b)$$

where the RWA has been made. This result applies to the case of a two-level atom interacting with a single frequency field, but in laser cooling it is often necessary to consider more complicated atoms and fields, such as multilevel atoms in a multifrequency field. This will be discussed in more detail in Chapter 4.

The solution of Eqs. 1.11 for the same initial conditions as on p. 4, namely, $c_g(0) = 1$ and $c_e(0) = 0$, are

$$c_g(t) = \left(\cos \frac{\Omega' t}{2} - i\frac{\delta}{\Omega'} \sin \frac{\Omega' t}{2} \right) e^{+i\delta t/2} \qquad (1.12a)$$

and

$$c_e(t) = -i\frac{\Omega}{\Omega'} \sin \frac{\Omega' t}{2} e^{-i\delta t/2}, \qquad (1.12b)$$

where

$$\Omega' \equiv \sqrt{\Omega^2 + \delta^2}. \qquad (1.12c)$$

Note that the probability for finding the atom in the initial state g or the excited state e, $|c_g(t)|^2$ or $|c_e(t)|^2$, oscillates at frequency Ω', and that increasing the detuning $|\delta|$ increases the frequency of the oscillation while decreasing its amplitude as shown in Fig. 1.1. The segment of the oscillation associated with the transition from the excited state down to the ground state corresponds exactly to stimulated emission, and the result here illustrates clearly why the Einstein coefficients B_{kj} and B_{jk} are equal. When $\sin^2(\Omega' t/2)$ is between its extreme values, the system may be driven toward either ground or excited state depending on the relative phase between the driving field $\vec{\mathcal{E}}(\vec{r}, t)$ and the oscillations of $\Psi(\vec{r}, t)$ (see Eq. 1.2).

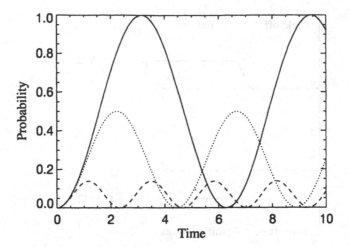

FIGURE 1.1. Probability $|c_e(t)|^2$ for the atom to be in the excited state for $\Omega = \gamma$ and $\delta = 0$ (solid line), $\delta = \gamma$ (dotted line), and $\delta = 2.5\gamma$ (dashed line). Time is in units of $1/\gamma$.

1.2.1 Light Shifts

In the presence of the off-diagonal Hamiltonian matrix elements of the operator $\mathcal{H}'(t)$, the energies E_n that are the eigenvalues of $\mathcal{H}_0$ are no longer the eigenvalues of the full Hamiltonian. The energy shifts are most readily found by first eliminating the time dependence associated with $\mathcal{H}'(t)$. An algebraic equivalent to the usual textbook approach of transforming to a rotating frame [14] is to replace the c's in Eqs. 1.6 by

$$c'_g(t) \equiv c_g(t) \tag{1.13a}$$

and

$$c'_e(t) \equiv c_e(t)e^{-i\delta t}. \tag{1.13b}$$

(Note that the rotating frame transformation is exact, as is this algebraic equivalent, and is completely different from the RWA.) Substituting these c's into Eqs. 1.6 and then making the RWA for $\mathcal{H}'_{ge}(t)$ gives

$$i\hbar\frac{dc'_g(t)}{dt} = c'_e(t)\frac{\hbar\Omega}{2} \tag{1.14a}$$

and

$$i\hbar\frac{dc'_e(t)}{dt} = c'_g(t)\frac{\hbar\Omega}{2} - c'_e(t)\hbar\delta. \tag{1.14b}$$

Now the oscillations of Eqs. 1.6 are gone. This new set of equations is exactly what would arise by evaluating Eq. 1.4 directly with both a time-independent perturbation $\mathcal{H}'_{ge}$ and the time dependence $e^{-i\omega_k t}$ absorbed directly into each of the $c_k(t)$'s so that the unperturbed energies of states g and e do not appear. This justifies the next step of diagonalizing the matrix formed from the coefficients of

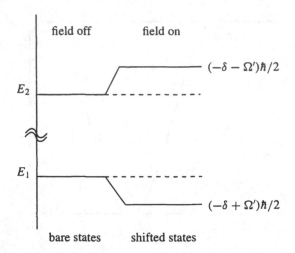

FIGURE 1.2. Energies of the two coupled states with the light field off and the light field on. The states are shifted due to the atom-light interaction, and the shift is called light shift.

Eqs. 1.14:

$$\mathcal{H} = \frac{\hbar}{2} \begin{bmatrix} -2\delta & \Omega \\ \Omega & 0 \end{bmatrix}. \tag{1.15}$$

The solutions show that the shifted energies are given by

$$E_{e,g} = \frac{\hbar}{2}(-\delta \mp \Omega'), \tag{1.16}$$

as shown in Fig. 1.2. In the limit where $\Omega \ll |\delta|$, the resulting energies are shifted by

$$\Delta E_g = \frac{\hbar \Omega^2}{4\delta} \tag{1.17a}$$

and

$$\Delta E_e = -\frac{\hbar \Omega^2}{4\delta}. \tag{1.17b}$$

Since the light intensity is proportional to Ω^2, $\Delta E_{g,e}$ as given above is appropriately called the light shift. In the limit $\Omega \gg |\delta|$, the solutions give $\Delta E_g = \mathrm{sgn}(\delta)\hbar\Omega/2$ and $\Delta E_e = -\mathrm{sgn}(\delta)\hbar\Omega/2$, where $\mathrm{sgn}(\delta) \equiv \delta/|\delta|$. The eigenstates corresponding to $\Delta E_{g,e}$ are called the dressed states of the atom and are calculated in the next section. Very often the light field is not homogeneous (e.g., in a standing wave) producing a spatially dependent light shift $\Delta E_{g,e}(\vec{r})$. The force that results from this gradient of energy is called the dipole force and is discussed in more detail in Chapter 9.

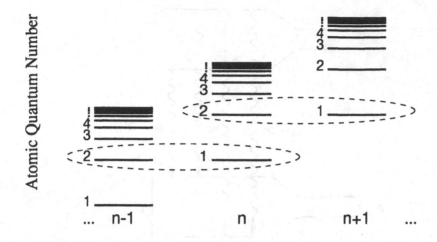

Atomic Quantum Number

Laser Field Quantum Number

FIGURE 1.3. Energy level diagram for the atom plus field Hamiltonian. In each vertical column there is the familiar level scheme of a typical atom, but the columns are vertically displaced by $\hbar\omega_\ell$ because of the addition of one laser photon per column. The nearly degenerate pairs are indicated.

1.2.2   The Dressed Atom Picture

The eigenfunctions of the Schrödinger equation for a two-level atom in a monochromatic field are best described in terms of the "dressed states" of the atom [5]. It begins with the total Hamiltonian

$$\mathcal{H} = \mathcal{H}_a + \mathcal{H}_{\text{rad}} + \mathcal{H}_{\text{int}}, \qquad (1.18)$$

where $\mathcal{H}_a$ is the usual atomic part denoted by $\mathcal{H}_0$ in Sec. 1.1 that gives the atomic energy levels, $\mathcal{H}_{\text{rad}} = \hbar\omega_\ell(a^\dagger a + \tfrac{1}{2})$ is the radiation part whose eigenvalues are $E_n = (n + \tfrac{1}{2})\hbar\omega_\ell$, and $\mathcal{H}_{\text{int}}$ is the atom-field interaction such as $\mathcal{H}'(t)$ in Sec. 1.1 that causes transitions as well as light shifts.

The energy level diagram of the first two terms in Eq. 1.18 consists of the ordinary atomic energies repeated for each value of n and vertically displaced by $\hbar\omega_\ell$ each time, as shown schematically in Fig. 1.3. Attention is focused on the two atomic states coupled by the laser light that form closely spaced pairs of one excited state and one ground state separated by $\hbar\delta$, as shown in Fig. 1.4. They are each mixtures of the ground and excited states, found by diagonalizing the matrix in Eq. 1.15.

The third term in the Hamiltonian, the interaction between the atom and the field embodied in $\mathcal{H}_{\text{int}}$, couples the ground and excited states that form each of these pairs through the off-diagonal matrix elements $\mathcal{H}'_{ge}(t)$. This splits the energy levels farther apart to $\hbar\Omega'$ as given in Eq. 1.12c. Ω' is independent of the sign of δ, and the shift $\hbar(\Omega' - |\delta|)/2$ is the light shift of each dressed state (see Eq. 1.16).

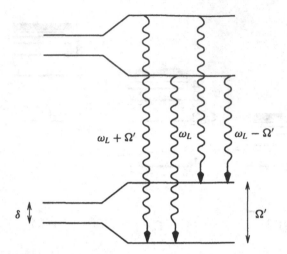

FIGURE 1.4. The nearly degenerate pairs of energy levels of Fig. 1.3. In the presence of the coupling interaction, each of these pairs is a mixture of ground and excited states, so each can decay by spontaneous emission as indicated. This figure is different from Fig. 1.2 because the energy levels here are separated only by δ, whereas in Fig. 1.2 their separation is ω_a.

The light also mixes the states by an amount expressed in terms of a mixing angle θ given by $\cos(2\theta) \equiv -\delta/\Omega'$, so that each ground state is mixed with a component of excited state and vice versa. These eigenstates of the Hamiltonian including this interaction are called the "dressed states" of the atom in the field [5]. The eigenfunctions are given by

$$|\phi_1\rangle = \cos\theta|g\rangle - \sin\theta|e\rangle \qquad (1.19a)$$

and

$$|\phi_2\rangle = \sin\theta|g\rangle + \cos\theta|e\rangle. \qquad (1.19b)$$

In a standing wave, the light shifts of these dressed states vary from zero at the nodes to a maximum at the antinodes. The spatially oscillating energies found from Eq. 1.16 are not sinusoidal, except in the limit of $\delta \gg \Omega$. This is apparent because these oscillatory terms will always be dominated by δ^2 in the vicinity of a node. Thus, for any value of $\Omega \gg \delta$, the expansion of Eq. 1.16 as $\Delta E \sim \hbar\Omega|\cos kz|/2$ will eventually fail near a node.

The spatial variation of the internal energy of the atoms results in a force related to the gradient of the energy. Although a more thorough and rigorous discussion of optical forces is given in later chapters, it is simply noted here that the spatial average over a wavelength of this force vanishes. However, the potential, and hence the force, is different for different atomic states, and spatially dependent optical pumping among various states of multilevel atoms can result in a non-vanishing force.

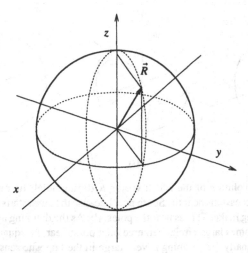

FIGURE 1.5. Graphical representation of the Bloch vector $\vec{R}$ on the Bloch sphere.

1.2.3   The Bloch Vector

Because the overall phase of the wavefunction has no physical meaning, there are really only three free parameters in the solutions of Eqs. 1.6 for the complex $c_k(t)$'s. In a classic paper, Feynman, Vernon, and Hellwarth [15] considered a transformation to a rotating frame where they then combined the real and imaginary parts of the $c_k(t)$'s to form the three real parameters

$$r_1 \equiv c_g c_e^* + c_g^* c_e, \tag{1.20a}$$

$$r_2 \equiv i(c_g c_e^* - c_g^* c_e), \tag{1.20b}$$

and

$$r_3 \equiv |c_e|^2 - |c_g|^2. \tag{1.20c}$$

The equations of motion 1.6 can be used to show that, in the rotating frame, a vector $\vec{R}$ whose components are the three r_i's given above, obeys

$$\frac{d\vec{R}}{dt} = \vec{\Omega} \times \vec{R}, \tag{1.21}$$

where the vector $\vec{\Omega}$ has the three components $\mathrm{Re}(\mathcal{H}'_{ge})$, $\mathrm{Im}(\mathcal{H}'_{ge})$, and $\hbar\delta$. Usually $\mathcal{H}'$ is taken to be real, so $\mathrm{Im}(\mathcal{H}'_{ge})$ vanishes and the components of $\vec{\Omega}$ become $\mathcal{H}'_{ge}$, 0, and $\hbar\delta$. This result is equivalent to the "Bloch vector" picture [16] and is graphically depicted in Fig. 1.5.

Equation 1.21 shows that the Bloch vector $\vec{R}$ precesses with time without changing length, and its motion is thus confined to the surface of a sphere. The south (north) poles of this sphere correspond to the ground (excited) states of the atom, and equatorial points correspond to equal superpositions with various phases.

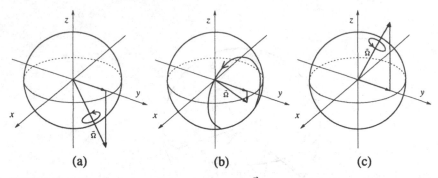

(a)   (b)   (c)

FIGURE 1.6. The evolution of the Bloch vector $\vec{R}$ during adiabatic rapid passage. (a) It begins with small precessions near the South pole because the atom starts in the ground state and the large detuning makes $\vec{\Omega}$ pass near the poles. (b) As the detuning approaches zero the precession of $\vec{R}$ becomes large circles centered on a point near the equator as shown in the center picture. (c) Finally the detuning is very large in the opposite sense to its beginning, resulting in small circular precession near the North pole. Thus the atom is left in the excited state. In all three pictures, the $\hat{x}$ component of $\vec{\Omega}$ is chosen to be zero, and the $\hat{y}$ component is constant as shown. The detuning is represented by the $\vec{z}$ vector pointing downward at the start (a) and upward at the end (c).

When $\hbar\delta \gg |\mathcal{H}'_{ge}|$, the precession axis passes very nearly through the poles. In this case, an atom initially in the ground state undergoes rapid precessions on a small circle near the south pole and thus has a small excitation probability, as shown in Eq. 1.12 and Fig. 1.1. By contrast, for $\delta = 0$, $\vec{\Omega}$ passes through the equator so an atom initially in the ground state is described by a Bloch vector $\vec{R}$ undergoing slow, full-circle oscillations through the poles. The response of an atom initially in an equal superposition of ground and excited states (on the equator) to a field tuned to resonance ($\delta = 0$) will therefore depend strongly on the components of $\vec{R}$ and thus on the mechanism that produced the superposition.

The steady state for the Bloch vector is given by $\vec{\Omega} \times \vec{R} = 0$. There are two such Bloch vectors, where one is parallel and the other antiparallel to $\vec{\Omega}$. It can easily be shown that these two vectors correspond to the eigenstates of Eq. 1.19.

1.2.4   Adiabatic Rapid Passage

The motion of $\vec{R}$ on the Bloch sphere allows a particularly graphic interpretation of a phenomenon called adiabatic rapid passage. If the frequency of the applied field is swept through resonance, an atom initially in the ground state is left in the excited state (and vice versa) with very high probability. At the beginning of the frequency sweep, $\vec{R}$ executes small, rapid orbits near the south pole, and these grow in size as δ sweeps toward 0 and the precession axis consequently approaches the equatorial plane. At $\delta = 0$, $\vec{R}$ undergoes polar oscillations because $\vec{\Omega}$ is now in the equatorial plane, but the continually shifting axis now moves the center of the orbit on the surface of the sphere toward the north pole. Near the end of the sweep,

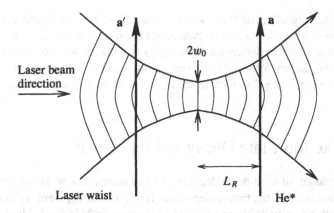

FIGURE 1.7. Schematic diagram of the adiabatic rapid passage experiment of Ekstrom *et al.* [17]. A metastable He beam crosses the axis of a focused laser beam at a distance L_R.

$\vec{R}$ executes small, rapid orbits near the north pole, and at the end of the sweep, $\vec{R}$ is left at the north pole, and the atom is left in the excited state (see Fig. 1.6).

The name *adiabatic rapid passage* may seem a bit enigmatic: how can something be both adiabatic and rapid? During the process of raising $\vec{R}$ from the south to the north pole of the Bloch sphere, there is always some component of the excited state present, with a corresponding probability of spontaneous decay. Thus this coherent excitation process can succeed only if it occurs in a time short compared with the natural lifetime of the excited state $\phi_e(\vec{r})$, so it must be fast. Needless to say, it must also be slow enough for the precessing Bloch vector $\vec{R}$ to follow the evolving axis of $\vec{\Omega}$ adiabatically. Thus there are boundaries determined by the atomic parameters on the rate of sweeping the detuning $d\delta/dt$. In practice, these limits can be satisfied with ordinary lasers and atoms, but it takes some effort.

In a very clever experiment [17] this has been accomplished in an atomic beam (see Fig. 1.7). The atoms traverse a focused laser beam along path **a**, well away from its waist, so they experience a significant part of the wavefront curvature (which is strongest at the Rayleigh length, L_R). As a result of the Doppler shift, the atoms first experience light whose frequency is shifted toward the blue, and then the frequency sweeps through $\delta = 0$ and toward the red as they leave the laser beam. (This sweep can be reversed by aligning the atomic beam along the alternative path **a'** shown in Fig. 1.7.) The Doppler shift is $\omega_D \equiv -\vec{k} \cdot \vec{v} = -kv \cos\theta$, where $\vec{v}$ is the atomic velocity, and $\vec{k}$ is the laser's wavevector whose magnitude is $2\pi/\lambda$. From geometry, $\cot\theta = vt/L_R$. Thus for small angles θ, the frequency sweep $\omega_D t = (-kv^2/L_R)t$ is linear in time t.

For the experiment discussed here [17], metastable He(2^3S_1) atoms ($v \sim 2000$ m/s) were excited to their 2^3P state (lifetime $\tau \approx 100$ ns) by $\lambda = 1.083$ μm light. For such atoms traversing a beam with a waist $w_0 \approx 10$ μm at a Raleigh length from it, the passage time $2\sqrt{2}w_0/v \approx 15$ ns, is considerably less than the excited-state lifetime. The waist size and wavelength determine the beam's angular divergence to be $\lambda/\pi w_0$, and hence a total frequency sweep of $2v/\pi w_0$, which is over 100 MHz.

Thus the intensity of the light is chosen to make $\Omega \ll 100$ MHz, and corresponds to a few mW/cm^2 at the waist. The total power required is thus only a few nW! The experimenters used the deflection of the highly collimated beam of atoms resulting from their excitation, followed by the spreading caused by spontaneous emission, to determine that more than 98% of the atoms were excited this way.

1.3   Excited-State Decay and its Effects

In the discussion of time-dependent response of atoms to a radiation field above, the eigenfunctions of the time-independent Hamiltonian $\mathcal{H}_0$ were written as time independent. The wavefunction of the total Hamiltonian $\mathcal{H}$ in Eq. 1.2 included only the purely oscillatory behavior associated with the eigenvalues and the radiation-induced time dependence of the $c_k(t)$'s. That discussion omitted the spontaneous decay of the excited states resulting from their interaction with the zero-point energy of the electromagnetic field. Spontaneous emission has played an important role in atomic physics since the conception of discrete atomic states by Bohr in 1913.

The problem of radiative transitions between discrete states in atoms was discussed by Einstein in 1917 [18], where he considered three radiative processes. In the first process, an amount of optical energy $\hbar\omega_\ell$ (a "photon") is absorbed from an applied radiation field of angular frequency ω_ℓ, and atoms make transitions from the ground to the excited state. The newly introduced second process is stimulated emission, where a photon is emitted into the applied radiation field and the atoms make a transition from the excited to the ground state. Note that in both of these processes the total energy of the system consisting of the applied radiation field and the atoms is conserved. The third process is spontaneous emission, where a photon is also emitted and the atoms also make transitions from the excited to the ground state. However, unlike stimulated emission, the photon is *not* emitted in the mode of the radiation field, but has a random direction or polarization. Since the photon is emitted into the vacuum field, there is no longer conservation of energy for the system of radiation field plus atoms, since the vacuum field is outside the system. Finally, from the distribution of black body radiation, Einstein deduced that the fourth process, spontaneous absorption, is not possible.

The discussion in this chapter so far has properly accounted for the two stimulated processes discussed above (see Eqs. 1.11 and 1.12). The combined action of these two processes causes the oscillation in both the excited and ground state probabilities (see Fig. 1.1). For atoms initially in the ground state, the probability for absorption is large and the probability for them to go into the excited state increases. Once the atoms have a large probability to be in the excited state, however, the probability for absorption decreases and the probability for stimulated emission increases, which leads to the oscillations.

Up to now, spontaneous emission has been left out of the discussion. Including it is very complicated, since it leads to loss of photons, and hence energy, from

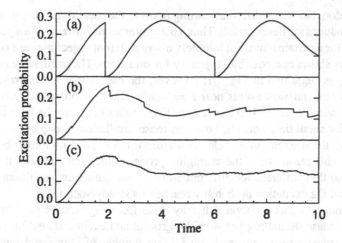

FIGURE 1.8. Trajectories for atoms is a radiation field with $\Omega = \gamma$ and $\delta = -\gamma$, where γ is the natural width. The number of atoms averaged over is 1 (a), 10 (b), and 100 (c).

the system of radiation field and atoms. One way to avoid the difficulty might be to include the vacuum field in the system, which would then be closed as before. However, the task of doing so is formidable because both the emission direction and the polarization direction are random in spontaneous emission. Thus it would be necessary to include the entire continuum of these parameters in the system, and such a description is beyond the scope of this book. Furthermore, in most cases the properties of the emitted photon are not of interest, and information on the atom and the applied radiation field suffices.

The usual way to treat this problem in quantum mechanics is to introduce the density matrix ρ and to discuss the excitation of the atoms in terms of populations and coherences instead of amplitudes. This follows in the next chapter. Here an alternative view of this problem is presented.

This view is called the Monte Carlo wavefunction method and was recently described anew [19]. It is a numerical simulation that treats the evolution of the system with the same coupled Eqs. 1.6. However, at each instant there is some probability that an atom will undergo spontaneous emission within a certain, small time interval. This probability is proportional to the probability of the atom being in the excited state, $|c_e|^2$. In this "gedanken" experiment the state of the system is observed by detecting the emitted photons with a photon counter. At each instant, the output of a random number generator is compared with the probability for a spontaneous emission, and if the random number is smaller, it is assumed that spontaneous emission has occurred (this is why this method is named after a city most famous for gambling). At that instant the evolution starts again from the values $c_g = 1$ and $c_e = 0$. Since there is no interest in the emitted photon, it is disregarded.

Numerical results from this method are shown in Fig. 1.8. Note that there is a random aspect of the description, which means that repeating the procedure for the same atom but with a different starting point in the pseudo random number sequence produces a different result. Thus a particular sequence results in a particular trajectory for a certain atom, but infinitely many different trajectories are possible. Figure 1.8a shows one possible trajectory for one atom. The oscillatory behavior is evident, as suggested in Fig. 1.1; however, the oscillations are interrupted by a spontaneous emission events near $t = 1.9/\gamma$ and $t = 6.0/\gamma$. Repeating the procedure with $N=10$ or 100 atoms (see Figs. 1.8b,c) still results in oscillatory behavior for small time periods; however, these oscillations damp out for longer times. Also the discrete jumps, clearly visible for $N=1$, can no longer be easily observed. This results from the averaging process, since the emission times are random and thus different for different atoms. This causes the oscillations to be damped and the excitation probability reaches its steady-state value.

One common misconception that may arise from Fig. 1.8c is that the atoms eventually cease oscillating between the ground and excited states. In most experiments, measurement are made on a large number of atoms and indeed the oscillations are damped. However, Fig. 1.8a clearly shows that each individual atom still oscillates, but that these oscillations are damped out by the averaging process. This topic will reappear in the density matrix approach that describes the evolution of an ensemble of atoms.

2
The Density Matrix

Chapter 1 presented the equations for the coherent evolution of the amplitudes of a two-level atom in a radiation field. However, the effects of spontaneous emission cannot be described in terms of such coherent evolution of the eigenstates of the system. Spontaneous emission is most readily handled by the density matrix, which is introduced in this chapter. Since this topic is covered by many textbooks in quantum mechanics, it is only briefly presented in the first section here (for instance, see Ref. 20). In the next section it is applied to the specific case of a two-level atom in a radiation field. The resulting equations are solved and discussed in terms of the effects of spontaneous emission on the interaction of atoms by radiation fields.

2.1 Basic Concepts

In quantum mechanics all information about a system in a pure state is stored in the wavefunction $|\Psi\rangle$. However, in an experiment $|\Psi\rangle$ cannot be measured directly. Instead, one can only determine the expectation values of a set of quantum mechanical operators A given by

$$\langle A \rangle = \langle \Psi | A | \Psi \rangle, \tag{2.1}$$

when Ψ is normalized according to $\langle \Psi | \Psi \rangle = 1$. By proper arrangement of the experiment the wavefunction can be determined completely, except for one unnecessary parameter, the overall phase.

Alternatively, the state of the system can be described by the density operator ρ, which is given by $\rho = |\Psi\rangle\langle\Psi|$. The density operator ρ can be written in terms of the $n \times n$ density matrix, where n is the number of wavefunctions that completely spans the Hilbert space. In general, the wavefunction Ψ can be expanded in a basis set $\{\phi_n\}$ as in Eq. 1.2,

$$\Psi = \sum_{i=1}^{n} c_i \phi_i, \qquad (2.2)$$

so that the elements of the density matrix become

$$\rho_{ij} = \langle \phi_i | \rho | \phi_j \rangle = \langle \phi_i | \Psi \rangle \langle \Psi | \phi_j \rangle = c_i c_j^* \qquad (2.3)$$

and the normalization of the wavefunction yields $\text{Tr}(\rho) = \langle \Psi | \Psi \rangle = 1$. In the case of a two-level atom in a radiation field, $n = 2$ so that ρ is a 2×2 matrix.

Clearly the elements ρ_{ij} depend on the basis states $\{\phi_n\}$. The diagonal elements are the probabilities $|c_i|^2$ for the atom to be in state i, which are all between 0 and 1. The off-diagonal elements $c_i c_j^*$ are called the coherences, since they depend on the phase difference between c_i and c_j.

The expectation value of an operator given in Eq. 2.1 can be written as

$$\langle A \rangle = \langle \sum_i c_i \phi_i | A | \sum_j c_j \phi_j \rangle = \sum_{i,j} c_i^* c_j \langle \phi_i | A | \phi_j \rangle \qquad (2.4)$$

$$= \sum_{i,j} \rho_{ji} A_{ij} = \sum_j (\rho A)_{jj} = \text{Tr}(\rho A).$$

Note that if the wavefunction Ψ is multiplied by an arbitrary phase factor $e^{i\alpha}$, there is no change of any observable of the system as shown by Eq. 2.4. Also ρ remains unchanged in this case, as required for an observable.

Since the density matrix contains n^2 complex elements, in principle it would have $2n^2$ real, independent parameters. Because ρ is Hermitian (see Eq. 2.3), $\rho_{ij} = \rho_{ji}^*$ and there remain n^2 independent elements. By contrast, the wavefunction Ψ is completely specified by the expansion coefficients c_i, which contain only $2n - 1$ independent parameters apart from its overall phase. This reduction in the number of parameters arises because the system under discussion here is in a pure state, which means that there is a fixed relation between the diagonal and off-diagonal elements. This relation is found from Eq. 2.3 to be $\rho_{ij}\rho_{ji} = \rho_{ii}\rho_{jj}$.

The alternative to such a pure state is a statistical mixture of several states $\{\Psi_n\}$ that can no longer be specified by just a single wavefunction. In that case the state is represented by a density operator of the form

$$\rho = \sum_i p_i |\Psi_i\rangle \langle \Psi_i|. \qquad (2.5)$$

This relation has the intuitive meaning that the system is in state i with a certain probability p_i. It can easily be checked that there is no longer a fixed relation between diagonal and non-diagonal elements, but instead $\rho_{ij}\rho_{ji} \leq \rho_{ii}\rho_{jj}$. The complete information on the system now requires n^2 independent elements of the density matrix.

The advantages of the density matrix formalism compared to the wavefunction approach can be summarized as follows: (1) It eliminates the arbitrary overall phase, (2) it establishes a more direct connection with observable quantities, and (3) it provides a powerful method for doing calculations. In addition, it can handle pure states as well as mixed states, the last one being of importance in the case of spontaneous emission.

The distinction between pure states and statistical mixtures is of fundamental importance in quantum mechanics. Suppose that for a certain quantum mechanical system there is a complete set of commuting operators. The question if a set of commuting operators is complete depends on the system under study. Then one measurement with each operator completely determines the state. Any subsequent measurement with one of the operators yields the same outcome as before, since all operators commute with each other. In this way the system has been prepared in a pure state, also referred to as a state of "maximum knowledge". If there is no measurement with one of the operators of this complete set, there is no information on the outcome of such a measurement. The system will then be in a statistical mixture of states $\{|\Psi_n\rangle\}$ with a probabilities p_i to be in a pure state Ψ_i, where i labels the eigenstates of the unmeasured operator.

Spontaneous emission results in a transition of the system from an initial to a final state and can convert a pure state to a statistical mixture. This can happen because statistical mixtures are not only a consequence of incomplete preparation of the system, but also occur if there is only partial detection of the final state. Suppose a system consists of two parts A and B, such as an atom and a radiation field that are coupled, but only part A is observed. Then information about part B is lost, and a statistical average over part B is necessary. Using the density matrix to describe the system, one has to take the trace over part B, or

$$\rho_A = \text{Tr}_B (\rho_{AB}). \tag{2.6}$$

If the system was initially in a pure state, the incomplete detection process causes the pure state to evolve into a statistical mixture.

As an example, consider a two-level atom in the excited state. After a short time the atom has a probability to remain in the excited state or it can make a transition to the ground state by spontaneous emission of a photon. The evolution of this system is given by

$$|\Psi\rangle = \alpha(t)|e; 0\rangle + \sum_S \beta_S(t)|g; 1_S\rangle, \tag{2.7}$$

where the state of the atom is indicated by e or g and the emitted photon by $S = (\vec{k}, \hat{\varepsilon})$ with its wavevector $\vec{k}$ and its polarization $\hat{\varepsilon}$. Note that the photon can be emitted in all directions with a certain polarization, so the sum runs over all possible values of S. If one only observes the state of the atom and not the emitted photon, then the atom will be found in either the excited state $|e\rangle$ or the ground state $|g\rangle$; however, it will no longer be in a pure state. The new state can be described by its density matrix ρ_{atom}:

$$\rho_{\text{atom}} = \text{Tr}_{\text{ph}} |\Psi\rangle\langle\Psi| = |\alpha(t)|^2 |e\rangle\langle e| + \sum_S |\beta_S(t)|^2 |g\rangle\langle g|. \qquad (2.8)$$

The pure state $|\Psi\rangle$ has evolved to a statistical mixture of $|e\rangle$ and $|g\rangle$ since the emitted photon has not been observed. Equation 2.8 shows that phase information has been lost from Eq. 2.7.

From the definition of the density matrix in Eq. 2.3, it is easy to show that for a pure, normalized state $\rho^2 = \rho$, whereas for a statistical mixture $\rho^2 \neq \rho$. In a pure state, one of the eigenvalues of the density matrix is unity and all the others are zero. In the case of a statistical mixture there are several eigenvalues between 0 and 1, which are the probabilities for the state to be in a particular eigenstate. These properties make it possible to determine from a given density matrix whether the system is in a pure state or not.

2.2   Spontaneous Emission

The previous section showed that spontaneous emission causes a pure state to evolve into a mixed state because only the atom and the laser field are considered (part A) and not the spontaneously emitted light (part B). This results in a huge simplification of the description because the spontaneously emitted light can travel in many different directions and have different polarizations. The number of modes is infinite and this complicates the situation enormously. Furthermore, spontaneous emission cannot be properly handled within the framework of a semiclassical description of the electromagnetic field as was done in Chapter 1, because it is induced by vacuum fluctuations of the field. There are various books describing the quantization of the field that produces such fluctuations, and these books should be consulted for details [21–23].

In his famous 1917 paper [18], Einstein not only showed that stimulated emission was necessary to explain Planck's blackbody spectrum, but also derived the spontaneous emission rate using detailed balancing between spontaneous and stimulated processes. Although his result is correct, his derivation does not show the true nature of the spontaneous emission process. Its properties emerge from the Wigner-Weisskopf theory that is summarized here [24]. In this theory it is shown that an atom in the excited state decays exponentially as a result of the fluctuations of the quantized vacuum field. The rate of this decay process is just the spontaneous emission rate.

Consider an atom in the excited state at $t = 0$ and no photons in the radiation field. The system is in a pure state $|e; 0\rangle$, where the first parameter in the ket describes the state of the atom and the zero indicates the absence of photons in the field. The system makes a transition from the excited to the ground state by spontaneous emission, emitting one photon into the radiation field. Then the state is denoted by $|g; 1_S\rangle$ with $S = (\vec{k}, \hat{\varepsilon})$ the mode of spontaneous emission, where the direction of the emitted photon is explicitly indicated by its wavevector $\vec{k}$ and its polarization by $\hat{\varepsilon}$. The state of the system can now be described analogously to Eq. 1.2 by

$$\Psi(t) = c_{e0}e^{-i\omega_e t}|e; 0\rangle + \sum_S c_{g1_S}e^{-i(\omega_g + \omega)t}|g; 1_S\rangle, \qquad (2.9)$$

where the sum is over all possible modes S. Note that the frequency ω in the exponent must be replaced by kc for the summation. Even though the summation runs over an infinite number of modes, this notation is sufficient for now.

To describe the evolution of the wavefunction in time, the Hamiltonian of the system has to be defined. This requires the quantization of the electromagnetic field, which will not be described here. However, the only part of the Hamiltonian that couples the two states in Eq. 2.9 is the atom-field interaction: the atomic and field parts play no role. This coupling is analogous to its semiclassical counterpart, and the result for the time evolution of the two states is

$$i\frac{dc_{e0}(t)}{dt} = \sum_S c_{g1_S}(t)\, \Omega_S\, e^{-i(\omega - \omega_a)t} \qquad (2.10a)$$

and

$$i\frac{dc_{g1_S}(t)}{dt} = c_{e0}(t)\, \Omega_S^*\, e^{i(\omega - \omega_a)t}. \qquad (2.10b)$$

These equations are similar to Eq. 1.4, where the coupling for each mode is given by $\hbar\Omega_S = -\vec{\mu} \cdot \vec{E}_\omega$ and Ω_S is called the vacuum Rabi frequency. The dipole moment is $\vec{\mu} = e\langle e|\vec{r}|g\rangle$ and the electric field per mode is found from the classical expression for the energy density to be

$$\vec{E}_\omega = \sqrt{\frac{\hbar\omega}{2\epsilon_0 V}}\hat{\varepsilon}. \qquad (2.11)$$

Here V is the volume used to quantize the field, and it will eventually drop out of the calculation. The total energy of the electromagnetic field in the volume V is given by $\hbar\omega/2$, corresponding to the zero point energy of the radiation field. By directly integrating Eq. 2.10b and substituting the result into Eq. 2.10a, the time evolution of $c_{e0}(t)$ is found to be

$$\frac{dc_{e0}(t)}{dt} = -\sum_S |\Omega_S|^2 \int_0^t dt'e^{-i(\omega - \omega_a)(t - t')}c_{e0}(t'). \qquad (2.12)$$

This represents an exponential decay of the excited state, and to evaluate the decay rate it is necessary to count the number of modes for the summation and then do the time integral.

To count the number of modes $S = (\vec{k}, \hat{\varepsilon})$, represent the field by the complete set of traveling waves in a cube of side L. Since the field is periodic with a periodicity L, the components of $\vec{k}$ are quantized as $k_i = 2\pi n_i/L$, with $i = x, y, z$. Then

$dn_i = (L/2\pi)dk_i$ and therefore $dn = (L/2\pi)^3 d^3k$. The frequency ω is given by $\omega = kc$, so

$$dn = 2 \times \frac{V\omega^2}{8\pi^3 c^3} \sin\theta \, d\omega d\theta d\phi. \qquad (2.13)$$

The factor of 2 on the right-hand side of Eq. 2.13 derives from the two independent polarizations $\hat{\varepsilon}$ of the fluorescent photons. Now replace the summation in Eq. 2.12 by an integration over all possible modes, insert the result of Eq. 2.13, and then integrate over the angles θ and ϕ to find

$$\frac{dc_{e0}(t)}{dt} = -\frac{1}{6\epsilon_0 \pi^2 \hbar c^3} \int d\omega \, \omega^3 \mu^2 \int_0^t dt' \, e^{-i(\omega-\omega_a)(t-t')} c_{e0}(t'), \qquad (2.14)$$

where the volume V has dropped out, since $|\Omega_S|^2 \propto 1/V$. In this result, the orientation of the atomic dipole with respect to the emission direction has been taken into account, which yields a reduction factor of $\frac{1}{3}$ for a random emission direction.

The remaining time integral can be evaluated by assuming that the dipole moment μ varies slowly over the frequency interval of interest, so it can be evaluated at $\omega = \omega_a$. Furthermore, the time integral is peaked around $t = t'$, so that the coefficient $c_{e0}(t)$ can be evaluated at time t and taken out of the integral. The upper boundary of the integral can be shifted toward infinity, and the result becomes

$$\lim_{t\to\infty} \int_0^t dt' \, e^{-i(\omega-\omega_a)(t-t')} = \pi\delta(\omega - \omega_a) - \mathcal{P}\left(\frac{i}{\omega - \omega_a}\right), \qquad (2.15)$$

where $\delta(x)$ is the delta function and $\mathcal{P}(x)$ is the principal value. The last term is purely imaginary and causes a shift of the transition frequency, which will not be discussed further. Substitution of the result of Eq. 2.15 into Eq. 2.14 yields the final result

$$\frac{dc_{e0}(t)}{dt} = -\frac{\gamma}{2}c_{e0}(t), \qquad (2.16a)$$

where

$$\gamma = \frac{\omega^3 \mu^2}{3\pi \epsilon_0 \hbar c^3}. \qquad (2.16b)$$

Since the amplitude of the excited state decays at a rate $\gamma/2$, the population of the state decays with γ and the lifetime of the excited state becomes $\tau \equiv 1/\gamma$.

The decay of the excited state is irreversible. In principle, the modes of the spontaneously emitted light also couple to the ground state in Eqs. 2.10, but there is an infinite number of modes in free space. The amplitude for the reverse process has to be summed over these modes. Since the different modes add destructively, the probability for the reverse process becomes zero. The situation can be changed by putting the atom in a reflecting cavity with dimensions of the order of the optical wavelength λ. Then the number of modes can be changed considerably compared to free space. In quantum optics, several experiments have been carried out where this effect has been detected.

2.3   The Optical Bloch Equations

It is straightforward to use Eq. 1.4 to show that the time dependence of the density matrix depends on the Hamiltonian simply as

$$i\hbar \frac{d\rho}{dt} = [\mathcal{H}, \rho]. \tag{2.17}$$

This relation points out the special role of ρ in quantum mechanics. Note that the sign on the right-hand side is *opposite* to the usual Heisenberg equation of motion for quantum mechanical operators. The rest of this section continues the analysis of the Rabi two-level problem using the density matrix, which is written for a pure state as

$$\rho = \begin{pmatrix} \rho_{ee} & \rho_{eg} \\ \rho_{ge} & \rho_{gg} \end{pmatrix} = \begin{pmatrix} c_e c_e^* & c_e c_g^* \\ c_g c_e^* & c_g c_g^* \end{pmatrix}. \tag{2.18}$$

The effects of the coupling to the light field and spontaneous emission can be added independently [25]. The evolution equation for the terms ρ_{ij} in the case of interaction with a laser can be found by applying the evolution equation for the amplitudes, given by Eq. 1.11. For instance, in the case of ρ_{gg} this is

$$\frac{d\rho_{gg}}{dt} = \frac{dc_g}{dt} c_g^* + c_g \frac{dc_g^*}{dt} = i\frac{\Omega^*}{2}\tilde{\rho}_{eg} - i\frac{\Omega}{2}\tilde{\rho}_{ge}, \tag{2.19}$$

where $\tilde{\rho}_{ge} \equiv \rho_{ge}e^{-i\delta t}$. In the same manner, equations for the time derivative of the other elements of the density matrix can be obtained. Solving these equations gives the same solutions as Eqs. 1.12. The identification of ρ_{ij} in terms of $c_i c_j^*$ is valid for a pure state, but loses its meaning for a statistical mixture.

Spontaneous emission can now be described by an exponential decay of the coefficient $\rho_{eg}(t)$ with a constant rate $\gamma/2$,

$$\left(\frac{d\rho_{eg}}{dt}\right)_{spon} = -\frac{\gamma}{2}\rho_{eg}. \tag{2.20}$$

The ground state is stable against spontaneous emission, but the population of the ground state still changes because of the spontaneous emission process, since the excited state decays to the ground state. The loss of population of the excited state leads to a gain of population in the ground state. This leads to the following equations for the two-level system, including spontaneous emission:

$$\frac{d\rho_{gg}}{dt} = +\gamma\rho_{ee} + \frac{i}{2}\left(\Omega^*\tilde{\rho}_{eg} - \Omega\tilde{\rho}_{ge}\right) \tag{2.21}$$

$$\frac{d\rho_{ee}}{dt} = -\gamma\rho_{ee} + \frac{i}{2}\left(\Omega\tilde{\rho}_{ge} - \Omega^*\tilde{\rho}_{eg}\right)$$

$$\frac{d\tilde{\rho}_{ge}}{dt} = -\left(\frac{\gamma}{2} + i\delta\right)\tilde{\rho}_{ge} + \frac{i}{2}\Omega^*\left(\rho_{ee} - \rho_{gg}\right)$$

$$\frac{d\tilde{\rho}_{eg}}{dt} = -\left(\frac{\gamma}{2} - i\delta\right)\tilde{\rho}_{eg} + \frac{i}{2}\Omega\left(\rho_{gg} - \rho_{ee}\right),$$

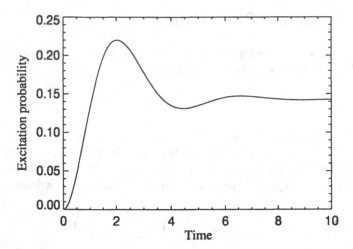

FIGURE 2.1. Probability $|c_e(t)|^2$ for the atom to be in the excited state for $\Omega = \gamma$ and $\delta = -\gamma$ by numerical integration of the OBEs. The solutions are identical to the Monte Carlo wavefunction method with an infinite number of atom trajectories. Time is in units of $1/\gamma$.

These equations are called the optical Bloch equations (OBE), in analogy to the Bloch equations for nuclear magnetic resonance. Note that $d\rho_{ee}/dt = -d\rho_{gg}/dt$, in accordance with the requirement of a closed two-level system where the total population $\rho_{gg} + \rho_{ee} = 1$ is conserved.

Furthermore, it is explicitly assumed that the decay of the coherences and the decay of the excited state are described by a single parameter γ. This will always be the case in the systems discussed within the framework of laser cooling. However, in cases where collisions between atoms play a role, the decay of the coherences and the populations are described by different decay parameters, and in those cases parameters T_1 and T_2 are introduced to account for this difference. For details regarding this issue the reader is referred to several books on this topic [7, 14].

The steady-state solutions of Eqs. 2.21 are discussed in the next section. However, the temporal behavior can be found by direct numerical integration. The results are shown in Fig. 2.1 for the case $\Omega = \gamma$ and $\delta = -\gamma$. This is identical to the result obtained with the Monte Carlo wavefunction method at the end of Chapter 1, if an infinite number of trajectories is used in the Monte Carlo wavefunction method.

2.4   Power Broadening and Saturation

The steady-state solutions of the OBE can be found by setting the time derivatives to zero and exploiting certain relationships among the $n^2 = 4$ real, independent parameters of ρ for a two-level system. The conservation of the population given by $\rho_{gg} + \rho_{ee} = 1$ eliminates one of these parameters, and two of the others are complex conjugates. Using the population difference $w \equiv \rho_{gg} - \rho_{ee}$ and the optical coherence $\rho_{eg} = \rho_{ge}^*$ in the OBE gives

$$\frac{d\rho_{eg}}{dt} = -\left(\frac{\gamma}{2} - i\delta\right)\rho_{eg} + \frac{iw\Omega}{2} \tag{2.22a}$$

and

$$\frac{dw}{dt} = -\gamma w - i(\Omega\rho_{eg}^* - \Omega^*\rho_{eg}) + \gamma. \tag{2.22b}$$

The steady-state case has $d\rho_{eg}/dt = dw/dt = 0$, and the resulting equations can be solved for w and ρ_{eg}:

$$w = \frac{1}{1+s} \tag{2.23a}$$

and

$$\rho_{eg} = \frac{i\Omega}{2(\gamma/2 - i\delta)(1+s)}. \tag{2.23b}$$

Here the saturation parameter s is given by

$$s \equiv \frac{|\Omega|^2}{2|(\gamma/2 - i\delta)|^2} = \frac{|\Omega|^2/2}{\delta^2 + \gamma^2/4} \equiv \frac{s_0}{1 + (2\delta/\gamma)^2}, \tag{2.24a}$$

where the last step defines the on-resonance saturation parameter

$$s_0 \equiv 2|\Omega|^2/\gamma^2 = I/I_s \tag{2.24b}$$

with the saturation intensity given by

$$I_s \equiv \pi hc/3\lambda^3\tau. \tag{2.24c}$$

For the case of a low saturation parameter, $s \ll 1$, the population is mostly in the ground state ($w = 1$), whereas in the case of high s the population is equally distributed between the ground and excited state ($w = 0$). The population ρ_{ee} of the excited state is given by

$$\rho_{ee} = \frac{1}{2}(1 - w) = \frac{s}{2(1+s)} = \frac{s_0/2}{1 + s_0 + (2\delta/\gamma)^2}, \tag{2.25}$$

and for $s \gg 1$, ρ_{ee} approaches $1/2$. Since the population in the excited state decays at a rate γ, and in steady state the excitation rate and the decay rate are equal, the total scattering rate γ_p of light from the laser field is given by

$$\gamma_p = \gamma\rho_{ee} = \frac{s_0\gamma/2}{1 + s_0 + (2\delta/\gamma)^2}. \tag{2.26}$$

At very high intensities, where $s_0 \gg 1$, γ_p saturates to $\gamma/2$. This equation can be rewritten as

$$\gamma_p = \left(\frac{s_0}{1+s_0}\right)\left(\frac{\gamma/2}{1 + (2\delta/\gamma')^2}\right), \tag{2.27a}$$

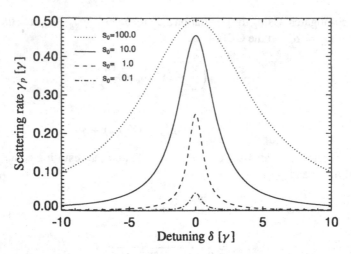

FIGURE 2.2. Excitation rate γ_p as a function of the detuning δ for several values of the saturation parameter s_0. Note that for $s_0 > 1$ the line profiles start to broaden substantially from power broadening.

where

$$\gamma' = \gamma\sqrt{1 + s_0} \qquad (2.27b)$$

is called the power-broadened linewidth of the transition. Because of saturation, the linewidth of the transition as observed in an experiment, where the absorption of light is detected while scanning its frequency, is broadened from its natural linewidth γ to its power-broadened value γ'.

Figure 2.2 shows a plot of γ_p as a function of the detuning δ for several values of the saturation parameter s_0. For large values of s_0 there is a significant power broadening of the spectral profile, which is a direct consequence of the fact that for large s_0, the absorption continues to increase with increasing intensity in the wings, whereas in the center half of the atoms are already in the excited state. The absorption in the center of the profile is therefore saturated, whereas in the wings it is not.

Note that other line-broadening mechanisms, such as the Doppler effect, pressure broadening, and others, have been left out of the present discussion. However, they might also play a significant role under certain conditions, and their convolution with power broadening has to be considered carefully because of the different line shapes.

The scattering of light from a laser beam results in intensity loss when the beam travels through a sample of resonant atoms. The amount of scattered power per unit of volume is given by $\hbar\omega\gamma_p n$, where n is the density of the atoms. Thus $dI/dz = -\hbar\omega\gamma_p n$ for a laser beam of intensity I traveling in the z-direction. For low intensity light tuned near the atomic resonance, the scattering rate is given by

$\gamma_p \approx s_0 \gamma / 2$, so the absorption rate is

$$\frac{\mathrm{d}I}{\mathrm{d}z} = -\sigma_{eg} n I, \tag{2.28a}$$

where the cross section σ_{eg} for scattering light out of the beam on resonance is given by

$$\sigma_{eg} = \frac{\hbar \omega \gamma}{2 I_s} = \frac{3\lambda^2}{2\pi}. \tag{2.28b}$$

Note that this cross section is of the order of λ^2, which is much larger than the cross section for atom-atom interactions, typically of the order of a_0^2.

The solution of Eq. 2.28a is $I(z) = I_0 \exp(-\sigma_{eg} n z)$, and the cross section of Eq. 2.28b allows for an estimate of the densities for which absorption becomes important. Using $\lambda \approx 500$ nm and an interaction length of 1 mm, the laser beam is appreciably absorbed if the density is of the order of 10^{10} atoms/cm³. Such densities can be achieved in optical traps, and so the total absorption of the light at the edge of the atomic cloud can severely diminish the trapping potential. Also, the reabsorption of spontaneously emitted light causes a repulsion between the atoms, which limits the obtainable density in optical traps.

3
Force on Two-Level Atoms

Laser cooling and trapping rely on the interaction between laser light and atoms to exert a controllable force on the atoms, and many sophisticated schemes have been developed using the special properties of the interaction. The outcome is a new field called laser cooling and trapping of atoms that has flourished over the last decade.

This chapter considers the simplest schemes for exerting optical forces on atoms, namely, a single-frequency light field interacting with a two-level atom. The description is one dimensional (the z-direction) and shows how the absorption and emission of light alters the velocity of the atoms. It is based on the interaction of two-level atoms with a laser field as discussed in Chapters 1 and 2. Although this is the simplest possible scheme, it is pedagogically valuable because it shows many of the features that will be encountered in the rest of the book.

3.1 Laser Light Pressure

The philosophy of the correspondence principle requires a smooth transition between quantum and classical mechanics. Clearly the orbits of the planets can be described with arbitrary accuracy using classical mechanics, but just as clearly, they must conform to the rules of quantum mechanics. The quantum version of Newton's laws is embodied in the Ehrenfest theorem [26], a simple statement that the expectation value of an operator must correspond to the behavior of its classical counterpart.

In this section the semiclassical description of the interaction of a light field with a two-level atom is used to derive the laser light pressure on an atom. The force F on an atom is defined as the expectation value of the quantum mechanical force operator $\mathcal{F}$, as defined by

$$F = \langle \mathcal{F} \rangle = \frac{d}{dt} \langle p \rangle. \tag{3.1}$$

The time evolution of the expectation value of a time-independent quantum mechanical operator $\mathcal{A}$ is given by [6]

$$\frac{d}{dt} \langle A \rangle = \frac{i}{\hbar} \langle [\mathcal{H}, \mathcal{A}] \rangle. \tag{3.2}$$

The commutator of $\mathcal{H}$ and p is given by

$$[\mathcal{H}, p] = i\hbar \frac{\partial \mathcal{H}}{\partial z}, \tag{3.3}$$

where the operator p has been replaced by $-i\hbar(\partial/\partial z)$. The force on an atom is thus given by

$$F = -\left\langle \frac{\partial \mathcal{H}}{\partial z} \right\rangle. \tag{3.4}$$

This relation is a specific example of the Ehrenfest theorem and forms the quantum mechanical analog of the classical expression that the force is the negative gradient of the potential.

Discussion of the force on atoms caused by light fields begins with the relevant part of the Hamiltonian of the system, $\mathcal{H}'(t)$ given in Eq. 1.8. Then the force is simply

$$\langle \mathcal{F} \rangle = F = e \left\langle \frac{\partial}{\partial z} \left(\vec{\mathcal{E}}(\vec{r}, t) \cdot \vec{r} \right) \right\rangle. \tag{3.5}$$

Using the electric dipole approximation, i.e., neglecting the spatial variation of the electric field over the size of an atom, allows the interchange of the gradient with the expectation value, and gives

$$F = e \frac{\partial}{\partial z} \left(\left\langle \vec{\mathcal{E}}(\vec{r}, t) \cdot \vec{r} \right\rangle \right), \tag{3.6}$$

whose matrix has only off-diagonal entries. The expectation value can be found using the definition of the Rabi frequency of Eq. 1.10 and the expectation value $\langle A \rangle = \text{Tr}(\rho A)$ from Eq. 2.4, resulting in

$$F = \hbar \left(\frac{\partial \Omega}{\partial z} \rho_{eg}^* + \frac{\partial \Omega^*}{\partial z} \rho_{eg} \right). \tag{3.7}$$

Deriving this result requires the RWA that neglects terms oscillating with the laser frequency. Note that the force depends on the state of the atom, and in particular, on the optical coherence between the ground and excited states, ρ_{eg}.

Although it may seem a bit artificial, it is instructive to split $\partial\Omega/\partial z$ into its real and imaginary parts (the matrix element that defines Ω in Eq. 1.10 can certainly be complex):

$$\frac{\partial\Omega}{\partial z} = (q_r + iq_i)\Omega. \tag{3.8}$$

Here $q_r + iq_i$ is the logarithmic derivative of Ω. In general, for a field $E(z) = E_0(z)\exp(i\phi(z)) + \text{c.c.}$ the real part of the logarithmic derivative corresponds to a gradient of the amplitude $E_0(z)$ and the imaginary part to a gradient of the phase $\phi(z)$. Then the expression for the force becomes

$$F = \hbar q_r(\Omega\rho_{eg}^* + \Omega^*\rho_{eg}) + i\hbar q_i(\Omega\rho_{eg}^* - \Omega^*\rho_{eg}). \tag{3.9}$$

Equation 3.9 is a very general result that can be used to find the force for any particular situation as long as the optical Bloch equations (OBE) for ρ_{eg} can be solved (see Eqs. 2.21). In spite of the chosen complex expression for Ω, it is important to note that the force itself is real, and that first term of the force is proportional to the real part of $\Omega\rho_{eg}^*$, whereas the second term is proportional to the imaginary part.

3.2 A Two-Level Atom at Rest

The remainder of this chapter will be devoted to two specific cases for the laser field. The first one is a traveling wave whose electric field is given by Eq. 1.9:

$$E(z) = \frac{E_0}{2}\left(e^{i(kz-\omega t)} + \text{c.c.}\right). \tag{3.10}$$

In calculating the Rabi frequency from this, the RWA causes the positive frequency component of $E(z)$ to drop out (see Eqs. 1.2 and 1.10). Then the gradient of the Rabi frequency becomes proportional to the gradient of the surviving negative frequency component, so that $q_r = 0$ and $q_i = k$. For such a traveling wave the amplitude is constant but the phase is not, and this leads to the nonzero value of q_i.

This is in direct contrast to the case of a standing wave, composed of two counterpropagating traveling waves so its amplitude is twice as large, for which the electric field is given by

$$E(z) = E_0\cos(kz)\left(e^{-i\omega t} + \text{c.c.}\right), \tag{3.11}$$

so that $q_r = -k\tan(kz)$ and $q_i = 0$. Again, only the negative frequency part survives the RWA, but the gradient does not depend on it. Thus a standing wave has an amplitude gradient, but not a phase gradient. The singularity in q_r from the tangent function for a standing wave does not lead to problems, since it occurs at the node of the field where the Rabi frequency Ω is zero.

The steady-state solutions of the OBE for an atom at rest are given in Eqs. 2.23. Substituting the solution for ρ_{eg} of Eq. 2.23b into Eq. 3.9 gives

$$F = \frac{\hbar s}{1 + s} \left(-\delta q_r + \frac{1}{2} \gamma q_i \right).$$

(3.12)

Note that the first term is proportional to the detuning δ, whereas the second term is proportional to the decay rate γ. For zero detuning, the force becomes $F = (\hbar k \gamma / 2)[s_0/(s_0 + 1)]$, a very satisfying result because it is simply the momentum per photon $\hbar k$, times the scattering rate γ_p of Eq. 2.26.

It is instructive to identify the origin of both of the terms in Eq. 3.12. Absorption of light leads to the transfer of momentum from the optical field to the atoms. If the atoms decay by spontaneous emission, the recoil associated with the spontaneous fluorescence is in a random direction, so its average over many emission events results in zero net effect on the atomic momentum. Thus the force from absorption followed by spontaneous emission can be written as

$$F_{sp} = \hbar k \, \gamma \, \rho_{ee},$$

(3.13)

where the first factor is the momentum transfer for each photon, the second factor is the rate for the process, and the last factor is the probability for the atoms to be in the excited state. Although it may seem natural for this expression to depend on the ground-state population ρ_{gg} and not the excited-state population ρ_{ee}, using ρ_{ee} simply builds in the dependence of absorption on detuning and intensity, including saturation. Using Eq. 2.26, the force resulting from absorption followed by spontaneous emission becomes

$$F_{sp} = \frac{\hbar k s_0 \gamma / 2}{1 + s_0 + (2\delta/\gamma)^2},$$

(3.14)

which saturates at large intensity as a result of the factor s_0 in the denominator. Increasing the rate of absorption by increasing the intensity does not increase the force without limit, since that would only increase the rate of stimulated emission, where the transfer of momentum is opposite in direction compared to the absorption. Thus the force saturates to a maximum value of $\hbar k \gamma / 2$, because ρ_{ee} has a maximum value of $1/2$ (see Eq. 2.25).

Examination of Eq. 3.13 shows that it clearly corresponds to the second term of Eq. 3.9. This term is called the light pressure force, radiation pressure force, scattering force, or dissipative force, since it relies on the scattering of light out of the laser beam. It vanishes for an atom at rest in a standing wave where $q_i = 0$, and this can be understood because atoms can absorb light from either of the two counterpropagating beams that make up the standing wave, and the average momentum transfer then vanishes. This force is dissipative because the reverse of spontaneous emission is not possible, and therefore the action of the force cannot be reversed. It plays a very important role in the slowing and cooling of atoms as discussed in Chapters 6 and 7.

By contrast, the first term in Eq. 3.9 derives from the light shifts of the ground and excited states, described in Sec. 1.2.1. Such light shifts depend on the strength of the optical electric field. A standing wave is composed of two counterpropagating laser beams, and their interference produces an amplitude gradient that is not present in a traveling wave. The resulting spatially modulated light shift produces a force that is different from that of Eq. 3.13. The force is proportional to the gradient of the light shift, and Eq. 1.17a can be used to find the force on ground-state atoms in low intensity light:

$$F_{dip} = -\frac{\partial(\Delta E_g)}{\partial z} = \frac{\hbar\Omega}{2\delta}\frac{\partial\Omega}{\partial z}. \qquad (3.15)$$

For an amplitude-gradient light field such as a standing wave, $\partial\Omega/\partial z = q_r\Omega$, and this force corresponds to the first term in Eq. 3.9 in the limit of low saturation ($s \ll 1$). The apparent difference in the dependence on δ is merely a consequence of the expansion of the radical as done in Eqs. 1.17.

For the case of a standing wave Eq. 3.12 becomes

$$F_{dip} = \frac{2\hbar k\delta s_0 \sin 2kz}{1 + 4s_0 \cos^2 kz + (2\delta/\gamma)^2}, \qquad (3.16)$$

where s_0 is the saturation parameter of each of the two beams that form the standing wave. For $\delta < 0$ the force drives the atoms to positions where the intensity has a maximum, whereas for $\delta > 0$ the atoms are attracted to the intensity minima. The force is conservative and can be written for an atom at rest as the gradient of a potential U_{dip} given by

$$U_{dip} = \frac{1}{2}\hbar\delta \log\left(\frac{1 + 4s_0\cos^2 kz + (2\delta/\gamma)^2}{1 + (2\delta/\gamma)^2}\right). \qquad (3.17)$$

The potential depth can be increased by increasing δ because of the first factor, but when $(\delta/\gamma)^2$ becomes much larger than s_0 the potential depth decreases because of the logarithmic term in Eq. 3.17. When $(\delta/\gamma)^2 \gg s_0$ the potential U_{dip} reduces to the light shift ΔE_g of Eq. 1.17a, corrected for the presence of two beams.

The force F_{dip} is called the dipole force, reactive force, gradient force, or redistribution force. It has the same origin as the force of an inhomogeneous dc electric field on a classical dipole, but relies on the redistribution of photons from one laser beam to the other. The entire Chapter 9 is devoted to dipole forces, and they play an important role in both cooling *and* trapping of atoms, as discussed in Chapters 8 and 11.

It needs to be emphasized that the forces of Eqs. 3.14 and 3.16 are two fundamentally different kinds of forces. For an atom at rest, the scattering force vanishes for a standing wave, whereas the dipole force vanishes for a traveling wave. The scattering force is dissipative, and can be used to cool, whereas the dipole force is conservative, and can be used to trap. Dipole forces can be made large by using high intensity light because they do not saturate. However, since the forces are conservative, they cannot be used to cool a sample of atoms. Nevertheless, they can be combined with the dissipative scattering force to enhance cooling in several different ways, as described in Chapters 8 and 9. By contrast, scattering forces are always limited by the rate of spontaneous emission γ and cannot be made arbitrarily strong, but they are dissipative and are required for cooling.

3.3   Atoms in Motion

Laser cooling requires velocity-dependent forces that cannot derive from the gradient of a potential. Instead, it depends upon dissipative forces that are velocity dependent. Including the velocity of the atoms in the OBE is possible, but the resulting equations are usually too hard to solve analytically.

Instead, the procedure will be to treat the velocity of the atoms as a small perturbation, and make first-order corrections to the solutions of the OBE obtained for atoms at rest [27]. It begins by adding drift terms in the expressions for the relevant quantities. Thus the Rabi frequency satisfies

$$\frac{d\Omega}{dt} = \frac{\partial\Omega}{\partial t} + v\frac{\partial\Omega}{\partial z} = \frac{\partial\Omega}{\partial t} + v(q_r + iq_i)\Omega, \qquad (3.18)$$

where Eq. 3.8 has been used to separate the gradient of Ω into real and imaginary parts. In the same way, differentiating Eq. 2.23a leads to

$$\frac{dw}{dt} = \frac{\partial w}{\partial t} + v\frac{\partial w}{\partial z} = \frac{\partial w}{\partial t} - \frac{2vq_rs}{(1+s)^2}, \qquad (3.19a)$$

since $s_0 = 2|\Omega|^2/\gamma^2$ and Ω depends on z. Similarly, differentiating Eq. 2.23b leads to

$$\frac{d\rho_{eg}}{dt} = \frac{\partial\rho_{eg}}{\partial t} + v\frac{\partial\rho_{eg}}{\partial z} = \frac{\partial\rho_{eg}}{\partial t} + \frac{iv\Omega}{2(\gamma/2 - i\delta)(1+s)}\left[q_r\left(\frac{1-s}{1+s}\right) + iq_i\right]. \qquad (3.19b)$$

In both of these calculations it must be remembered that Ω is complex, so differentiating s_0 results in two terms that give $\partial s_0/\partial z = 2q_r s_0$. In Eqs. 3.19 the value of w in $\partial w/\partial z$ has been taken from its steady-state value given by Eq. 2.23a, and similarly for ρ_{eg}. Since neither w nor ρ_{eg} is explicitly time dependent, both $\partial w/\partial t$ and $\partial\rho_{eg}/\partial t$ vanish. The Eqs. 3.19 are still difficult to solve analytically for a general optical field, and the results are not very instructive. However, the solution for the two special cases of the standing and traveling waves provide considerable insight.

3.3.1   Traveling Wave

For a traveling wave $q_r = 0$, and the velocity-dependent force can be found by combining Eqs. 3.19 with Eqs. 2.22 to eliminate the time derivatives dw/dt and

$d\rho_{eg}/dt$. The resulting coupled equations for w and ρ_{eg} can be separated and substituted into Eq. 3.9 for the force to find, after considerable algebra,

$$F = \hbar q_i \frac{s\gamma/2}{1+s}\left(1 + \frac{2\delta v q_i}{(1+s)(\delta^2 + \gamma^2/4)}\right) \equiv F_0 - \beta v. \qquad (3.20)$$

The first term is the velocity-independent force F_0 for an atom at rest given by Eq. 3.12. The second term is velocity-dependent and can lead to compression of the velocity distribution. For a traveling wave $q_i = k$ and thus the damping coefficient β is given by

$$\beta = -\hbar k^2 \frac{4s_0(\delta/\gamma)}{(1 + s_0 + (2\delta/\gamma)^2)^2}. \qquad (3.21)$$

Note that such a force would compress the velocity distribution of an atomic sample for negative values of δ, i.e., for red detuned light. For small detuning and low intensity the damping coefficient β is linear in both parameters. However, for detunings much larger than γ and intensities much larger than I_s, β saturates and even decreases as a result of the dominance of δ in the denominator of Eq. 3.21. This behavior can be seen in Fig. 3.1, where the damping coefficient β has been plotted as a function of detuning for different saturation parameters. The decrease of β for large detunings and intensities is caused by saturation of the transition, in which case the absorption rate becomes only weakly dependent on the velocity. The maximum value of β is obtained for $\delta = -\gamma/2$ and $s_0 = 2$, and is given by

$$\beta_{\text{max}} = \hbar k^2/4. \qquad (3.22)$$

The damping rate Γ is given by $\Gamma \equiv \beta/M$, and its maximum value is

$$\Gamma_{\text{max}} = \frac{\hbar k^2}{4M} = \frac{\omega_r}{2}, \qquad (3.23)$$

where ω_r is the recoil frequency discussed near the end of Sec. 5.1. For the alkalis this rate is of the order of 10^4–10^5 s^{-1}, indicating that atomic velocity distributions can be compressed on the order of 10-100 μs. Furthermore, F_0 in Eq. 3.20 is always present and so the atoms are *not* damped toward any constant velocity.

3.3.2 Standing Wave

For a standing wave $q_i = 0$, and just as above in Sec. 3.3.1, the velocity-dependent force can be found by combining Eqs. 3.19 with Eqs. 2.22 to eliminate the time derivatives. The resulting coupled equations for w and ρ_{eg} can again be separated and substituted into Eq. 3.9 for the force to find

$$F = -\hbar q_r \frac{s\delta}{1+s}\left(1 - vq_r \frac{(1-s)\gamma^2 - 2s^2(\delta^2 + \gamma^2/4)}{(\delta^2 + \gamma^2/4)(1+s)^2\gamma}\right), \qquad (3.24)$$

where $q_r = -k\tan(kz)$. In the limit of $s \ll 1$, this force is

$$F = \hbar k \frac{s_0\delta\gamma^2}{2(\delta^2 + \gamma^2/4)}\left(\sin 2kz + kv\frac{\gamma}{(\delta^2 + \gamma^2/4)}(1 - \cos 2kz)\right). \qquad (3.25)$$

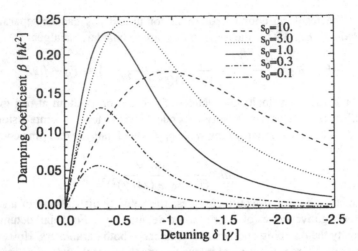

FIGURE 3.1. The damping coefficient β for an atom in a traveling wave as a function of the detuning for different values of the saturation parameter s_0. The damping coefficient is maximum for intermediate detunings and intensities.

Here s_0 is the saturation parameter of each of the two beams that compose the standing wave. The first term is the velocity-independent part of Eq. 3.12 and is sinusoidal in space, with a period of $\lambda/2$. Thus its spatial average vanishes. The force remaining after such averaging is $F_{av} = -\beta v$, where the damping coefficient β is given by

$$\beta = -\hbar k^2 \frac{8s_0(\delta/\gamma)}{(1 + (2\delta/\gamma)^2)^2}. \tag{3.26}$$

In contrast to the traveling-wave case, this is a true damping force because there is no F_0, so atoms are slowed toward $v = 0$ independent of their initial velocities. Note that this expression for β is valid only for $s \ll 1$ because it depends on spontaneous emission to return excited atoms to their ground state. By contrast, the value of β for a traveling wave given in Eq. 3.21 is valid for all values of s by virtue of its saturation, as discussed below. The standing-wave value of β is twice as large as the traveling-wave value, since a standing wave is the sum of two traveling waves, and their damping coefficients add constructively.

There is an appealing description of the mechanism for this kind of cooling in a standing wave. With light detuned below resonance, atoms traveling toward one laser beam see it Doppler shifted upward, closer to resonance. Since such atoms are traveling away from the other laser beam, they see its light Doppler shifted further downward, hence further out of resonance. Atoms therefore scatter more light from the beam counterpropagating to their velocity, and thus their velocity is lowered. This is the damping mechanism called optical molasses, discussed in detail in Chapter 7. It is one of the most important tools of laser cooling.

Needless to say, such a pure damping force would reduce the atomic velocities, and hence the absolute temperature, to zero. Since this violates thermodynamics, there must be something left out of the description. It is the discreteness of the momentum changes in each case, $\Delta p = \hbar k$ that results in a minimum velocity change. The consequences of this discreteness can be described as a diffusion of the atomic momenta in momentum space by finite steps, and is discussed in Secs. 5.3 and 7.2.

The damping coefficient β for a traveling wave given in Eq. 3.21 can easily include the effects of high values of s because the momentum change from stimulated emission exactly cancels that of absorption, and the saturation of the absorption described by Eq. 2.26 accounts for the high-intensity effects. By contrast, the damping coefficient β for a high-intensity standing wave is more complicated because there can be absorption from one beam followed by stimulated emission from the other. Since the order of these processes can be random, this constitutes a totally different form of momentum diffusion. It is related to the spatially sinusoidal dipole force deriving from the light shift, which itself is the result of absorption followed by stimulated emission. Thus this "dipole force contribution to the diffusion" may be viewed as momentum impulses arising from atoms located at different positions on the sinusoidal potential of the light shift in the standing wave, whose amplitude increases with intensity.

4
Multilevel Atoms

The discussion up to here has focused on the two-level atom problem where the light field couples a single ground and excited state. In practice atoms have many levels, and in general the light field couples more than two levels at the same time. Two-level atoms are often discussed in the literature because it is straightforward to obtain analytical results. Such solutions provide much insight and understanding that cannot be obtained from the numerical solutions required for more complicated atoms.

However, in laser cooling one must deal with the coupling of large numbers of states by light. This chapter discusses the nature of these states and shows their origin for specific atoms (alkali-metal and metastable rare gas atoms). The discussion is generally restricted to the ground and first excited states, since these are the only ones that play a significant role in laser cooling.

4.1   Alkali-Metal Atoms

Alkali-metal atoms were the first ones to be cooled and trapped. Their popularity stems from multiple origins. Most important is that the excitation frequency from the lowest to the first excited state is in the visible region, which makes it relatively simple to generate light for the optical transitions. Another reason for their popularity is that it is easy to generate an atomic beam for the alkalis, which have a large vapor pressure at a modest temperature of only a few hundred degrees Centigrade. Heating alkali-metals in an oven with a small opening produces an effusive beam of atoms that can be readily manipulated by laser light.

The ground states of all the alkali-metal atoms have a closed shell with one valence electron. For sodium (Na), which is often used in laser cooling experiments, the electron configuration is given by ^{23}Na $(1s)^2$ $(2s)^2$ $(2p)^6$ $(3s)$. Since the core is a closed shell, it does not contribute to the orbital angular momentum of the atom, and there remains only the outer, valence electron. The state of this electron is completely determined by its orbital angular momentum ℓ and spin angular momentum s. These two momenta couple in the usual way to form the total angular momentum j of the electron:

$$|\ell - s| \le j \le \ell + s. \tag{4.1}$$

Since the only contribution to the total angular momentum of the atom comes from the valence electron, the total orbital angular momentum is $\vec{L} = \vec{\ell}$, spin angular momentum $\vec{S} = \vec{s}$, and total angular momentum $\vec{J} = \vec{j}$ for all electrons.[1] Different values of $\vec{J}$ lead to different energies of the states, since the spin-orbit interaction $V_{so} = A\vec{L} \cdot \vec{S}$ depends on the orientation of $\vec{S}$ with respect to $\vec{L}$. This splitting of the states by the spin-orbit interaction is called the fine structure of the atom. The LS-coupling discussed above is therefore only valid if this spin-orbit interaction is small compared to the level separation of the states.

For the alkali-metal atoms the electronic states are fully specified in the Russell-Saunders notation as n $^{(2S+1)}L_J$, where n is the principal quantum number of the valence electron. The lowest state for Na is the $3^2S_{1/2}$ state, whereas the first excited states are the $3^2P_{1/2,3/2}$ states, where the valence electron is excited to the (3p)-state. In this case the angular momentum $L = 1$ can couple with the total spin $S = {}^1\!/_2$ to form either $J_e = {}^1\!/_2$ or $J_e = {}^3\!/_2$. The fine structure splitting between these two states is ≈ 515 GHz in Na, and other values for the alkali-metal atoms are given in Table C.4 of Appendix C.

The structure of the alkali-metal atoms becomes somewhat more complicated when the interaction of the nuclear spin $\vec{I}$ with the total angular momentum of the electron $\vec{J}$ is included. These angular momenta couple in the usual way to form the total angular momentum as $\vec{F} = \vec{I} + \vec{J}$. Different values of $\vec{F}$ for the same values of both $\vec{I}$ and $\vec{J}$ are split by the $A\vec{I} \cdot \vec{J}$ interaction between the nuclear spin and the electronic angular momentum. The resulting energy structure is called the hyperfine structure (hfs). This hfs is generally much smaller than the fine structure because of the much smaller size of the nuclear magnetic moment. For Na, with a nuclear spin of $I = {}^3\!/_2$, the ground state has $F_g = 1$ and 2, and the hfs is $\approx$ 1.77 GHz. The excited state has $F_e = 0$, 1, 2, and 3, and the resulting hfs is only on the order of 100 MHz. In general, the shift of the energy levels because of the hyperfine interaction can be written as [11, 28]

$$\Delta E_{\text{hfs}} = \frac{1}{2}hAK + hB\frac{{}^3\!/_2K(K+1) - 2I(I+1)J(J+1)}{2I(2I-1)2J(2J-1)}, \tag{4.2}$$

---

[1] Here the convention is that the angular momentum of one electron is indicated in lower case, whereas the angular momentum of the atom is indicated in capitals.

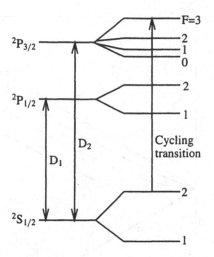

FIGURE 4.1. The ground S state and lowest lying P states of atomic Na, showing the hfs schematically (see Table C.4 of Appendix C for numerical values). These transitions are near $\lambda = 590$ nm (see Table C.1) in the orange-yellow region of the spectrum, and are accessible with dye laser light.

where $K = F(F+1) - I(I+1) - J(J+1)$ and A and B are two parameters, that are adjusted using experimental data [28]. The splitting between adjacent levels becomes

$$\Delta E_{\text{hfs}}(F) - \Delta E_{\text{hfs}}(F-1) = hAF + 3hBF\frac{F^2 - I(I+1) - J(J+1) + \frac{1}{2}}{2I(2I-1)J(2J-1)},$$

(4.3)

where F denotes the highest value of the total angular momentum of the two adjacent levels. A schematic diagram for the fine and hyperfine structure of Na, or other alkalis with $I = \frac{3}{2}$, is given in Fig. 4.1. More detailed information on specific values of the hfs of the alkali-metal atoms is given in Table C.4 of Appendix C.

Each of these states of alkali-metal atoms is further split into $(2I+1) \times (2J+1)$ Zeeman sublevels. In the case of Na with $I = \frac{3}{2}$, this leads to 8 Zeeman sublevels in the ground state ($J_g = \frac{1}{2}$), 8 sublevels in the first excited state ($J_e = \frac{1}{2}$), and 16 sublevels in the next excited state ($J_e = \frac{3}{2}$). In principle, the light can drive all transitions between ground and excited sublevels. However, certain selection rules have to be obeyed, and these limit the number of transitions considerably. These selection rules are discussed in more detail in Sec. 4.4.

In the absence of any perturbations, many of these Zeeman sublevels are degenerate, but application of an external field lifts the degeneracy. It has already been shown in Sec. 1.2.1 that the presence of a light field not only induces transitions, but also shifts the energy levels. Later in this chapter in Sec. 4.5 it is shown that the transition strengths vary among the Zeeman sublevels, and thus a laser field can lift the degeneracy through the different light shifts. In fact, this feature is at the heart of the sub-Doppler cooling schemes described in Chapter 8.

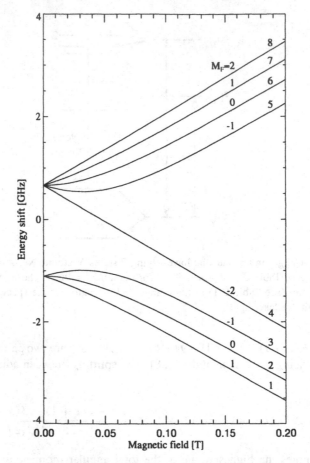

FIGURE 4.2. Energies of the ground hyperfine states of Na, where the states are numbered 1–8 and M_F is the projection of the total angular momentum of the atom on the magnetic field axis.

An applied magnetic field B can also lift these degeneracies, producing the well-known Zeeman effect, as shown in Fig. 4.2. At low fields the energy level shifts ΔE are proportional to the field strengths according to $\Delta E = g\mu_B M B$, where $\mu_B \equiv e\hbar/2m_e c$ is the Bohr magneton, M is the projection of the angular momentum along B, and g is the Landé g-factor (here m_e is the electron mass). The presence of the nuclear spin changes the g-factor from its usual g_J value given by

$$g_J = 1 + \frac{J(J+1) + S(S+1) - L(L+1)}{2J(J+1)} \tag{4.4a}$$

to

$$g_F = g_J \frac{F(F+1) + J(J+1) - I(I+1)}{2F(F+1)}. \tag{4.4b}$$

Here L, S, and J refer to the electron's angular momenta, I is the nuclear spin, and F is the total atomic angular momentum that ranges from $F = |J - I|$ to $F = J + I$ in integer steps. Thus the different manifolds of Fig. 4.2 have different slopes at small field values.

4.2 Metastable Noble Gas Atoms

The metastable noble gases are next in popularity for laser cooling and trapping. With the development of metastable noble gas sources, where a discharge is run through a supersonic expansion of the noble gas, a beam of metastable noble gas atoms can be formed. The efficiency of such sources is low ($\approx 10^{-5}$–10^{-4}) because electron impact in the discharge is the only way to excite the atoms, since the excitation energy of 10–20 eV is much greater than the photon energy of ordinary lasers. Some of the low-lying excited states of the noble gases cannot decay to the ground state by a dipole transition because of the selection rules, and therefore can have lifetimes of more than 100 s and are thus called metastable states. Since the noble gas atoms in the ground state are inert, they do not interfere with experiments on the metastables in spite of the dominance of ground-state atoms.

One advantage of the metastable noble gases over the alkali metals is that most of them do not have a nuclear spin and therefore they do not show any hfs. This reduces the number of states by an appreciable amount. Since the outer electron is highly excited and the state is close to the ionization level, metastable noble gas atoms can be treated effectively as one-electron atoms. One important difference between them and the alkali-metal atoms is that the last shell of the core is not closed and therefore possesses both orbital and spin angular momentum. Together with the orbital and spin angular momentum of the outer electron, these have to be coupled to form the total angular momentum of the atom.

The scheme most often used to couple these four angular momenta, referred to as $j\ell$ coupling,[2] is first to couple the momenta L and S of the core to a total momentum j of the core, which is subsequently coupled with the orbital angular momentum ℓ of the valence electron to form the angular momentum K, which is finally coupled to the spin s of the valence electron to form the total angular momentum J. The notation for the states is then given by $^{2S+1}L_j n\ell[K]_J$, with n the principal quantum number of the outer electron. However, in order to appreciate the correspondence between the alkali-metal atoms and the metastable noble gas atoms, the Russell-Saunders notation $n\ ^{2S'+1}L_J$ will be used for the metastable noble gases as well. Note that here S' is the total spin of the core plus the valence electron and therefore not identical to S.

---

[2] The conventional spectroscopic notation for the rare gases is different from that of the alkalis that was introduced earlier.

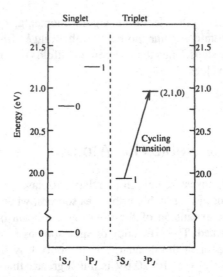

FIGURE 4.3. The $n = 1$ and 2 states of He, showing the metastable 2^3S state that constitutes He* and serves as the ground state for the triplet system. The first excited P state is truly a triplet, its $J = 0$ sublevel lies highest, 29.62 GHz above $J = 1$, which is 2.29 GHz above $J = 2$. These transitions are near $\lambda = 1.083\ \mu$m in the near infrared (see Table C.1 of Appendix C) and are accessible with a solid state laser called LNA, Ti:Sapphire, and rather special diode lasers.

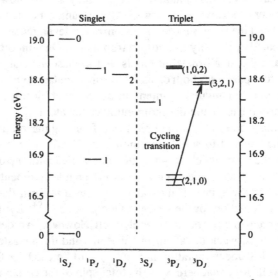

FIGURE 4.4. The low-lying levels of Ne, showing the metastable 3^3P sublevels that constitute Ne*. The $J = 1$ state is not truly metastable. These transitions are near $\lambda = 640$ nm in the red (see Table C.1 of Appendix C) and are accessible with dye laser light. Similar transitions in the other noble gases Ar and Kr are near $\lambda = 810$ nm (see Table C.1) and are more conveniently accessible with Ti:Sapphire and common diode lasers.

Metastable He (He*) is the noble gas with the lowest mass and also with the easiest level structure. This arises because the electron excited from the core is an s-electron and therefore the total orbital angular momentum of the core is also an S-state. The spin of the core and of the valence electron couple to form either singlet or triplet states, and since optical transitions between triplet and singlet states are forbidden, the triplet states have the longest lifetimes. Thus only the triplet states are discussed in connection with laser cooling. The lowest metastable triplet state, which for all practical purposes can be considered as another ground state of the atom, is the 2^3S_1-state, whereas the first excited triplets are the $2^3P_{0,1,2}$-states (see Fig. 4.3). Note that $J = 0$ has the highest energy and $J = 2$ has the lowest.

The situation is considerably more complicated for the other metastable noble gas atoms, of which only neon (Ne*) will be considered in more detail. Since the configuration for the lowest, metastable state is ^{20}Ne $(1s)^2$ $(2s)^2$ $(2p)^5$ $(3s)$, the core is missing a (2p)-electron and the resulting core state is a $^2P_{1/2,3/2}$-state, depending on how the orbital and spin angular momenta of the core are coupled. As in the case of He*, only the triplet system will be discussed, and the lowest triplet states are $3^3P_{0,1,2}$, of which only the $(J = 0)$- and $(J = 2)$-state are truly metastable. For the first excited metastable triplet states, there are total orbital angular momentum values of 0, 1, and 2, so there are the following states: 3^3S_1, $3^3P_{0,1,2}$, and $3^3D_{1,2,3}$ (see Fig. 4.4).

4.3   Polarization and Interference

In the treatment of the interaction of two-level atoms and a laser field, the discussion of the polarization has been deferred. In the case of multilevel atoms, this is no longer possible because the orientation of the dipole moment of the atoms with respect to the polarization of the light is important. Since the atoms can be in different ground states, their coupling to the light field in these states will in general be different. Another aspect is that the interference of two laser beams depends on their mutual polarization. Since the light field used in laser cooling may consist of many laser beams, their polarizations often play a key role.

A laser beam has a high degree of polarization. Although its polarization is in general elliptical, only the extreme cases of linear and circular polarization will be considered here. Because of the transverse nature of the electromagnetic field of a laser beam, the unit polarization vector $\hat{\varepsilon}$ of the field is always perpendicular to the propagation direction $\vec{k}$.

Consider the light field of two counterpropagating plane-wave laser beams with the same frequency ω_ℓ. If the polarizations of the two laser beams are identical, then the polarization of the resulting light field is everywhere the same as that of the incoming laser beams. However, the two plane waves interfere and produce a standing wave. The resulting electric field for a linear polarization $\hat{\varepsilon}$ can be written as (see Eq. 3.11)

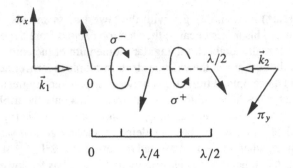

FIGURE 4.5. Polarization gradient field for the lin $\perp$ lin configuration (see also Chapter 8).

$$\vec{E} = E_0\,\hat{\varepsilon}\,\cos(\omega_\ell t - kz) + E_0\,\hat{\varepsilon}\,\cos(\omega_\ell t + kz) \qquad (4.5)$$
$$= 2E_0\,\hat{\varepsilon}\cos kz\,\cos\omega_\ell t.$$

The intensity of the light field has a $\cos^2 kz$ spatial dependence with a period of $\lambda/2$. This situation of a standing wave is very common in laser cooling, and it will reappear in the discussion of optical traps and lattices.

If the polarization of the laser beams is not identical, then the situation becomes rather complicated. Only the two special cases that play important roles in laser cooling will be considered here. The first is where the two counterpropagating laser beams are both linearly polarized, but their $\hat{\varepsilon}$ vectors are perpendicular (*e.g.*, $\hat{x}$ and $\hat{y}$, which is called lin $\perp$ lin or lin-perp-lin). Then the total field is the sum of the two counterpropagating beams given by

$$\vec{E} = E_0\,\hat{x}\,\cos(\omega_\ell t - kz) + E_0\,\hat{y}\,\cos(\omega_\ell t + kz) \qquad (4.6)$$
$$= E_0\left[(\hat{x} + \hat{y})\cos\omega_\ell t\,\cos kz + (\hat{x} - \hat{y})\sin\omega_\ell t\,\sin kz\right].$$

At the origin, where $z = 0$, this becomes

$$\vec{E} = E_0(\hat{x} + \hat{y})\cos\omega_\ell t, \qquad (4.7)$$

which corresponds to linearly polarized light at an angle $+\pi/4$ to the x-axis. The amplitude of this field is $\sqrt{2}E_0$. Similarly, for $z = \lambda/4$, where $kz = \pi/2$, the field is also linearly polarized but at an angle $-\pi/4$ to the x-axis.

Between these two points, at $z = \lambda/8$, where $kz = \pi/4$, the total field is

$$\vec{E} = E_0\left[\hat{x}\,\sin(\omega_\ell t + \pi/4) + \hat{y}\,\cos(\omega_\ell t + \pi/4)\right]. \qquad (4.8)$$

Since the $\hat{x}$ and $\hat{y}$ components have sine and cosine dependence, they are $\pi/2$ out of phase, and so Eq. 4.8 represents circularly polarized light rotating about the z-axis in the negative sense. Similarly, at $z = 3\lambda/8$ where $kz = 3\pi/4$, the polarization is circular but in the positive sense. Thus in this lin $\perp$ lin scheme the polarization cycles from linear to circular to orthogonal linear to opposite circular in the space of only half a wavelength of light, as shown in Fig. 4.5. It truly has a very strong polarization gradient.

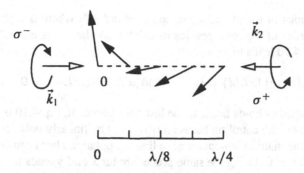

FIGURE 4.6. Polarization gradient field for the σ^+-σ^- configuration (see also Chapter 8).

The other important polarization configuration is that of counterpropagating, oppositely circularly polarized light beams. The total electric field is

$$\vec{E} = E_0 \left[\hat{x} \, \cos(\omega_\ell t - kz) + \hat{y} \, \sin(\omega_\ell t - kz) \right] \tag{4.9}$$
$$+ E_0 \left[\hat{x} \, \cos(\omega_\ell t + kz) - \hat{y} \, \sin(\omega_\ell t + kz) \right]$$
$$= 2 E_0 \, \cos \omega_\ell t \left[\hat{x} \cos kz + \hat{y} \sin kz \right].$$

Since there is no temporal phase difference between the two polarization directions $\hat{x}$ and $\hat{y}$ at any position, this represents a linearly polarized field whose $\hat{\varepsilon}$ vector is fixed in time but rotates uniformly in space along z, rotating through 180° as z changes by $\lambda/2$ (see Fig. 4.6). This arrangement is called the σ^+-σ^- polarization scheme.

These two cases of lin $\perp$ lin and σ^+-σ^- polarization schemes play an important role in laser cooling. Since the coupling of the atoms to the light field depends on the polarization of the field, atoms moving in a polarization gradient will be coupled differently at different positions as discussed in Sec. 4.5. Furthermore, since in a multilevel atom different states are coupled differently to the light field depending on the polarization, this will have important consequences for the laser cooling, as described in Chapter 8.

4.4   Angular Momentum and Selection Rules

For optical transitions the coupling between the atomic states is given by the dipole moment, and selection rules exist for such transitions. Selection rules can be inferred from the equations derived in the next section, but they can also be quite simply calculated from the commutation relations [9]. For the z-component of the orbital angular momentum L of the atom, the following commutation rules apply:

$$[L_z, x] = i\hbar y, \qquad [L_z, y] = -i\hbar x, \qquad [L_z, z] = 0. \tag{4.10}$$

The eigenfunctions of the atoms are denoted by $|\alpha L M\rangle$, where α represents all the other properties of the state besides its orbital angular momentum. The third relation of Eqs. 4.10 leads to

$$\langle \alpha' L' M'| [L_z, z] |\alpha L M\rangle = (M' - M)\hbar\langle \alpha' L' M'|z|\alpha L M\rangle = 0, \qquad (4.11)$$

where the last equality holds because the last commutator in Eq. 4.10 is 0. As the next section shows, the coupling between two states by linearly polarized light is proportional to the matrix element for z, so linearly polarized light can couple two states only if $\Delta M = 0$. Using the same procedure for x and y leads to

$$\langle \alpha' L' M'| [L_z, x] |\alpha L M\rangle = (M' - M)\hbar\langle \alpha' L' M'|x|\alpha L M\rangle \qquad (4.12)$$
$$= i\hbar\langle \alpha' L' M'|y|\alpha L M\rangle$$

and

$$\langle \alpha' L' M'| [L_z, y] |\alpha L M\rangle = (M' - M)\hbar\langle \alpha' L' M'|y|\alpha L M\rangle \qquad (4.13)$$
$$= -i\hbar\langle \alpha' L' M'|x|\alpha L M\rangle.$$

The combination of these two relations requires that either $\Delta M = \pm 1$ or that the matrix element for x or for y must vanish. Again, the next section shows that for circularly polarized light the appropriate matrix element is a combination of x and y. The selection rules for circularly polarized light are thus $\Delta M = \pm 1$, where the $(+)$-sign is for right-handed and the $(-)$-sign for left-handed circular polarization.

Note that these selection rules reflect the conservation of angular momentum. Since each photon carries an angular momentum $\hbar$, the projection of this angular momentum on the z-axis can be $0, \pm 1$. Conservation of angular momentum requires that absorption of a photon be accompanied by a corresponding change of the projection of the angular momentum of an atom. In the case of fine or hyperfine interaction, the orbital angular momentum L can be replaced by the total angular momentum J of the electron or F of the atom, respectively. The same selection rules thus apply for M_J or M_F.

For the selection rules for L, consider the commutation relation

$$\left[L^2, [L^2, \vec{r}]\right] = 2\hbar^2(\vec{r}L^2 + L^2\vec{r}), \qquad (4.14)$$

which can be obtained from the usual algebra for commutators [9]. (Equation 4.14 explicitly depends on the fact that $\vec{L} \equiv \vec{r} \times \vec{p}$ is the orbital angular momentum of the atom, and this relation cannot be generalized for either J or F.) Calculating the matrix element for both sides of Eq. 4.14 results in

$$\langle \alpha' L' M'| \left[L^2, [L^2, \vec{r}]\right] |\alpha L M\rangle \qquad (4.15)$$
$$= 2\hbar^4[L(L + 1) + L'(L' + 1)]\langle \alpha' L' M'|\vec{r}|\alpha L M\rangle$$
$$= \hbar^4[L'(L' + 1) - L(L + 1)]^2 \langle \alpha' L' M'|\vec{r}|\alpha L M\rangle.$$

Thus the coupling between two states is zero for any polarization, unless the two factors in front of the matrix elements in Eq. 4.15 are equal. Rearrangement of this requirement leads to [9]

$$[(L' + L + 1)^2 - 1][(L' - L)^2 - 1] = 0. \tag{4.16}$$

The first term can only be zero if $L = -L'$, e.g., $L = L' = 0$, but this is prohibited since L' is the vector sum of L and $L_{ph} = 1$ for the photon, and thus cannot be zero. The second term is zero only if $\Delta L = \pm 1$, so this is the selection rule for L. Again, this selection rule reflects the conservation of angular momentum for absorption of one photon.

Also for $\Delta L = 0$ the final state angular momentum L' can be the vector sum of L and L_{ph}. But the parity of the state for a one-electron system is given by $(-1)^L$ and $\vec{r}$ is antisymmetric, so symmetry demands that the matrix element be zero between states where L and L' are both either odd or even.

The selection rules for J and F are $\Delta J = 0, \pm 1$ and $\Delta F = 0, \pm 1$. In contrast with the case for ΔL, $\Delta J = 0$ is allowed since L and S couple to J, so $\Delta J = 0$ does not imply $\Delta L = 0$. Only for $J = J' = 0$ is $\Delta L = 0$ a necessary consequence, and therefore transitions with $J = 0 \rightarrow J' = 0$ are forbidden. The same rule applies to F, namely, $F = 0 \rightarrow F' = 0$ is also forbidden.

In laser cooling, selection rules play a very important role. In order to slow atoms from their thermal velocity down to zero velocity, a large number of photons have to be scattered. Therefore, the coupling strength between the two levels involved in the laser cooling has to be sufficiently high. Furthermore, since the atoms have to undergo a very large number of cycles, the decay from the excited to the ground state must be to only the sublevel coupled by the light. This restricts the number of possible cooling transitions. The selection rules can be used to determine whether two states are coupled by the laser light without extensive calculations.

For the alkali-metal atoms, the hfs complicates the level structure and most of the optically accessible transitions do not meet these criteria. Since the same selection rules for excitation are valid for spontaneous emission, the $\Delta F = 0, \pm 1$ selection rule allows the decay of one excited state to many ground states, and some of these may not be coupled by the laser to an excited state. This is because the laser's spectral width is generally much smaller than the ground-state hfs splitting. However, for the states with $J = L + \frac{1}{2}$, the decay from the highest F_e-state can only occur to the highest F_g-state, since the other ground state has $F_g = F_e - 2$ (see Fig. 4.1). Therefore these two states form a closed two-level system. A similar system exists between the lowest F_e and F_g states. However, since the hfs splitting between the two lowest excited states is usually very small, exciting the lowest F_e state can often also partially excite the next F_e state, which can then decay to the other hyperfine ground-state sublevels. Laser cooling in the alkalis is therefore usually carried out on the highest F_g and F_e states.

These complications do not appear in the metastable noble gas atoms where the splitting between the states is caused by the spin-orbit interaction instead of the hyperfine interaction. For Ne* only the $^3P_{0,2}$ states are truly metastable. The only closed system can be formed by the $^3P_2 \rightarrow \,^3D_3$ transition, which is the one most often used for laser cooling. Similar transitions exist for the other metastable noble gases. For He* the situation is very simple since there is only the $^3S_1$-state (see Sec. 4.2).

4.5 Optical Transitions in Multilevel Atoms

4.5.1 Introduction

The optical transitions considered in Chapter 1 were restricted to the particularly simple case of a two-level atom, and these transitions can be described by a single Rabi frequency. Real atoms have more than two levels that can be coupled by the optical field, and furthermore, the relative strengths of their multiple transitions depend on the orientation of the atomic dipole moment with respect to the polarization of the light. The single Rabi frequency of Chapter 1 that describes the coupling is given by $\hbar\Omega = -\mu_{eg}E_0$ (see Eq. 1.10), where

$$\mu_{eg} = e\langle e|\hat{\varepsilon} \cdot \vec{r}|g\rangle \tag{4.17}$$

and $\hat{\varepsilon}$ represents the polarization of the light. The value of the dipole moment of Eq. 4.17 depends on the wavefunctions of the ground and excited states, and is generally complicated to calculate.

It is often convenient to introduce the spherical unit vectors [29] given by

$$\hat{u}_{-1} = \left(\hat{x} - i\hat{y}\right)/\sqrt{2}, \qquad \hat{u}_0 = \hat{z}, \qquad \hat{u}_{+1} = -\left(\hat{x} + i\hat{y}\right)/\sqrt{2} \tag{4.18}$$

and to expand the polarization vector $\hat{\varepsilon}$ in terms of these vectors. Note that $\hat{u}_{\pm 1}$ corresponds to circularly polarized light, whereas $\hat{u}_0$ corresponds to linearly polarized light. For simplicity, only cases where the polarization of the light field is given by just one of these vectors will be considered, and this will be indicated by the symbol q ($q = 0, \pm 1$ is the subscript of $\hat{u}_q$). In this notation the components of the dipole moment can be written as

$$\hat{\varepsilon} \cdot \vec{r} = \hat{u}_q \cdot \vec{r} = \sqrt{\frac{4\pi}{3}} r Y_{1q}(\theta, \phi) \tag{4.19}$$

where the Y_{1q}'s represent the simplest of the spherical harmonic functions.

The matrix element of Eq. 4.17 can be broken up into two parts, one depending on all the various quantum numbers of the coupled states and the other completely independent of M, the projection of $\vec{L}$ on the quantization axis. This separation is embodied in the well-known Wigner-Eckart theorem discussed in many quantum mechanics texts [29]. Here, the treatment will be somewhat different, since this section treats the simplest case, namely, that fine and hyperfine structure are absent. The more general case will be treated in Sec. 4.5.4. Thus the hydrogenic wavefunctions for the ground and excited state can be used:

$$|g\rangle = |nlm\rangle = R_{nl}(r)Y_{lm}(\theta, \phi) \tag{4.20a}$$

and

$$|e\rangle = |n'l'm'\rangle = R_{n'l'}(r)Y_{l'm'}(\theta, \phi). \tag{4.20b}$$

Substitution of Eqs. 4.19 and 4.20 into Eq. 4.17 leads to

$$\mu_{eg} = e \langle n'l'm'|\hat{\varepsilon}\cdot\vec{r}|nlm\rangle \tag{4.21}$$

$$= e \langle n'l'||r||nl\rangle \langle l'm'|\sqrt{\frac{4\pi}{3}}Y_{1q}|lm\rangle \equiv e \, \mathcal{R}_{n'l',nl} \, \mathcal{A}_{l'm',lm}.$$

The following sections first treat the radial or physical part $\mathcal{R}_{n'l',nl}$, also known as the reduced or double-bar matrix element, and then the angular or geometric part $\mathcal{A}_{l'm',lm}$.

4.5.2 Radial Part

The radial part of the matrix element is generally less important in laser cooling because experiments typically use an optical transition joining a set of states that all share the same ground- and excited-state radial wavefunctions. Therefore it becomes an overall multiplicative factor that determines only the magnitude of the coupling (e.g., the overall Rabi frequency). It is given by

$$\mathcal{R}_{n'\ell',n\ell} = \langle R_{n'\ell'}(r)|r|R_{n\ell}(r)\rangle = \int_0^\infty r^2 dr \, R_{n'\ell'}(r) \, r \, R_{n\ell}(r), \tag{4.22}$$

with R_{nl} the radial wavefunction of the state. Here the $r^2 dr$ term in the integral originates from the radial part of d^3r.

The radial part can be evaluated if the eigenfunctions are known. For all atoms except hydrogen, the eigenfunctions can only be calculated approximately and therefore only approximate values for the radial part can be found. However, for the hydrogen atom the eigenfunctions of the bound states are known and the radial matrix elements can be calculated exactly [30]. For instance, for the first optical allowed transition in H, the $1s \rightarrow 2p$ transition, the radial wavefunctions involved are $R_{1s}(r) = 2\exp(-r/a_0)/a_0^{3/2}$ and $R_{2p}(r) = (r/a_0)\exp(-r/2a_0)/\sqrt{3}(2a_0)^{3/2}$. Thus the integral becomes

$$\mathcal{R}_{2p,1s} = \int_0^\infty R_{2p}(r) \, r \, R_{1s}(r) \, r^2 dr = 2^7\sqrt{6}a_0/3^5 \approx 1.290 \, a_0. \tag{4.23}$$

For other transitions in hydrogen similar integrals can be evaluated.

The hydrogenic wavefunctions $R_{n\ell}(r)$ are given by [30]

$$R_{n\ell}(r) = N_{n\ell}\rho^\ell \exp(-\rho/2)L_{n-\ell-1}^{2\ell+1}(\rho), \tag{4.24}$$

where $\rho = 2r/na_0$, $L_n^m(r)$ are the Laguerre polynomials, and $N_{n\ell}$ is a normalization constant. These Laguerre polynomials can be expanded in a power series:

$$L_n^m = \sum_{k=0}^n c_k r^k, \tag{4.25}$$

$n\ell$	$2(\ell+1)$	$3(\ell+1)$	$4(\ell+1)$	$5(\ell+1)$
1s	1.2902	0.5166	0.3045	0.2087
2s	−5.1961	3.0648	1.2822	0.7739
2p	-	4.7479	1.7097	0.9750
3s	0.9384	−12.7279	5.4693	2.25957
3p	-	−10.0623	7.5654	2.9683
3d	-	-	10.2303	3.3186
4s	0.3823	2.4435	−23.2379	8.5178
4p	-	1.3022	−20.7846	11.0389
4d	-	-	−15.8745	14.0652
4f	-	-	-	17.7206
5s	0.2280	0.9696	4.6002	−36.7423
5p	-	0.4827	3.0453	−34.3693
5d	-	-	1.6613	−30.0000
5f	-	-	-	−22.5000

TABLE 4.1. Radial matrix elements $\mathcal{R}_{n\ell,n'\ell+1}$ in units of a_0 for hydrogen for a transition $n\ell \to n'(\ell+1)$. Note that $\mathcal{R}_{n\ell,n'\ell'}$ is symmetric with respect to interchange of n and ℓ, i.e., $\mathcal{R}_{n\ell,n'\ell'} = \mathcal{R}_{n'\ell',n\ell}$.

where c_k are the coefficients, for which a simple recurrence relation exists [30,31]. Substitution of Eq. 4.24 into Eq. 4.22 and integrating over r with the help of standard integrals, the matrix element for any transition can be found. The results for $n \leq 5$ are given in Table 4.1. Note that the radial matrix elements increase with increasing n, since the radius of the electron orbit increases with n.

For all other atoms, the situation is more complicated. In the case of alkali-metal atoms with only one active electron, the matrix elements can be quite accurately expressed in terms of the effective principal quantum number $n_\ell^* = n - \delta_\ell$ of the valence electron, where δ_ℓ is called the quantum defect and depends on the orbital quantum number ℓ [6]. The same analysis as in the hydrogen case can be applied for the alkali-metal atoms; however, in the summation n is now replaced by n^* [32]. Table 4.2 shows the matrix elements for the first optically allowed transition for the alkali-metal atoms. As the table shows, the agreement between the calculated elements and the values derived from experiments is reasonable.

4.5.3 Angular Part of the Dipole Matrix Element

The angular part $\mathcal{A}_{l'm',lm}$ of the dipole moment for atoms with $S = 0 = I$ is defined by Eq. 4.21:

$$\mathcal{A}_{l'm',lm} = \sqrt{\frac{4\pi}{3}} \langle Y_{l'm'} | Y_{1q} | Y_{lm} \rangle \tag{4.26}$$

El.	Transition			Theory		Experiment
		n_s^*	n_p^*	μ (a.u.)	Γ (MHz)	Γ (MHz)
H	1s→ 2p	1.000	2.000	0.745	99.52	99.47
He*	2s→ 2p	1.689	1.938	2.540	1.64	1.62
Li	2s→ 2p	1.589	1.959	2.352	5.93	5.92
Na	3s→ 3p	1.627	2.117	2.445	9.43	10.01
K	4s→ 4p	1.770	2.235	2.842	5.78	6.09
Rb	5s→ 5p	1.805	2.293	2.917	5.78	5.56
Cs	6s→ 6p	1.869	2.362	3.093	4.99	5.18

TABLE 4.2. Matrix element $\langle ns|r|np \rangle$ for the first optically allowed transition in the al-kali-metal atoms. The theoretical value is calculated using the procedure of Bates and Damgaard [32], whereas the experimental value is derived from the lifetimes of the np-states (see Table C.1 in Appendix C).

$$= \sqrt{\frac{4\pi}{3}} \int \sin\theta d\theta d\phi \; Y_{l'm'}(\theta, \phi) \; Y_{1q}(\theta, \phi) \; Y_{lm}(\theta, \phi),$$

where the integration limits are over 4π. The integral can be expressed in terms of the $3j$-symbols as

$$A_{l'm',lm} = (-1)^{\ell'-m'} \sqrt{\max(\ell, \ell')} \begin{pmatrix} \ell' & 1 & \ell \\ -m' & q & m \end{pmatrix}. \tag{4.27}$$

The $3j$-symbols are related to the Clebsch-Gordan coefficients and are tabulated in [33] (see Eq. 4.30). The symmetry of the $3j$-symbols dictates that they are only nonzero when the sum of the entries in the bottom row is zero, which means $m + q = m'$. Thus circularly polarized light only couples states that differ in m by ±1, whereas linearly polarized light only couples states that have equal m's. This result is thus identical to the result obtained in Sec. 4.4. Table 4.3 shows tabulated the values of $A_{l'm',lm}$ for optical transitions.

4.5.4   Fine and Hyperfine Interactions

In case of fine and hyperfine interaction the situation changes considerably. For the fine structure, the energy levels are split by the spin-orbit interaction and $\vec{L}$ is no longer a good quantum number. Here $\vec{\ell}$ is replaced with $\vec{L}$ to be more general. The states are now specified by $\vec{J}$, the vector sum of $\vec{L}$ and $\vec{S}$. However, the optical electric field still couples only to the orbital angular momentum $\vec{L} = \vec{r} \times \vec{p}$ of the states. In this situation the Wigner-Eckart theorem could also be applied to calculate the transition strength [29], but again this section will follow a different route that provides more insight in the problem. Although the formulas below may appear rather complicated, the principle is simple.

ℓ'	$q = \pm 1$	$q = 0$
$\ell + 1$	$\left[\dfrac{(\ell \pm m')(\ell \pm m' + 1)}{2(2\ell + 1)(2\ell + 3)} \right]^{1/2}$	$\left[\dfrac{(\ell - m' + 1)(\ell + m' + 1)}{(2\ell + 1)(2\ell + 3)} \right]^{1/2}$
$\ell - 1$	$-\left[\dfrac{(\ell \mp m')(\ell \mp m' + 1)}{2(2\ell - 1)(2\ell + 1)} \right]^{1/2}$	$\left[\dfrac{(\ell - m')(\ell + m')}{(2\ell - 1)(2\ell + 1)} \right]^{1/2}$

TABLE 4.3. The angular part $\mathcal{A}$ for optical transitions $(\ell, m) \rightarrow (\ell', m')$ with the polarization of the light indicated by q, with $q = 0$ for linear and $q = \pm 1$ for right- and left-circular polarized light. Because of the selection rules, $\ell' = \ell \pm 1$ and $m' = m + q$.

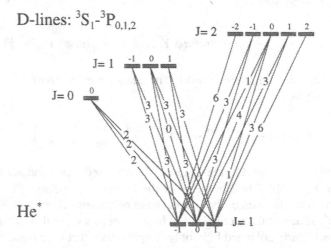

FIGURE 4.7. Transition strength for the D-lines in He*. The strength is normalized to the weakest allowed transition.

The atomic eigenstates are denoted by $|\alpha J M_J\rangle$ in the J-basis, and M_J explicitly indicates for which angular momentum the magnetic quantum number M is the projection. In most cases, this is obvious from the notation, but in this section it is not. The dipole transition matrix element is therefore given by

$$\mu_{eg} = e \, \langle \alpha' J' M'_J | \hat{\varepsilon} \cdot \vec{r} | \alpha J M_J \rangle. \qquad (4.28)$$

Since the optical electric field only couples the $\vec{L}$ component of these $\vec{J}$ states, these eigenfunctions must be first expanded in terms of the L and S wavefunctions:

$$|\alpha J M_J\rangle = \sum_i C_i \, |\alpha L M_L\rangle |S M_S\rangle, \qquad (4.29)$$

where i represents an appropriate set of angular momentum quantum numbers. The C_i's are Clebsch-Gordan coefficients that can also be expressed in terms of the more symmetrical $3j$ symbols as

$$C_i = \langle LM_L; SM_S|JM_J\rangle = (-1)^{-L+S-M_J}\sqrt{2J+1}\begin{pmatrix} L & S & J \\ M_L & M_S & -M_J \end{pmatrix}.$$
(4.30)

The fact that Eq. 4.27 for the integral of the product of three spherical harmonics and Eq. 4.30 both contain the $3j$ symbols is a result of the important connection between the Y_{lm}'s and atomic angular momenta.

Substitution of Eq. 4.29 in Eq. 4.28 twice leads to a double summation, which contains matrix elements in the (L, S) basis of the form

$$\langle \alpha'L'M'_L|\langle S'M'_S|r|\alpha LM_L\rangle|SM_S\rangle = \langle \alpha'L'M'_L|r|\alpha LM_L\rangle\delta_{SS'}\delta_{M_S M'_S}.$$
(4.31)

The first term on the right-hand side is the matrix element that has been evaluated before (see Eq. 4.21). The δ-functions reflect the notion that the light couples the orbital angular momenta of the states, and not the spin. The spin and its projection are not changed by the transition. Substitution of Eq. 4.31 into Eq. 4.28, expansion of the matrix elements in the L-basis, and recoupling of all the Clebsch-Gordan coefficients leads to

$$\mu_{eg} = e(-1)^{L'+S-M'_J}\sqrt{(2J+1)(2J'+1)}$$
(4.32)

$$\times \begin{Bmatrix} L' & J' & S \\ J & L & 1 \end{Bmatrix}\begin{pmatrix} J & 1 & J' \\ M_J & q & -M'_J \end{pmatrix}\langle \alpha'L'||r||\alpha L\rangle.$$

The array of quantum numbers in the curly braces is not a $3j$ symbol, but is called a $6j$ symbol. It summarizes the recoupling of six angular momenta. Values for the $6j$ symbols are also tabulated in Ref. 33.

Note that the radial part of the dipole moment has remained unchanged, and so the results of the previous section can still be used. For metastable helium the transition strengths for the triplet system are shown in Fig. 4.7. Triplet metastable helium only has one "ground" state, so that decay out of the excited states is always to this state.

In case of hyperfine interactions the situation becomes even more complicated. However, the procedure is the same. First the eigenfunctions in the F-basis are expanded in the (J, I)-basis, where I is the nuclear spin, and a $6j$ symbol involving I, J, and F appears. Then the eigenfunctions of the J-basis are further reduced into the (L, S)-basis. Since the procedure is similar to the procedure for the fine structure interaction, only the result is shown:

$$\mu_{eg} = e(-1)^{1+L'+S+J+J'+I-M'_F}\langle \alpha'L'||r||\alpha L\rangle$$
(4.33)

$$\times \sqrt{(2J+1)(2J'+1)(2F+1)(2F'+1)}$$

$$\times \begin{Bmatrix} L' & J' & S \\ J & L & 1 \end{Bmatrix}\begin{Bmatrix} J' & F' & I \\ F & J & 1 \end{Bmatrix}\begin{pmatrix} F & 1 & F' \\ M_F & q & -M'_F \end{pmatrix}.$$

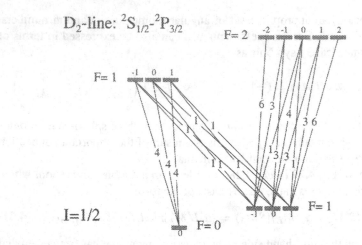

FIGURE 4.8. Transition strength for the first optical transition in an alkali system with a nuclear spin of $1/2$, for example, H. The strength is normalized to the weakest transition. For the D_1 lines see Appendix D.

The hyperfine interaction is important for the alkalis. For a system with nuclear spin $1/2$, such as H, the result is given in Fig. 4.8. Since S can be parallel or anti-parallel to L, $J' = 1/2, 3/2$ and the fine-structure interaction is usually large compared to the hyperfine interaction. Results for transitions important for the alkalis are given in Appendix D.

5
General Properties Concerning Laser Cooling

This chapter presents some of the general ideas regarding laser cooling. One of the characteristics of optical control of atomic motion is that the speed of atoms can be reduced by a considerable amount. Since the spread of velocities of a sample of atoms is directly related to its temperature, the field has been dubbed laser cooling, and this name has persisted throughout the years. Laser cooling has much in common with the field of optics. In laser cooling, light is used to manipulate atoms, whereas in optics matter is used to manipulate light. The more proper identification for the field would therefore be "atom optics" or "optical control of atomic motion". The similarities between atom optics and electromagnetic optics will be pointed out.

These experiments almost always involve the use of nearly resonant light, which can populate the atomic excited state and hence result in spontaneous emission. As discussed in Chapter 2, such events produce unpredictable changes in atomic momenta. Hence, the discussion here begins with a "random walk" model, which provides the background on a microscopic scale for how the rapid exchange of momenta between the light field and the atoms influences their velocity distribution. This leads to the Fokker-Planck equation, which can be used for a more formal treatment of the laser cooling process. Solutions of the Fokker-Planck equation in a limiting case can ultimately be used to relate the velocity distribution of the atoms with their temperature.

5.1 Temperature and Thermodynamics in Laser Cooling

The idea of "temperature" in laser cooling requires some careful discussion and disclaimers. In thermodynamics, temperature is carefully defined as a parameter of the state of a closed system in thermal equilibrium with its surroundings. This, of course, requires that there be thermal contact, *i.e.* heat exchange, with the environment. In laser cooling this is clearly not the case because a sample of atoms is always absorbing and scattering light, making major changes to its environment. Furthermore, there is essentially no heat exchange (the light cannot be considered as heat even though it is indeed a form of energy). Thus the system may very well be in a steady-state situation, but certainly not in thermal equilibrium, so that the assignment of a thermodynamic "temperature" is completely inappropriate.

Nevertheless, it is convenient to use the label of temperature to describe an atomic sample whose average kinetic energy $\langle E_k \rangle$ in one dimension has been reduced by the laser light, and this is written simply as

$$\frac{1}{2} k_B T = \langle E_k \rangle, \tag{5.1}$$

where k_B is Boltzmann's constant. It must be remembered that this temperature assignment is absolutely inadequate for atomic samples that do not have a well-defined velocity distribution, whether or not they are in thermal equilibrium: there are infinitely many velocity distributions that have the same value of $\langle E_k \rangle$ but are so different from one another that characterizing them by the same "temperature" is a severe error.

With these ideas in mind, it is useful to define a few rather special values of temperatures associated with laser cooling. Each of these quantities appear elsewhere in this book in connection with the special domain of their applications. Their place on the energy scale is shown in Fig. 5.1.

The highest of these temperatures corresponds to the energy associated with atoms whose speed and concomitant Doppler shift puts them just at the boundary of absorption of light. This velocity is $v_c \equiv \gamma/k \sim 1$ m/s, and the corresponding temperature is

$$k_B T_c \equiv \frac{M \gamma^2}{k^2}, \tag{5.2}$$

and is typically several mK.

The next characteristic temperature corresponds to the energy associated with the natural width of atomic transitions, and is called the Doppler temperature. It is given by

$$k_B T_D \equiv \frac{\hbar \gamma}{2}. \tag{5.3}$$

Because it corresponds to the limit of certain laser cooling processes, it is often called the Doppler limit, and is typically several hundred μK (see Sec. 7.2). Associated with this temperature is the one-dimensional velocity $v_D = \sqrt{k_B T_D / M} \sim$ 30 cm/s.

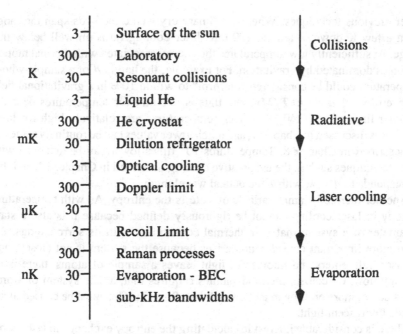

	3 —	Surface of the sun	Collisions
	300 —	Laboratory	
K	30 —	Resonant collisions	↓
	3 —	Liquid He	
	300 —	He cryostat	Radiative
mK	30 —	Dilution refrigerator	↓
	3 —	Optical cooling	
	300 —	Doppler limit	
μK	30 —	-	Laser cooling
	3 —	Recoil Limit	↓
	300 —	Raman processes	
nK	30 —	Evaporation - BEC	Evaporation
	3 —	sub-kHz bandwidths	↓

FIGURE 5.1. Temperature scale.

The last of these three characteristic temperatures corresponds to the energy associated with a single photon recoil. In the absorption or emission process of a single photon, the atoms obtain a recoil velocity $v_r = \hbar k/M$. The corresponding energy change can be related to a temperature, the recoil limit, defined as

$$k_B T_r \equiv \frac{\hbar^2 k^2}{M}, \tag{5.4}$$

and is generally regarded as the lower limit for optical cooling processes, although there are a few clever schemes that cool below it. It is typically a few μK, and corresponds to speeds of $v_r \sim 1$ cm/s.

These three temperatures are related to one another through a single parameter ε that is ubiquitous in describing laser cooling. It corresponds to the ratio of the recoil frequency $\omega_r \equiv \hbar k^2/2M$ to the natural width γ, and as such embodies most of the important information that characterize laser cooling on a particular atomic transition. Typically $\varepsilon \sim 10^{-3} - 10^{-2}$, and is given by

$$\varepsilon \equiv \omega_r/\gamma = \frac{\hbar k^2}{2M\gamma}. \tag{5.5}$$

From this it is clear that

$$T_r = 4\varepsilon T_D = 4\varepsilon^2 T_c. \tag{5.6}$$

It is instructive to put these temperatures on a scale to compare with others as shown in Fig. 5.1. Clearly laser cooling is in a temperature domain far below any other previous techniques. Whereas ordinary cryogenic methods span the range from a few K down to around 100 mK, laser cooling turns on well below this range. At sufficiently low temperatures the energy associated with thermal motion becomes dominated by gravitation. For example, the height d of a sample whose temperature could be considered "uniform" to within 10% in a gravitational field is given by 10 $d \simeq k_B T / Mg$, and thus is $\simeq 1$ mm for temperatures near the Doppler limit of Na (240 μK). Such temperatures are relatively high for laser cooling as discussed in Chapter 7, and much lower values can be routinely achieved as described in Chapter 8. Temperatures 10^4 times lower can be achieved with other techniques such as the evaporative cooling described in Chapter 12, and this corresponds to $d < \lambda$, with λ the optical wavelength.

Another thermodynamic variable of state is the entropy. As with temperature, entropy in laser cooling cannot be rigorously defined because it is also a state parameter of a system that is in thermal equilibrium with its surroundings. An even more important consideration arises because the system is not closed, and as laser light enters and fluorescent light leaves a sample of atoms, there is an entropy flow. Of course, thermodynamics requires that, as the system of atoms cools down, more entropy must flow out than flows in. This must be carried away by the fluorescent light.

There is considerable interest in calculating the entropy exchange in laser cooling, but there are serious difficulties in doing so. Part of the problem arises because it's much more difficult to achieve a working definition for entropy in a non-equilibrium situation than the "average kinetic energy" definition used for temperature earlier in this section. In fact, there is little agreement among the experts in statistical mechanics about a usable definition: every one of the several choices presents some difficulty. Another part of the problem arises because the energy change of the light field is a small fraction of its total energy because the total number of photons is conserved, and the frequency shift of the fluoresced photons is small. By contrast, the relative entropy change of the light is huge because the light in the incoming, well-defined laser beams of very low entropy is converted into disorganized fluorescence having very high entropy. Unlike the energy exchange between the atoms and the light field, the entropy exchange cannot be treated as a small perturbation.

Probably the simplest way to begin is to count the number of states accessible to the system, and use the von Neumann "maximum entropy" approach. For an incoming single-mode laser beam, this can only be applied in a quantum mechanical description of the light field (coherent state) because a classical description would lead to a contradiction. Needless to say, the emitted fluorescence can occupy states of various frequencies, polarizations, and directions, so the outgoing entropy flow is huge. In general, the outflow calculated this way is orders of magnitude larger than the entropy lost by cooling the atomic sample, so laser cooling is really a quite poor refrigerator [34].

In spite of these difficulties, it is both interesting and challenging to think about the entropy flow in the various laser cooling schemes that are described in Part II of this book. In every case, spontaneous emission is the necessary dissipative process that provides for the increase of entropy of the light field. In several examples where optical forces based on stimulated emission can lower the average kinetic energy, it must always be remembered that this is *not* the only criterion for cooling. The width of the velocity distribution must be narrowed for cooling, and any scheme purported to accomplish this without spontaneous emission violates thermodynamics.

5.2   Kinetic Theory and the Maxwell-Boltzmann Distribution

The modern era of kinetic theory began with Bernoulli, Clausius, Maxwell, and Boltzmann, who showed that the experimental "laws" of Boyle, Charles, and Gay-Lussac could be derived from considerations of molecular motion. The underlying assumptions are that a gas is composed of a large number of small particles having no intrinsic properties other than their mass. They undergo collisions that redistribute their kinetic energies and momenta without dissipation or bias toward any particular energy or direction. They occupy negligible volume, although there are simple approximate corrections for small volume, and their density is sufficiently low that all collisions are binary because three-body collisions are too improbable.

From these assumptions it is straightforward to calculate the equation of state of a confined gas, and the perfect gas law follows directly from it as $PV = Nk_BT$, where P, V, and T are the pressure, volume, and temperature of the gas of N particles, and k_B is Boltzmann's constant. It is readily shown that all sets of particles in the sample have the same average kinetic energy, even if there are different kinds of particles (*i.e.*, a mixture of different masses). This is not to say that they all have the same energy, but only that the energy distribution of those particles of one mass is the same as those of another mass if they are mixed, so that collisional redistribution occurs.

The system can approach true thermal equilibrium at a defined rate, which is generally quite fast on the human scale, and the laws of thermodynamics are readily applied. At equilibrium the velocity, or momentum, distribution is the most likely one of the infinitely many possibilities, and this is the Maxwell-Boltmann (MB) distribution derived in many standard texts. The MB distribution is characterized by a Gaussian shape and is given by

$$f(v) = \frac{1}{\sqrt{2\pi}\tilde{v}} \exp\left(-\frac{v^2}{2\tilde{v}^2}\right), \tag{5.7}$$

where $\tilde{v} \equiv \sqrt{k_BT/M}$. The distribution only depends on the speed of the atoms, and therefore it is spherically symmetric. Although the MB distribution of Eq. 5.7 is remarkably simple, its use in atomic physics can easily lead to confusion if no proper distinction is made between different cases.

Theoretical descriptions of laser cooling are often done in only 1D. In order to test the outcome of such models, experiments are carried out in 1D as well. In

Distribution $f(v)$	Range	v_{mp}	v_{ave}	v_{rms}
Gas (1D) $\dfrac{1}{\sqrt{2\pi}\,\tilde{v}}\exp\left(-\dfrac{v_x^2}{2\tilde{v}^2}\right)$	$(-\infty,\infty)$	0	0	$\tilde{v}$
Gas (3D) $\sqrt{\dfrac{2}{\pi}}\dfrac{v^2}{\tilde{v}^3}\exp\left(-\dfrac{v^2}{2\tilde{v}^2}\right)$	$(0,\infty)$	$\sqrt{2}\,\tilde{v}$	$\sqrt{\dfrac{8}{\pi}}\,\tilde{v}$	$\sqrt{3}\,\tilde{v}$
Beam $\dfrac{v_z^3}{2\tilde{v}^4}\exp\left(-\dfrac{v_z^2}{2\tilde{v}^2}\right)$	$(0,\infty)$	$\sqrt{3}\,\tilde{v}$	$\sqrt{\dfrac{9\pi}{8}}\,\tilde{v}$	$2\,\tilde{v}$

TABLE 5.1. Quantities appropriate for a gas of particles (1D and 3D) and a thermal beam. Note that the rms velocity is not the same as the average velocity for the distribution. Here v_{mp} is the most probable velocity and v_{ave} is the average velocity. For a 3D gas the velocity $\tilde{v} \equiv v_{rms}$ is defined as the velocity characteristic for the temperature.

both cases the results can be related to a "temperature", although strictly speaking temperature is not defined when only one velocity component is considered. Furthermore, laser cooling experiments are often done in atomic beams, and it is important to point out that the various averages and distributions of particles in a confined gas are not the same as those in a beam formed by letting that confined gas expand into a vacuum. The detailed nature of the expansion, ranging from thermal to highly supersonic, can result in a wide variety of distributions. The simplest case is thermal expansion, and occurs when the size of the aperture between the source volume (for example, an oven for metals) and the vacuum system is small compared with the mean free path of the particles in the oven.

The first case to consider is the distribution of velocities in 1D, for instance, the x-direction. Using $v^2 = v_x^2 + v_y^2 + v_z^2$, $d^3v = dv_x dv_y dv_z$, and integration of Eq. 5.7 over v_y and v_z yields the result shown in the first row of Table 5.1. The distribution is Gaussian with a maximum at $v_x = 0$, and has a width $\tilde{v}$. The second case is the distribution of speeds in a gas in 3D, where d^3v can be replaced by $v^2 \sin\theta\, dv d\theta d\phi$. Since the MB distribution is spherically symmetric, the integration over the angles can easily be performed and the result is shown in the second row of Table 5.1. This distribution is only defined for positive values of the speed v. Finally, consider the case of a thermal beam in the z-direction. Since the flux of atoms is proportional to v_z, the distribution is peaked toward higher v compared to that of a thermal gas. The result is shown in the last row of Table 5.1 (see Ref. 11 for a detailed derivation, where $\alpha \equiv \sqrt{2}\tilde{v}$).

Various moments of the velocity distributions are readily calculated from

$$\langle v^n \rangle = \int v^n f(v)\, d^3v. \tag{5.8}$$

The lowest moments of the distribution are the average velocity v_{ave} ($n = 1$) and the root mean square (rms) velocity $v_{rms} = \sqrt{\langle v^2 \rangle}$ ($n = 2$). The various moments can easily be found by solving the definite integral by using standard integrals (see e.g., Eqs. 2 and 3 in Sec. 3.461 of Ref. 35). The results for $n = 1, 2$ are shown in Table 5.1 together with the most probable velocity v_{mp}, for which the distribution has its maximum. Since the primary characteristic denoting the temperature of a 3D gas is the mean kinetic energy, the appropriate choice from Table 5.1 is the rms velocity $\bar{v}$ given by

$$\bar{v} = \sqrt{3}\, \tilde{v} \equiv \sqrt{\frac{3k_B T}{M}}. \tag{5.9}$$

For both the 1D and 3D cases the average kinetic energy $M\bar{v}^2/2$ is given by $k_B T/2$ times the number of degrees of freedom for the system, as required by thermodynamics.

5.3 Random Walks

In laser cooling and related aspects of optical control of atomic motion, the forces arise because of the exchange of momentum between the atoms and the laser field. Since the energy and momentum exchange is necessarily in discrete quanta rather than continuous, the interaction is characterized by finite momentum "kicks". This is often described in terms of "steps" in a fictitious space whose axes are momentum rather than position. These steps in momentum space are of size $\hbar k$ and thus are generally small compared to the magnitude of the atomic momenta at thermal velocities. This is easily seen by comparing $\hbar k$ with $M\bar{v}$,

$$\frac{\hbar k}{M\bar{v}} = \sqrt{\frac{T_r}{T}} \ll 1. \tag{5.10}$$

Thus the scattering of a single photon has a negligibly small effect on the motion of thermal atoms, but repeated cycles of absorption and emission can cause a large change of the atomic momenta and velocities.

Before delving into the details of these processes, it is helpful to discuss a simple model to provide some background. Consider an atom that is confined to motion in 1D and the effect of a 1D light field such as a traveling or standing plane wave. The atomic motion would be related to a "random walk" in a 1D momentum space whose step sizes are equal to the momentum of a photon, $\hbar k$. The randomness arises from spontaneous emission from the excited state, and uncertainty of the absorption direction in the case of a standing wave. At a certain instant t, an atom with momentum p has a probability $\epsilon_+(p)$ to make a step $\hbar k$ and a probability $\epsilon_-(p)$ to make a step $-\hbar k$. The dependence of $\epsilon_+(p)$ and $\epsilon_-(p)$ on momentum p can be understood by remembering that the force on an atom may depend on its velocity.

Figure 5.2 shows a simulation of this process. This simulation uses a pseudo-random number generator that produces numbers α uniformly distributed between

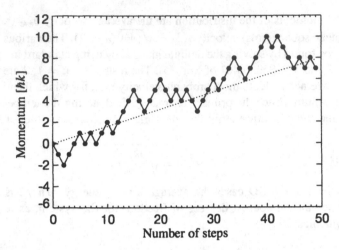

FIGURE 5.2. Random walk process for α_0=0.42, where a step in the positive direction is taken if the random number is larger than α_0 and in the opposite direction otherwise. This leads to a drift in the positive direction indicated by the dotted line. The fluctuations around this line is caused by the randomness of the process.

0 and 1. When α is below a certain value α_0, the step is negative, and otherwise it is positive. From Fig. 5.2 it is apparent that there is a drift in the positive direction indicated by a dotted line, because in this case α_0 has been chosen to be smaller than 0.5. In addition, the trajectory of the atom does not follow the dotted line exactly since there is randomness involved in this process, and at each instant the atom might undergo a large number of steps in one direction only (see for instance around 14 and 38 steps).

If this simulation were to be repeated under the same initial conditions for a large number of atoms, each trajectory would be different, but there would still be the same drift. Such random walks form the basis of many processes in physics, such as the Brownian motion of particles in a liquid, the current through an electrical circuit, and the electric field in a laser. Several techniques have been developed over the last century to model these kinds of processes, one of which, the Fokker-Planck equation, will be discussed in the next section.

The distribution of the momenta of the atoms is described by a function $W(p, t)$. As a result of the random walk process, this distribution is changed in time according to

$$W(p, t + \Delta t) - W(p, t) = -\left[\epsilon_+(p) + \epsilon_-(p)\right] W(p, t) \qquad (5.11)$$
$$+ \epsilon_+(p - \hbar k)W(p - \hbar k, t) + \epsilon_-(p + \hbar k)W(p + \hbar k, t).$$

The first term on the right-hand side is the probability $\epsilon_\pm$ to jump away from the momentum p in the $+$ or $-$ direction, multiplied by $W(p, t)$, and this product is the rate. The second and third terms give the rates of jumping toward momentum p. These last two terms can each be Taylor expanded as

$$\epsilon_{\pm}(p \mp \hbar k)W(p \mp \hbar k, t) = \epsilon_{\pm}(p)W(p, t) \tag{5.12}$$

$$\mp \hbar k \frac{\partial}{\partial p}[\epsilon_{\pm}(p)W(p, t)] + \frac{(\hbar k)^2}{2}\frac{\partial^2}{\partial p^2}[\epsilon_{\pm}(p)W(p, t)] + \mathcal{O}\left(\frac{\hbar k}{M\bar{v}}\right)^3,$$

where the expansion is truncated after the term of the order $(\hbar k/M\bar{v})^2$. When Eq. 5.12 is inserted into Eq. 5.11, the first term $\epsilon_{\pm}(p)W(p, t)$ is cancelled by its negative in Eq. 5.11. As long as Eq. 5.10 is satisfied, so that higher-order terms in Eq. 5.12 can be safely neglected, combining Eqs. 5.11 and 5.12 leads directly to

$$\frac{\partial W(p, t)}{\partial t} = -\frac{\partial[M_1 W(p, t)]}{\partial p} + \frac{1}{2}\frac{\partial^2[M_2 W(p, t)]}{\partial p^2} + \dots, \tag{5.13}$$

with

$$M_1 = [\epsilon_+(p) - \epsilon_-(p)]\frac{\hbar k}{\Delta t} \tag{5.14a}$$

and

$$M_2 = [\epsilon_+(p) + \epsilon_-(p)]\frac{(\hbar k)^2}{\Delta t}. \tag{5.14b}$$

The expressions on the right-hand side of Eq. 5.13 are called the drift and diffusion terms respectively.

For the case of Doppler cooling as discussed in Chapter 7, the absorption parameters $\epsilon_+(p)$ and $\epsilon_-(p)$ of the two laser beams coming from the right and the left depend on the Doppler shift kv. Since the linewidth for absorption is of the order of γ, the difference in absorption $[\epsilon_+(p) - \epsilon_-(p)]$ depends on the ratio kv/γ. Since the scattering rate of one beam is proportional to $s\gamma$ for low intensity, M_1 and M_2 can be written as

$$M_1 = -\left(\frac{kv}{\gamma}\right)s\gamma\hbar k = -\hbar k^2 sv \equiv -\beta v \tag{5.15a}$$

and

$$M_2 = s\gamma(\hbar k)^2 = 2D \tag{5.15b}$$

with β the damping coefficient and D the diffusion coefficient.

The stationary-state distribution $\overline{W}(p)$ is found by setting $\partial W(p, t)/\partial t = 0$ in Eq. 5.13, and using M_1 and M_2 as defined as in Eqs. 5.15. Then the simplified partial differential equation can be directly integrated twice to give

$$\overline{W}(p) \propto e^{-\beta p^2/2MD}. \tag{5.16}$$

This is a Maxwell-Boltzmann distribution with a characteristic temperature of $k_B T = D/\beta$. Using the values of β and D in Eqs. 5.15 gives $k_B T = \hbar\gamma/2$, which is the usual Doppler cooling limit given in Eq. 5.3, and is derived more carefully in Sec. 7.2.

5.4   The Fokker-Planck Equation and Cooling Limits

The random walk process discussed above is simply a particular case for the more general case of a force on an atom when the force can be written in two parts:

$$F(p, t) = F_c(p) + F_v(p, t), \qquad (5.17)$$

where $F_c(p)$ is a continuous force that damps the atomic motion and $F_v(p, t)$ is a random force that fluctuates in time and has an ensemble time average of zero:

$$\langle F_v(p, t) \rangle = 0. \qquad (5.18)$$

Now the ensemble averages for the moments M_i of such a force can be calculated as before for the 1D random walk. The first and second moments are given by

$$M_1 = \langle F_c(p) \rangle \qquad (5.19a)$$

and

$$M_2 = \langle F_v(p, t') F_v(p, t'') \rangle = 2D(p, t)\delta(t' - t''), \qquad (5.19b)$$

where the second equality with the δ-function holds if $F_v(p, t)$ is Markovian. Note that M_1 is determined by the continuous force, whereas M_2 is determined by the fluctuating force. When the correlation time of the force vanishes (Eq. 5.19b), it can be shown that all higher-order moments also vanish [36]. Using only these first two moments results in the Fokker-Planck equation [36]:

$$\frac{\partial W(p, t)}{\partial t} = -\frac{\partial [F(p, t)W(p, t)]}{\partial p} + \frac{\partial^2 [D(p, t)W(p, t)]}{\partial p^2} \qquad (5.20)$$

For the special case when both the force and the diffusion are independent of time, the formal stationary solution is

$$\overline{W}(p) = \frac{C}{D(p)} \exp\left(\int_0^p \frac{F(p')}{D(p')} dp' \right), \qquad (5.21)$$

where C is an integration constant. Once the force and diffusion are known, the stationary solution of the Fokker-Planck equation emerges easily. The fact that this discussion closely parallels that in Sec. 5.3 shows that the random walk picture in 1D is not only appealing, but is also a close approximation to atomic behavior.

In the simplest and most common case in laser cooling the force is proportional to the velocity (a true damping force as in Chapter 7) and the diffusion is independent of velocity:

$$F(v) = -\beta v \qquad (5.22a)$$

and

$$D(v) = D_0. \qquad (5.22b)$$

Then the stationary solution of Eq. 5.20 for $\overline{W}(v)$ is

$$\overline{W}(p) \propto e^{-\beta p^2 / 2MD_0}. \tag{5.23}$$

This distribution is the Maxwell-Boltzmann distribution with a characteristic temperature of $k_B T = D_0/\beta$. The fact that the conditions of Eqs. 5.22 for the force and diffusion are often approximately correct explains why the notion of temperature often appears as a description of a laser-cooled sample.

In kinetic theory, a Maxwell-Boltzmann distribution is the result of elastic collision of molecules of a gas with the walls of the container that holds the gas (see Sec. 5.2). After a while the gas is in thermal equilibrium with the walls and its temperature equals that of the walls. In laser cooling quite the opposite is the case, because the atoms have no interaction with the walls. Their only contact with the exterior is through the light field, and it has no definable temperature for the atoms to equilibrate with. Instead there is a competition between cooling and heating effects described by the damping force and the diffusion in momentum space. Thus the conditions of Eq. 5.22 are fulfilled, so the stationary distribution is also a Maxwell-Boltzmann distribution. It should be emphasized as in Sec. 5.2, however, that this stationary state is *not* an equilibrium state, and thus there is no thermodynamically definable temperature.

In general, laser cooling forces can act over only a limited range of velocity $\pm v_c \ll \bar{v}$ for atoms at room temperature. This happens because the frequency of the laser light in the rest frame of a moving atom is modified by the Doppler effect, so the absorption only takes place for a small range of velocities. This can be described by introducing a capture velocity v_c that characterizes the velocity range where an appreciable force can be generated by writing

$$F(v) = \frac{-\beta v}{1 + (v/v_c)^2}. \tag{5.24}$$

In this case the stationary distribution can also be calculated directly from Eq. 5.20:

$$\overline{W}(p) \propto \left(1 + \left(\frac{p}{Mv_c}\right)^2\right)^{(-M\beta v_c^2 / 2D_0)}. \tag{5.25}$$

This is a not a Maxwell-Boltzmann distribution. Figure 5.3 shows plots of the distribution function for a constant ratio of D_0/β but different values of v_c. The spread of the distribution becomes large when $v_c \ll \sqrt{D_0/M\beta}$, and it does not make sense to define $\langle E_k \rangle$ or a temperature. However, using v_{rms} as the velocity for which $\overline{W}(p)$ has decreased by a factor $1/\sqrt{e}$ with respect to its maximum, and taking $Mv_{\text{rms}}^2 \equiv k_B T$, then the temperature can be written as

$$k_B T = Mv_c^2 \left(\exp\left(\frac{D_0}{M\beta v_c^2}\right) - 1\right). \tag{5.26}$$

For the case $D_0/M\beta \ll v_c^2$ this reduces to a Maxwell-Boltzmann distribution with a temperature given by Eq. 5.23. For $D_0/M\beta \gg v_c^2$ the temperature increases exponentially with D_0/β.

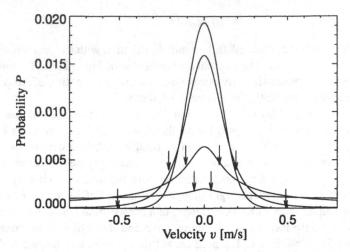

FIGURE 5.3. Stationary velocity distribution of the atoms for a constant value of D_0/β, but different values of the capture velocity v_c. The value of v_c is indicated with arrows.

5.5 Phase Space and Liouville's Theorem

One of the most important properties of laser cooling is its ability to change the phase space density of an atomic sample. Changing the phase space density provides a most important distinction between light optics and atom optics (see Chapters 13 and 15).

The phase space density $\rho(\vec{r}, \vec{p}, t)$ can be defined in terms of the probability that a single particle is at position $\vec{r}$ and has a momentum $\vec{p}$ at time t. In classical mechanics it is possible to know position and momentum of a particle with absolute certainty, and $\rho(\vec{r}, \vec{p}, t)$ is peaked for just these values. Then the phase space density for a system of N particles is the sum divided by N of the single-particle phase space densities of all the particles in the system, and the position and momentum of the *ensemble* are of interest.

Since $\rho(\vec{r}, \vec{p}, t)$ is a probability, it is always positive and normalized according to

$$\int \rho(\vec{r}, \vec{p}, t)\, \mathrm{d}^3r\, \mathrm{d}^3p = 1. \qquad (5.27)$$

Integrating $\rho(\vec{r}, \vec{p}, t)$ only over position yields the velocity distribution function $f(\vec{v})$, which becomes the Maxwell-Boltzmann function for a gas in free space as discussed in Sec. 5.2. Integrating $\rho(\vec{r}, \vec{p}, t)$ only over momentum yields the density $n(\vec{r})$ divided by the total number of atoms N. However, the aim of laser cooling is to increase the phase space density, not just the density in one or the other parameter.

For the discussion of the phase space density, which is defined in a six-dimensional space of $\vec{r}$ and $\vec{p}$, it is convenient to introduce the 6D vector $\vec{q} = (\vec{r}, \vec{p})$. The probability $P(\mathcal{V})$ to find one particle in a subspace $\mathcal{V}$ is then simply

$$P(\mathcal{V}) = \int_{\mathcal{V}} \rho(\vec{q}, t) d^6 q. \tag{5.28}$$

This probability can change only if probability "flows" out of the subspace $\mathcal{V}$ through its surface S, and the rate of change is given by

$$\frac{dP(\mathcal{V})}{dt} = \frac{\partial}{\partial t} \int_{\mathcal{V}} \rho(\vec{q}, t) d^6 q = -\oint_S \rho(\vec{q}, t) \, \dot{\vec{q}} \cdot d\vec{S}. \tag{5.29}$$

Here the 6D flux is given by $\dot{\vec{q}} \equiv (d\vec{r}/dt, d\vec{p}/dt)$, having three components of $\vec{r}$ and three components of $\vec{p}$, and $d\vec{S}$ is a differential surface element of $\vec{S}$. This is similar to fluid mechanics, where the change of fluid in a volume is just the flow of fluid through the surface out of the volume. Using Gauss' theorem, the surface integral of Eq. 5.29 can be converted in a volume integral, and this leads to

$$\frac{\partial}{\partial t} \int_{\mathcal{V}} \rho(\vec{q}, t) d^6 q = -\int_{\mathcal{V}} \vec{\nabla}_q \left(\rho(\vec{q}, t) \, \dot{\vec{q}} \right) d^6 q, \tag{5.30}$$

where the 6D gradient operator is defined by $\vec{\nabla}_q = (\partial/\partial\vec{r}, \partial/\partial\vec{p})$. Because the volume $\mathcal{V}$ can be defined arbitrarily, the integrands in Eq. 5.30 must be equal, and this leads to

$$\frac{\partial \rho(\vec{q}, t)}{\partial t} + \vec{\nabla}_q \left(\rho(\vec{q}, t) \, \dot{\vec{q}} \right) = 0. \tag{5.31}$$

The motion of an ensemble of classical particles in phase space can be described by Hamilton's equations. It is sufficient to describe only a single particle because the interaction between the particles does not play an important role in laser cooling. The classical Hamiltonian is $H(\vec{r}, \vec{p})$, and Hamilton's equations for the time dependence are

$$\frac{d\vec{p}}{dt} = -\frac{\partial H}{\partial \vec{r}} \quad \text{and} \quad \frac{d\vec{r}}{dt} = \frac{\partial H}{\partial \vec{p}}. \tag{5.32}$$

Here the Hamiltonian is chosen to be time-independent so the total energy of the system is conserved, and this leads to the total energy $H(\vec{r}, \vec{p}) \equiv E$.

The gradient operator in Eq. 5.31 acts on both ρ and $\dot{\vec{q}}$. However, for the $\dot{\vec{q}}$ part, the equations of motion 5.32 demand that

$$\vec{\nabla}_q \cdot \dot{\vec{q}} = \frac{\partial}{\partial \vec{r}} \left(\frac{\partial \vec{r}}{\partial t} \right) + \frac{\partial}{\partial \vec{p}} \left(\frac{\partial \vec{p}}{\partial t} \right) = \frac{\partial}{\partial \vec{r}} \left(\frac{\partial H}{\partial \vec{p}} \right) - \frac{\partial}{\partial \vec{p}} \left(\frac{\partial H}{\partial \vec{r}} \right) = 0. \tag{5.33}$$

Note that the order of differentiation can be changed, since $\vec{r}$ and $\vec{p}$ are independent coordinates. Using this result in Eq. 5.31 leads to

$$\frac{\partial \rho(\vec{q}, t)}{\partial t} + \dot{\vec{q}} \cdot \vec{\nabla}_q \left(\rho(\vec{q}, t) \right) \equiv \frac{d\rho(\vec{q}, t)}{dt} = 0. \tag{5.34}$$

Equation 5.34 is the key result of this section, and is called the Liouville theorem. It requires that the *phase space density cannot be changed*. However, it does not mean that the phase space density cannot be distributed differently over the degrees of freedom. Or, to put it into the context of fluid dynamics, the shape of a volume of flowing fluid can change in time, but the fluid volume is fixed. Traveling with the flow of the fluid, the density does not change.

The Liouville theorem depends on the fact that the motion of the particles can be described by a Hamiltonian. This is *not* the case when the forces depend not only on position, but also on velocity. Then a Hamiltonian description for the system can no longer be used and Eq. 5.32 no longer applies. For instance, consider the system in Sec. 5.4, where the force is directly proportional to the momentum, namely, $\vec{F} = -\Gamma \vec{p}$, with $\Gamma = \beta/M$. Inserting this force in Eq. 5.33 leads to

$$\frac{d\rho(\vec{q}, t)}{dt} = 3\Gamma\rho(\vec{q}, t), \tag{5.35}$$

where the factor of 3 stems from the compression in 3D. This leads to an exponential gain in the phase space density with a time constant $1/3\Gamma$, so using velocity dependent forces allows phase space compression to be obtained.

The Hamiltonian description of geometrical optics leads to a similar theorem to that of Eq. 5.34, called the brightness theorem, that can be found in many optics books. Thus bundles of light rays obey a similar phase space density conservation. But there is a fundamental difference between light and atom optics. In the first case, the "forces" that determine the behavior of bundles of rays are "conservative" and phase space density is conserved. For instance, a lens can be used to focus a light beam to a small spot; however, at the same time the divergence of the beam must be increased, thus conserving phase space density. By contrast, in atom optics dissipative forces that are velocity dependent can be used, and thus phase space density is no longer conserved. Optical elements corresponding to such forces can not exist for light, but in addition to the atom optic elements of lenses, collimators and others described in Chapter 13, phase space compressors can also be built. Such compression is essential in a large number of cases, but most importantly for the achievement of Bose-Einstein condensation.

Part II

Cooling & Trapping

6
Deceleration of an Atomic Beam

6.1 Introduction

The origin of optical forces on atoms has been discussed in Chapter 3, and here a specific application is introduced. The use of electromagnetic forces to influence the motion of neutral atoms has been a subject of interest for some years, and several review articles and books on the subject are listed in Appendix B. The force caused by radiation, particularly by light at or near the resonance frequencies of atomic transitions, originates from the momentum associated with light. In addition to energy $E = \hbar\omega$, each photon carries momentum $\hbar k$ and angular momentum $\hbar$. When an atom absorbs light, it stores the energy by going into an excited state; it stores the momentum by recoiling from the light source with a momentum $\hbar k$; and it stores the angular momentum in the form of internal motion of its electrons. The converse applies for emission, whether it is stimulated or spontaneous. It is the velocity change of the atoms, $v_r = \hbar k/M \simeq$ few cm/s, that is of special interest here, and although it is very small compared with thermal velocity, multiple absorptions can be used to produce a large total velocity change. Proper control of this velocity change constitutes a radiative force that can be used to decelerate and/or to cool free atoms.

Although there are many ways to decelerate and cool atoms from room temperature or higher, the one that has received the most attention by far depends on the scattering force that uses this momentum transfer between the atoms and a radiation field resonant with an atomic transition. By making a careful choice of geometry and of the light frequency one can exploit the Doppler shift to make the momentum exchange (hence the force) velocity dependent. Because the force is velocity dependent, it can not only be used for deceleration, but also for cooling that results in increased phase space density (see Sec. 5.5).

6.2 Techniques of Beam Deceleration

The idea that the radiation scattering force on free atoms could be velocity dependent and therefore be used for cooling a gas was suggested by Wineland and Dehmelt [37], Hansch and Schawlow [38], and Wineland and Itano [39], although Kastler, Landau, and others had made allusions to it in earlier years. The possibility for cooling stems from the fact that atomic absorption of light near a resonance is strongly frequency dependent, and is therefore velocity dependent because of the Doppler shift of the laser frequency seen by the atoms moving relative to the laboratory-fixed laser. Of course, a velocity-dependent dissipative force is needed for cooling.

The simplest form of this force to study, that from a low-intensity single plane wave of light, has been exploited for cooling of an atomic beam. Early experiments in several laboratories [40–43] have used this force, along with a variety of methods to overcome technical problems [44–46], to decelerate and cool thermal atomic beams to only a few hundredths of a Kelvin.

One very obvious implementation of radiative deceleration and cooling is to direct a laser beam opposite to an atomic beam as shown in Fig. 6.1 [42, 43]. In this case each atom can absorb light very many times along its path through the apparatus. Of course, excited-state atoms cannot absorb light efficiently from the laser that excited them, so between absorptions they must return to the ground state by spontaneous decay, accompanied by emission of fluorescent light. The emitted fluorescent light will also change the momentum of the atoms, but its spatial symmetry results in an average of zero net momentum transfer after many such fluorescence events. So the net deceleration of the atoms is in the direction of the laser beam, and the maximum deceleration is limited by the spontaneous fluorescence rate.

The maximum attainable deceleration is obtained for very high light intensities, and is limited because the atom must then divide its time equally between ground and excited states. High-intensity light can produce faster absorption, but it also causes equally fast stimulated emission; the combination produces neither deceleration nor cooling because the momentum transfer to the atom in emission is then in the opposite direction to what it was in absorption. The deceleration therefore saturates at a value $\vec{a}_{max} = \hbar \vec{k} \gamma / 2M$, where the factor of 2 arises because the atoms spend half of their time in each state (see the discussion on page 25).

The Doppler shifted laser frequency in the moving atoms' reference frame should match that of the atomic transition to maximize the light absorption and scattering rate. This rate γ_p is given by the Lorentzian (see Eq. 2.26)

$$\gamma_p = \frac{s_0 \gamma / 2}{1 + s_0 + \left[2(\delta + \omega_D)/\gamma \right]^2}, \tag{6.1}$$

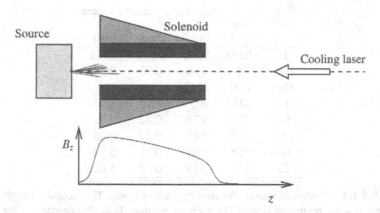

FIGURE 6.1. Schematic diagram of apparatus for beam slowing. The tapered magnetic field is produced by layers of varying length on the solenoid. A plot of B_z vs. z is also shown.

where $s_0 = I/I_s$ is the ratio of the light intensity I to the saturation intensity I_s, which is a few mW/cm^2 for typical atomic transitions (see Table C.2). Also $\delta = \omega_\ell - \omega_a$ is the laser detuning from resonance, ω_ℓ is the laser frequency and ω_a is the atomic resonance frequency. The Doppler shift seen by the moving atoms is $\omega_D = -\vec{k} \cdot \vec{v}$ (note that $\vec{k}$ opposite to $\vec{v}$ produces a positive Doppler shift). Maximum deceleration requires $(\delta + \omega_D) \ll \gamma$, so that the laser light is nearly resonant with the atoms in their rest frame. The net force $\vec{F}$ on the atoms is (see Eq. 3.14)

$$\vec{F} = \hbar\vec{k}\gamma_p, \tag{6.2}$$

which saturates at large s_0 to $M\vec{a}_{max} = \vec{F}_{max} \equiv \hbar\vec{k}\gamma/2$.

In Table 6.1 are some of the parameters for slowing a few atomic species of interest from the peak of the thermal velocity distribution. Since the maximum deceleration $\vec{a}_{max}$ is fixed by atomic parameters, it is straightforward to calculate the minimum stopping length L_{min} and time t_{min} for the rms velocity of atoms $\bar{v} = 2\sqrt{k_B T/M}$ at the chosen temperature. The result is

$$L_{min} = \bar{v}^2/2a_{max} \tag{6.3a}$$

and

$$t_{min} = \bar{v}/a_{max}. \tag{6.3b}$$

It is comforting to note that $|\vec{F}_{max}|L_{min}$ is just the atomic kinetic energy and that L_{min} is just $t_{min}\bar{v}/2$.

If the light source is spectrally narrow, then as the atoms in the beam slow down, their changing Doppler shift will take them out of resonance. They will eventually cease deceleration after their Doppler shift has been decreased by a few times the power-broadened width $\gamma' = \gamma\sqrt{1 + s_0}$ as given in Eq. 2.27b, corresponding to Δv of a few times γ/k. Although this Δv of a few m/s is considerably larger than the typical atomic recoil velocity v_r of a few cm/s, it is still only a small

atom	T_{oven} (K)	$\bar{v}$ (m/s)	L_{min} (m)	t_{min} (ms)
H	1000	5000	0.012	0.005
He*	4	158	0.03	0.34
He*	650	2013	4.4	4.4
Li	1017	2051	1.15	1.12
Na	712	876	0.42	0.96
K	617	626	0.77	2.45
Rb	568	402	0.75	3.72
Cs	544	319	0.93	5.82

TABLE 6.1. Parameters of interest for slowing various atoms. The stopping length L_{min} and time t_{min} are minimum values. The oven temperature T_{oven} that determines the peak velocity is chosen to give a vapor pressure of 1 Torr. Special cases are H at 1000 K and He in the metastable triplet state, for which two rows are shown: one for a 4 K source and another for the typical discharge temperature.

fraction of the atoms' average thermal velocity, so that significant further cooling or deceleration cannot be accomplished.

In order to accomplish deceleration that changes the atomic speeds by hundreds of m/s, it is necessary to maintain $(\delta + \omega_D) \ll \gamma$ by compensating such changes of the Doppler shift. This can be done by changing ω_D, or δ via either ω_ℓ or ω_a. The two most common methods for overcoming this problem are sweeping the laser frequency ω_ℓ to keep it in resonance with the decelerating atoms [47–49], and spatially varying the atomic resonance frequency with an inhomogeneous dc magnetic field to keep the decelerating atoms in resonance with the fixed frequency laser [42, 50]. Other methods that have also worked are discussed below.

6.2.1 Laser Frequency Sweep

In the method of changing ω_ℓ, the laser frequency is swept upward at rate $\dot{\omega}_\ell$ to compensate the decreasing Doppler shift as the atoms slow down. Of course, $(\delta + \omega_D)$ must be kept $\ll \gamma$ in order to maintain atomic resonance, and $a < a_{max}$ must always be satisfied. This requires that $-\dot{\omega}_\ell \approx \dot{\omega}_D = \vec{k} \cdot \vec{a} < \hbar k^2 \gamma / 2M = \omega_r \gamma$.

This method of Doppler compensation has several distinct advantages and disadvantages, and choosing it depends on the ultimate purpose for slowing the atoms. Although it was first implemented using a dye laser for Na [47], it is especially easy to use with semiconductor laser diodes [49] because of their fast and simple electronic tunability. A few mA sweep changes their frequency by several GHz, and this is easily enough to compensate for a Doppler shift corresponding to $v = \omega_D / k$ of a few km/s. The most obvious disadvantage is the time structure it imposes on the production of slow atoms. They arrive in pulses separated by a few times t_{min} given in Eq. 6.3b. This may be desired, of no importance, or undesired, depending on the nature of the experiments.

6.2.2   Varying the Atomic Frequency: Magnetic Field Case

The use of a spatially varying magnetic field to tune the atomic levels along the beam path was the first method to succeed in slowing atoms [42]. It works as long as the Zeeman shifts of the ground and excited states are different so that the resonant frequency is shifted (see p. 42). The field can be tailored to provide the appropriate Doppler shift along the moving atom's path. For uniform deceleration $a \equiv \eta a_{max}$ from initial velocity v_0, the appropriate field profile is

$$B(z) = B_0\sqrt{1 - z/z_0},   \tag{6.4}$$

where $z_0 \equiv M v_0^2/\eta \hbar k \gamma$ is the length of the magnet, $B_0 = \hbar k v_0/\mu'$, $\mu' \equiv (g_e M_e - g_g M_g)\mu_B$, subscripts g and e refer to ground and excited states, $g_{g,e}$ is the Landé g-factor, μ_B is the Bohr magneton, and $M_{g,e}$ is the magnetic quantum number. The design parameter $\eta < 1$ determines the length of the magnet z_0. A solenoid that can produce such a spatially varying field has layers of decreasing lengths as shown schematically in Fig. 6.1. The technical problem of extracting the beam of slow atoms from the end of the solenoid can be simplified by reversing the field gradient and choosing a transition whose frequency decreases with increasing field [44].

The equation of motion of an atom in the magnet cannot be easily solved in general because of the velocity-dependent force, but by transforming to a decelerating frame $\mathcal{R}$ [51] the problem can be addressed. For the special case of uniform deceleration the velocity of this frame in the lab is $v_{\mathcal{R}} = v_0\sqrt{1 - z/z_0}$, and the Doppler shift associated with this velocity is compensated by the position-dependent Zeeman shift in the magnet. The resulting equation of motion for the velocity of atoms $v' \equiv v - v_{\mathcal{R}}$ relative to this frame is given by

$$M\frac{d\vec{v}'}{dt} = -\vec{F}_{max}\left[\frac{s_0}{1 + s_0 + \left(2(\delta - \vec{k}\cdot\vec{v}')/\gamma\right)^2} - \eta\right],   \tag{6.5}$$

where $F_{max} = \hbar k \gamma/2$. For $dv'/dt = 0$ the steady-state velocity v'_{ss} is given by

$$kv'_{ss} = \delta \pm \frac{\gamma}{2}\sqrt{s_0\frac{1 - \eta}{\eta} - 1}.   \tag{6.6}$$

There are two values of v'_{ss} but the one with the $(+)$ sign is unstable. The magnitude of v'_{ss} is typically of order δ/k. This velocity is approximately constant as atoms decelerate along their paths through the magnet so the decreasing Doppler shift is compensated by the decreasing Zeeman shifts.

6.2.3   Varying the Atomic Frequency: Electric Field Case

The changing Doppler shift can also be compensated by a Stark shift using an inhomogeneous dc electric field, and this has been demonstrated in both Na [52] and Cs [53]. There are special problems with this technique that arise because adequate deceleration requires excited states with $\gamma = 1/\tau \simeq 10^7$/s or larger. However, only low-lying atomic states have such large values of γ, and the Stark shifts of such states are relatively small. Thus the method requires rather large electric fields.

The field profile for the Zeeman-compensated method has the form of Eq. 6.4 because the Zeeman shift is linear in field, but Stark shifts of eigenstates of parity (usual low-lying atomic states) are not linear. Their Stark shifts ΔE_S are typically quadratic in field, given by

$$\Delta E_S = \frac{1}{2}\alpha|\mathcal{E}|^2, \tag{6.7}$$

where α is the atomic polarizability and $\mathcal{E}$ is the applied dc electric field. Typical values of the Stark shift difference between ground and excited states are about $100\,\text{kHz} \times |\mathcal{E}|^2$, where $\mathcal{E}$ is given in kV/cm. Thus compensation of typical Doppler shifts of about 700 MHz requires $\mathcal{E} \simeq 80\,\text{kV/cm}$. Unlike the Zeeman-compensation method where the g-factors are nearly the same for many atoms, α can vary by factors of 3 – 5 among the alkalis. In order to achieve constant acceleration, the resulting field profile is

$$\mathcal{E}(z) = \mathcal{E}_0\sqrt{1 - \sqrt{1 - z/z_0}}, \tag{6.8}$$

where z_0 is the length of the field region. Needless to say, the condition $a < a_{max}$ must be maintained, and so z_0 must still conform to Eq. 6.3a.

The geometry for this experiment has two quite long, oppositely charged plates, typically made of highly polished stainless steel, separated by a tapered gap ranging from one to a few cm, and charged to a few tens of kV. Since the z-dependence of $\mathcal{E}$ is much weaker than that for $B(z)$ in Eq. 6.4, a linearly tapered gap provides an adequate approximation to Eq. 6.8. Like the Zeeman-compensating method, a slowing laser opposes the atomic velocity. However, unlike the Zeeman method, the open geometry allows lateral access to the beam because it's not enclosed in a solenoid, and transverse cooling and/or collimation can easily be applied in one direction [53].

6.2.4   Varying the Doppler Shift: Diffuse Light

It is also possible to compensate the changing Doppler shift of decelerating atoms by exploiting the angular dependence embodied in $\omega_D = -\vec{k} \cdot \vec{v}$ [54–56]. Atoms moving through diffuse monochromatic light see a range of frequencies that vary with the angle between the velocity and the light direction. The resonance frequency ω_a of an atom moving at a velocity v will be matched by the Doppler-shifted laser frequency when the angle θ between the wavevector $\vec{k}$ of the light and the atomic velocity $\vec{v}$ satisfies

$$\delta = \omega_\ell - \omega_a = kv\cos\theta = -\omega_D, \tag{6.9}$$

corresponding to $\delta + \omega_D = 0$. For red detuned light ($\delta < 0$) this resonance condition requires $\theta > \pi/2$, meaning that the recoil caused by absorbing light opposes the atomic motion and slows the atoms. As in the other cases, subsequent spontaneous emission does not exert a force on the atoms on average, but does provide the dissipative process needed for cooling. As the atoms are decelerated (v decreases), they absorb light from an increasingly forward angle θ until the maximum value of $\theta = \pi$ is reached.

When the light is tuned below resonance in the lab frame by an amount δ, then the Doppler shift will be toward the blue, closer to resonance, as long as a component of the light's propagation direction is antiparallel to the atomic velocity. (Light propagating nearly parallel to the atomic velocity is shifted further to the red, further out of resonance.)

Atoms can be efficiently slowed by scattering the counterpropagating light if the incident angle required for the Doppler effect to shift it close to atomic resonance is not too close to $\pi/2$ so that there remains a considerable component of the momentum vector $\hbar k$ antiparallel to $\vec{v}$. Of course, as the atoms slow down, δ doesn't change, but the smaller $\vec{v}$ requires a larger contribution from the angular part of $\vec{k} \cdot \vec{v}$. Thus atoms will interact with counterpropagating light from a cone of decreasing angle, closer to opposing the velocity direction, until they have decelerated to nearly $v = \delta/k$. Below this velocity, there is no angle for which the Doppler effect can shift the light into resonance, and so the deceleration ends.

Because the light is isotropic, the atomic motion in any direction is directly opposed. This is in contrast to the methods described above in Secs. 6.2.1, 6.2.2, and 6.2.3 where only the longitudinal velocity component of the atomic motion is opposed by the light because the force is determined by the $\vec{k}$-vector of the single laser beam. This major advantage helps to prevent the atomic beam from being transversely expanded (apart from spontaneous emissions).

6.2.5 Broadband Light

Still another method of deceleration uses light that is not spectrally narrow, but is white over a spectral region from ω_a to $(\omega_a - \vec{k} \cdot \vec{v}_0)$. Then Doppler compensation is not necessary because atoms of any velocity below v_0 will find resonant light in a counterpropagating beam. Such white light slowing has been considered in the past [57,58] and has also been demonstrated [59–62]. One important disadvantage is that saturation of the atomic transition for all velocities below v_0 requires much more light power because the spectral density must be at least I_s/γ, and since $\vec{k} \cdot \vec{v}_0$ is typically 100γ, the overall power must be at least 100 times larger than in the other Doppler-compensation techniques discussed above.

6.2.6 Rydberg Atoms

A quite different slowing scheme using the large Stark shifts of Rydberg states was proposed in 1981 [63]. In this method, the force on the atoms does not come from the momentum of the light but from the energy gradient associated with their

Stark shifts in an inhomogeneous dc electric field. Atoms are optically excited to a Rydberg state whose Stark-shifted energy is downward-going in a region of strong dc electric field. If the field is very inhomogeneous, produced, for example, by a pair of small electrodes of few mm size and separation, then atoms gain potential energy and thus must lose kinetic energy as they leave the region between the electrodes and travel to a region of zero field. When they entered such a region of strong field in the ground state, they did not gain as much energy as they lose when they leave it in an excited state.

The lifetime of the selected Rydberg state is chosen so that the atoms will decay to the ground state outside the field region. Thus the size scale of the experiment is determined by the Rydberg state lifetime and the atomic speed. Travel through regions of alternately small and large fields, coupled with proper excitation by well-focused laser beams, causes repeated kinetic energy loss. The atoms always climb up bigger hills than they fall down, and radiate higher frequency light from the tops of those hills than they absorb at the bottoms. Thus their kinetic energy is converted into potential energy and then radiated away.

This method has the advantage that the slowing distance depends on the properties of the Rydberg states, not on the atoms' kinetic energy or speed. Therefore the slowing distance can be much less than L_{min} for optical forces as given in Eq. 6.3a, and possibly more useful for fast, light atoms such as He* at discharge temperatures. Such large forces might also be used to control and deflect atomic velocities, and possibly even reverse them, thereby making a Rydberg atom mirror [64].

6.3   Measurements and Results

This section presents some results of experiments that used the Zeeman-tuning technique to compensate the changing Doppler shift. The most common way to measure the slowed velocity distribution is to detect the fluorescence from atoms excited by a second laser beam propagating at a small angle to the atomic beam [42]. Because of the Doppler shift, the frequency dependence of this fluorescence provides a measure of the atomic velocity distribution. In this method, the velocity resolution Δv is limited by the natural width of the excited state to $\Delta v = \gamma/k$ (≈ 6 m/s for Na).

In 1997 a new time-of-flight (TOF) method to accomplish the same result was reported, however, with a much improved resolution [65]. In addition, it provided a much more powerful diagnostic of the deceleration process. The TOF method has the capability to map out the velocity distribution for both hyperfine ground states of alkali atoms along their entire path through the solenoid. The experimental arrangement is shown in Fig. 6.2. The atoms emerge through an aperture of 1 mm^2 from an effusive Na source heated to approximately 300°C. During their subsequent flight through a solenoid, they are slowed by the counterpropagating laser light from laser 2, and the changing Doppler shift is compensated with a field that is well described by Eq. 6.4.

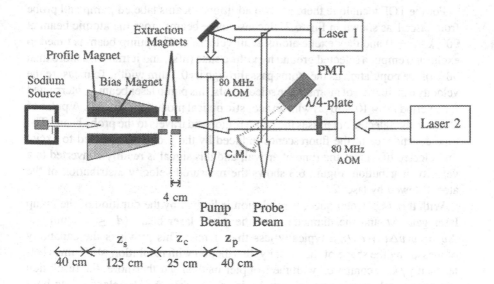

FIGURE 6.2. The TOF apparatus, showing the solenoid magnet and the location of the two laser beams used as the pump and probe. The resolution of the technique is ultimately determined by the flight path z_p (figure from Ref. 65).

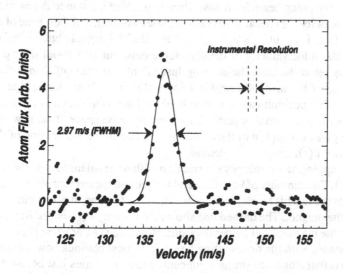

FIGURE 6.3. The velocity distribution measured with the TOF method. The experimental width of approximately $\frac{1}{6}(\gamma/k)$ is shown by the dashed vertical lines between the arrows. The Gaussian fit through the data yields a FWHM of 2.97 m/s (figure from Ref. 65).

For the TOF technique there are two additional beams labeled pump and probe from laser 1 as shown in Fig. 6.2. Because these beams cross the atomic beam at $90°$, $\vec{k} \cdot \vec{v} = 0$ and they excite atoms at all velocities. The pump beam is tuned to excite and empty a selected ground hyperfine state (hfs), and it transfers more than 98% of the population as the atoms pass through its 0.5 mm width. To measure the velocity distribution of atoms in the selected hfs, this pump laser beam is interrupted for a period $\Delta t = 10 - 50 \ \mu s$ with an acoustic optical modulator (AOM). A pulse of atoms in the selected hfs passes the pump region and travels to the probe beam. The time dependence of the fluorescence induced by the probe laser, tuned to excite the selected hfs, gives the time of arrival, and this signal is readily converted to a velocity distribution. Figure 6.3 shows the measured velocity distribution of the atoms slowed by laser 2.

With this TOF technique, the resolution is limited by the duration of the pump laser gate Δt and the diameter d of the probe laser beam ($d \leq 1.0$ mm) to $\Delta v = v(v\Delta t + d)/z_p$, typically less than 1 m/s. This provides the capability of measuring the shape of the velocity distribution with resolution ≈ 10 times better than γ / k as compared with the Doppler method. Furthermore, the resolution improves for decreasing velocity v; Δv is smaller than the Doppler cooling limit of $\sqrt{\hbar \gamma / 2M} \approx 30$ cm/s for $v \approx 80$ m/s and Na atoms. Figure 6.3 shows the final velocity distribution for such a measurement giving a FWHM of 3.0 m/s at a central velocity of 138 m/s. The width is about one half of γ / k.

The method of shutting off the slowing laser beam a variable time τ_{off} *before* the short shut-off of the pump beam offers a much more informative scheme of data acquisition. The atoms that pass through the pump region during the short time when the pump beam is off have already traveled a distance $\Delta z = v(z)\tau_{\text{off}}$ (at constant velocity $v(z)$ because the slowing laser was off), and their time of arrival at the probe laser is $z_p/v(z) = z_p \tau_{\text{off}}/\Delta z$. Thus the TOF signal contains information not only about the velocity of the detected atoms, but also about their position z in the magnet at the time the slowing laser light was shut off. Since the spatial dependence of the magnetic field is known (Fig. 6.1), both the field and atomic velocity at that position can be determined, and the TOF signal is proportional to the number of atoms in that particular region of phase space. This new technique therefore gives a mapping of the atomic population in the z-direction of the phase space, z and $v(z)$, within the solenoid.

Such mapping of the velocity distribution within the solenoid is a powerful diagnostic tool. The contours of Figs. 6.4a and b represent the strength of the TOF signal for each of the two hfs levels, and thus the density of atoms, at each velocity and position in the magnet. The dashed line shows the velocity $v(z) = (\mu' B(z)/\hbar - \delta)/k$ for which the magnetic field tunes the atomic transition $(F, M_F) = (2, 2) \to (3, 3)$ into resonance with the decelerating beam. The most obvious new information in Fig. 6.4a is that atoms are strongly concentrated at velocities just below that of the resonance condition. This corresponds to the strong peak of slow atoms shown in Fig. 6.3. Additional information about optical pumping among the hfs sublevels is also present, and discussed in Sec. 6.4.

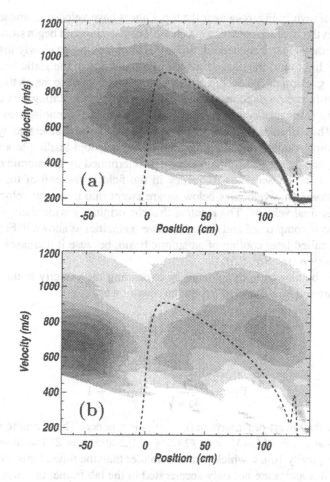

FIGURE 6.4. Contour map of the measured velocity and position of atoms in the solenoid, (a) for $F_g = 2$ atoms and (b) for $F_g = 1$ atoms. The dashed line indicates the resonance frequency for the $(F, M_F) = (2, 2) \to (3, 3)$ cycling transition. The density of atoms per unit phase space area $\Delta v \Delta z$ has been indicated with different gray levels (figure from Ref. 65).

6.4   Further Considerations

6.4.1   Cooling During Deceleration

It is important to stress that deceleration is *not* the same as cooling: cooling requires a compression of the velocity distribution in phase space as shown in Figs. 6.3 and 6.4. To see how this notion applies to laser deceleration of an atomic beam, consider again the example of Zeeman compensation of the Doppler shift [65].

Some atoms emerging from the oven are moving too fast to be decelerated at all because, for them, the laser frequency is Doppler shifted too far out of resonance to absorb light, even where the magnetic field is strongest at the solenoid entrance.

These are shown in Fig. 6.4a near the top. Others have velocities whose Doppler shift causes the laser frequency to match the Zeeman shift and begin slowing down as soon as they enter the solenoid. Still others are moving so slowly that they do not absorb light until they have traveled to a point where the static but spatially varying magnetic field has decreased to the appropriate value to match their smaller Doppler shift and produce resonance. These begin accumulating just below the curve of the resonance condition along the length of the solenoid, also shown in Fig. 6.4a. Thus *all atoms* with velocity lower than v_0 can be decelerated by the laser beam to some smaller velocity at the end of the solenoid leading to an increase in phase space density. This final velocity is determined by the atomic resonance condition at the chosen laser frequency in the field at the end of the solenoid. Thus *all atoms* with velocities below v_0 are swept into a narrow velocity group around this final velocity. The result is that the originally wide thermal velocity distribution is compressed and shifted to lower velocities as shown in Fig. 6.3: this process is called laser cooling of an atomic beam, because it increases the phase space density ρ (see Sec. 5.5).

This can be viewed in more detail by expanding the velocity in the reference frame $\mathcal{R}$ around v'_{ss}. Then Eq. 6.5 can be rewritten as

$$\frac{dv'}{dt} = -\Gamma_D(v' - v'_{ss}), \tag{6.10a}$$

where

$$\Gamma_D = \frac{4\omega_r\eta^2}{s_0}\sqrt{\frac{(1-\eta)s_0}{\eta} - 1} \tag{6.10b}$$

when only the lowest-order term in $(v' - v'_{ss})$ is retained. The damping rate Γ_D is maximum at $\Gamma_D = \hbar k^2/4M = \omega_r/2$ for $\eta = 0.5$ and $s_0 = 2$. The damping time $1/\Gamma_D$ is typically 10 μs, which is much smaller than the time atoms spend in the magnet. Thus atoms are not only decelerated in the lab frame, but cooled toward this velocity. With $\beta = M\Gamma_D = \hbar k^2/4$ and $D = s\gamma(\hbar)^2/2$ as given in Eq. 5.15b, it is straightforward to show that the final temperature is related to the Doppler temperature T_D.

6.4.2 Non-Uniformity of Deceleration

It is important to realize that even the ideal magnetic field profile does not produce a constant deceleration [44,65]. This is easily seen by differentiating the resonance condition found from Eq. 6.4 to find the deceleration satisfying the resonance condition:

$$a = -\eta a_{\max}\left(1 + \frac{1 - \hbar\delta/\mu'B_0}{\sqrt{\tilde{z}}}\right), \tag{6.11}$$

where $\tilde{z} \equiv (1-z/z_0)$, using $d\sqrt{\tilde{z}}/dt = v\, d\sqrt{\tilde{z}}/dz$. In the usual operating condition to extract slow atoms from the solenoid, $\hbar\delta/\mu'B_0 < 1$. Only for $\hbar\delta = \mu'B_0$ is the deceleration constant.

Furthermore, it is clear that as the atoms progress through the solenoid, $\bar{z}$ eventually becomes small enough that maintaining the resonance condition requires the magnitude of the deceleration to exceed a_{max}. The atoms then drop out of the deceleration process and emerge from the solenoid as desired.

6.4.3 Transverse Motion During Deceleration

As the longitudinal motion of atoms in a beam is slowed by counterpropagating laser light, their transverse motion becomes more important if it, too, is not compensated. For example, if an atomic beam of mean longitudinal velocity of 1000 m/s and angular spread 0.01 radian is decelerated to 50 m/s, its angular divergence expands to 0.2 radian at the end of the slowing region. It is hardly recognizable as a beam! In order to compensate this effect, the decelerating laser beam is chosen not to be parallel, but is focused so that it converges toward the atomic beam oven. Thus there is a small transverse component of the optical force that opposes the transverse velocity just enough to maintain the angular spread of the original atomic beam [42, 43]. Note that for the case of diffuse light slowing described in Sec. 6.2.4 such considerations are not necessary because the optical force is always directly opposite to the atomic velocity [54, 55].

There is another source of transverse motion that is not as easily controlled. Each time an atom spontaneously decays to the ground state, it receives an impulse $\hbar\vec{k}$ in a random direction. Although the average value of these impulses is zero, the rms value is not. Thus the atoms diffuse transversely as they move along their paths. The resulting distance from the axis Δx as a result of this diffusion is $\Delta x = \int v(t)dt$, where $v(t) = \hbar k\sqrt{\gamma_p t}/M$ ($\gamma_p t$ is the number of photons scattered). Integrating from $t = 0$ to $v_0/\eta a_{max}$ gives $\Delta x = (2\hbar k/3M)\sqrt{\gamma_p(v_0/\eta a_{max})^3}$, which is several mm for typical experiments. Thus the focused laser beam must be about 1 cm in diameter at the end of the deceleration region if too many atoms are not to escape out the sides as a result of this transverse diffusion. In the case of Stark compensation discussed in Sec. 6.2.3, the transverse motion in one of the two directions can be largely compensated with additional laser beams.

As atoms move down the magnet, their velocities are determined by s_0, δ, and $B(z)$ as discussed above. However the light intensity distribution across the counterpropagating laser beam is not uniform, but is given by $I(x, z) = I_0 e^{-2x^2/w^2(z)}$, where $w(z)$ is the Gaussian beam width. If the beam is focused, $w(z)$ is given by the usual expression for Gaussian beam propagation. Thus atoms in a plane perpendicular to the laser beam axis have different velocities, and the shape of the surface containing atoms of the same velocity is determined by the Gaussian intensity profile. Using $v_R = \sqrt{v_0 - 2az}$ and Eq. 6.2 with the Gaussian spatial profile in γ_p, this surface has the form $z(x) = z_1 e^{2x^2/w^2(z)}$ for small x, and is undefined where x is large enough for the intensity to fall below ηI_0, the minimum intensity required to maintain deceleration. Furthermore, fluctuations of the light intensity can cause instabilities on this surface [66] .

6.4.4   Optical Pumping During Deceleration

In Sec. 6.1 it was pointed out that very many absorptions and emissions are required to have a significant effect on the velocity of a thermal atom because $v_r \ll \bar{v}$ (see Table 6.1). Since alkali atoms have two well-separated hfs ground states, optical pumping can readily populate the one not in resonance with the slowing laser beam. Even in the first beam-slowing experiment [42], this was taken into account by choosing the laser beam's polarization to operate on a cycling transition that connects the state ($F_g = 2$, $M_g = 2$) with ($F_e = 3$, $M_e = 3$) of Na (see Fig. 4.1). However, even this could be insufficient unless the polarization is very pure. As an added precaution, the authors of Ref. 42 therefore applied a large homogeneous magnetic field in addition to the tapered one that compensated the changing Doppler shift. This homogeneous field altered the coupling among the hyperfine sublevels of the excited state to inhibit the undesired spontaneous decay channel.

The TOF diagnostic technique described in Sec. 6.3 has the capability of measuring such optical pumping processes along the entire length of the solenoid. Figure 6.4a shows a strong buildup of $F_g = 2$ atoms over a wide velocity range centered near 700 m/s at the entrance of the solenoid, along with a corresponding depletion of $F_g = 1$ atoms (Fig. 6.4b). This strong optical pumping occurs between the oven and the magnet where the slowing laser light is very intense because it is focused on the oven. For typical values of 30 mW laser power and a focal spot size of $\approx 100 \; \mu$m, the intensity of 3×10^5 mW/cm^2 broadens the absorption line from its natural width of 10 MHz to about 2 GHz (see Eq. 2.27), more than enough to compensate the Doppler shift of the entire velocity distribution. Since the detuning is close to the resonance for the $F_g = 2 \to F_e = 3$ transition, the optical pumping is most effective on $F_g = 1$ atoms because they are strongly Doppler-shifted by their velocities near 1000 m/s and not far from the peak of the distribution at 700 m/s. Of course, atoms in $F_g = 2$ can also be excited, but they are Doppler shifted further from resonance, so that the net transfer of population is from $F_g = 1 \to F_g = 2$.

As the atoms enter the solenoid, this effect is partially reversed. Figure 6.4a shows a decrease of the $F_g = 2$ population near 700 m/s in the rising edge of the magnetic field, while Fig. 6.4b shows a corresponding increase of $F_g = 1$ atoms. In this field region, the excitations from the $F_g = 2$, $M_g = -1$, 0, and $+1$ levels to the states appropriate for circularly polarized light come into resonance with the laser [65]. Various other optical pumping schemes can be used to explain the appearance and disappearance of other population islands in Fig. 6.4.

7
Optical Molasses

7.1 Introduction

Chapter 6 presented a discussion of the radiative force on atoms moving in a single laser beam. Here this notion is extended to include the radiative force from more than just one beam. For example, if two low-intensity laser beams of the same frequency, intensity, and polarization are directed opposite to one another (*e.g.*, by retroreflection of a single beam from a mirror), the net force found by adding the radiative forces given in Eq. 6.2 from each of the two beams obviously vanishes for atoms at rest because $\vec{k}$ is opposite for the two beams. However, atoms moving slowly along the light beams experience a net force proportional to their velocity whose sign depends on the laser frequency. If the laser is tuned below atomic resonance, the frequency of the light in the beam opposing the atomic motion is Doppler shifted toward the blue in the atomic rest frame, and is therefore closer to resonance; similarly, the light in the beam moving parallel to the atom will be shifted toward the red, further out of resonance. Atoms will therefore interact more strongly with the laser beam that opposes their velocity and they will slow down. This is illustrated in Fig. 7.1.

The slowing force is proportional to velocity for small enough velocities, resulting in viscous damping [67, 68] as shown in Fig. 3.1 on p. 36 that gives this technique the name "optical molasses"(OM). By using three intersecting orthogonal pairs of oppositely directed beams, the movement of atoms in the intersection region can be severely restricted in all three dimensions, and many atoms can thereby be collected and cooled in a small volume. OM has been demonstrated at several laboratories [69], often with the use of low cost diode lasers [70].

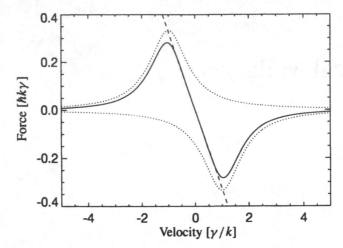

FIGURE 7.1. Velocity dependence of the optical damping forces for one-dimensional optical molasses. The two dotted traces show the force from each beam, and the solid curve is their sum. The straight line shows how this force mimics a pure damping force over a restricted velocity range. These are calculated for $s_0 = 2$ and $\delta = -\gamma$ so there is some power broadening evident (see Sec. 2.4).

Note that OM is not a trap for neutral atoms because there is no restoring force on atoms that have been displaced from the center. Still, the detainment times of atoms caught in OM of several mm diameter can be remarkably long.

7.2 Low-Intensity Theory for a Two-Level Atom in One Dimension

It is straightforward to estimate the force on atoms in OM from Eq. 3.14. The discussion here is limited to the case where the light intensity is low enough so that stimulated emission is not important. This eliminates consideration of excitation of an atom by light from one beam and stimulated emission by light from the other, a sequence that can lead to very large, velocity-independent changes in the atom's speed. In this low intensity case the forces from the two light beams are simply added to give $\vec{F}_{OM} = \vec{F}_+ + \vec{F}_-$, where

$$\vec{F}_{\pm} = \pm \frac{\hbar \vec{k} \gamma}{2} \frac{s_0}{1 + s_0 + \left[2(\delta \mp |\omega_D|)/\gamma\right]^2}. \qquad (7.1)$$

Then the sum of the two forces is

$$\vec{F}_{OM} \cong \frac{8\hbar k^2 \delta s_0 \vec{v}}{\gamma(1 + s_0 + (2\delta/\gamma)^2)^2} \equiv -\beta \vec{v}, \qquad (7.2)$$

where terms of order $(kv/\gamma)^4$ and higher have been neglected (see Eq. 3.26).

For $\delta < 0$, this force opposes the velocity and therefore viscously damps the atomic motion. $\vec{F}_{OM}$ has maxima near $v = \pm(\gamma'/2k)(X/\sqrt{3})$ and decreases rapidly for larger velocities. Here $\gamma' \equiv \gamma\sqrt{s_0 + 1}$ is the power broadened linewidth (see Eq. 2.27b), X is the numerical factor given by $\sqrt{x - 1 + 2\sqrt{x^2 + x + 1}}$, and $x \equiv (2\delta/\gamma')^2$. For $x \gg 1$ these maxima appear at $v = \pm\delta/k$ as expected, but for the usual realm of OM, $x \sim 1$ and $X \sim \sqrt{3}$. Since $\vec{F}_{OM}$ is nearly linear with velocity for $|v| < \gamma/k$ when $x \sim 1$ (i.e., $s_0 \leq 1$ and $\delta = -\gamma/2$), it is convenient to define a capture velocity $v_c = \gamma/k$. However, the range of this damping force can be increased considerably beyond v_c by using curved wavefronts [71], a properly arranged inhomogeneous magnetic field [72], high light intensity [73], or a variety of other tricks [74, 75], some of which are discussed in Sec. 13.5.

If there were no other influence on the atomic motion, all atoms would quickly decelerate to $v = 0$ and the sample would reach $T = 0$, a clearly unphysical result. There is also some heating caused by the light beams that must be considered, and it derives from the discrete size of the momentum steps the atoms undergo with each emission or absorption. Since the atomic momentum changes by $\hbar k$, their kinetic energy changes on the average by at least the recoil energy $E_r = \hbar^2 k^2/2M = \hbar\omega_r$. This means that the average frequency of each absorption is $\omega_{abs} = \omega_a + \omega_r$ and the average frequency of each emission is $\omega_{emit} = \omega_a - \omega_r$. Thus the light field loses an average energy of $\hbar(\omega_{abs} - \omega_{emit}) = 2\hbar\omega_r$ for each scattering. This loss occurs at a rate $2\gamma_p$ (two beams), and the energy becomes atomic kinetic energy because the atoms recoil from each event. The atomic sample is thereby heated because these recoils are in random directions, as discussed on p. 37.

The competition between this heating with the damping force of Eq. 7.2, results in a nonzero kinetic energy in steady state. At steady state, the rates of heating and cooling for atoms in OM are equal. Equating the cooling rate, $\vec{F} \cdot \vec{v}$, to the heating rate, $4\hbar\omega_r\gamma_p$, the steady-state kinetic energy is found to be $(\hbar\gamma/8)(2|\delta|/\gamma + \gamma/2|\delta|)$. This result is dependent on $|\delta|$, and it has a minimum at $2|\delta|/\gamma = 1$, whence $\delta = -\gamma/2$. The temperature found from the kinetic energy is then $T_D = \hbar\gamma/2k_B$, where k_B is Boltzmann's constant and T_D is called the Doppler temperature or the Doppler cooling limit (see p. 58). For ordinary atomic transitions T_D is below 1 mK, and several typical values are given in Table C.3 (see Appendix C).

Another instructive way to determine T_D is to note that the average momentum transfer of many spontaneous emissions is zero, but the rms scatter of these about zero is finite. One can imagine these decays as causing a random walk in momentum space with step size $\hbar k$ and step frequency $2\gamma_p$, where the factor of 2 arises because of the two beams. The random walk results in diffusion in momentum space with diffusion coefficient $D_0 \equiv 2(\Delta p)^2/\Delta t = 4\gamma_p(\hbar k)^2$. Then Brownian motion theory (see Sec. 5.3) gives the steady-state temperature in terms of the damping coefficient β to be $k_B T = D_0/\beta$. This turns out to be $\hbar\gamma/2$ as above for the case $s_0 \ll 1$ when $\delta = -\gamma/2$. There are many other independent ways to derive this remarkable result that predicts that the final temperature of atoms in OM is independent of the optical wavelength, atomic mass, and laser intensity (as long as it is not too large).

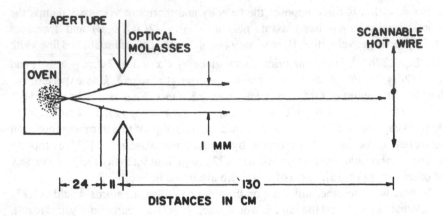

FIGURE 7.2. Overall schematic of the apparatus used for one-dimensional transverse cooling (figure from Ref. 76).

7.3   Atomic Beam Collimation

When an atomic beam crosses a one-dimensional OM as shown in Fig. 7.2, the transverse motion of the atoms is quickly damped while the longitudinal component is essentially unchanged. This transverse cooling of an atomic beam is an example of a method that can actually increase its brightness (atoms/sec-sr-cm^2) because such active collimation uses dissipative forces to compress the phase space volume occupied by the atoms. By contrast, the usual realm of beam focusing or collimation techniques for light beams and most particle beams, is restricted to selection by apertures or conservative forces that preserve the phase space density of atoms in the beam (see Sec. 5.5).

7.3.1   Low-Intensity Case

At low intensity the velocity dependence of the optical force that collimates atomic beams using transverse OM derives from the Doppler shift of the transverse velocity $v_t \equiv v \sin \theta$, where typically $v = \bar{v} \sim 500$ m/s. The damping coefficient β is maximum for $\delta = -\gamma/2$ and $s_0 = 2$, and for $v_x < v_c$ the damping force is approximately $\hbar k^2 v_x/2$ (see Eq. 7.2 and Fig. 7.1). For $v_x > v_c$, this force decreases approximately as $1/v_x$ from its maximum value of typically $\hbar k \gamma/4$, just as in OM. By contrast, for faster atoms the force from one of the beams dominates because of the Doppler shift, and for high enough speeds (v_x a few times larger than v_c) the Doppler shift can take an atom almost completely out of resonance with both beams.

This velocity compression at low intensity in one dimension can be estimated for two-level atoms as follows. The narrowest momentum distribution after cooling has an energy half width determined by the Doppler limit $M v_D^2/2 = \hbar \gamma/4$, corresponding to $v_D = \sqrt{\hbar \gamma/2M}$. The relevant transverse velocity width before optical collimation is about twice the capture range v_c. Therefore the one-dimensional mo-

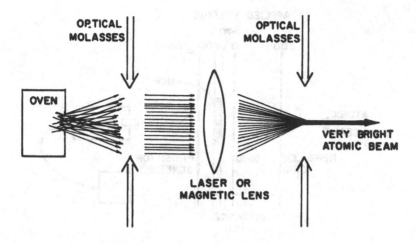

FIGURE 7.3. Scheme for optical brightening of an atomic beam. First the transverse velocity components of the atoms are damped out by an optical molasses, then the atoms are focused to a spot, and finally the atoms are recollimated in a second optical molasses (figure from Ref. 76).

mentum space compression is about $v_c/v_D = \sqrt{\gamma/\omega_r} = \sqrt{1/\varepsilon}$, with ε as defined in Eq. 5.5. For an atomic beam, this compression can be done in the two transverse directions, reducing the occupied volume of transverse momentum space by $\gamma/\omega_r = 1/\varepsilon$. Furthermore, longitudinal cooling of an atomic beam as described in Chapter 6 can compress the width of the longitudinal velocity distribution from its original thermal value of $\bar{v}$ to about the same v_D. For Rb $v_D = 12$ cm/s, $v_c \simeq 4.6$ m/s, $\omega_r \simeq 2\pi \times 3.8$ kHz, and $1/\varepsilon \simeq 1600$. Thus the decrease in phase space volume from the momentum contribution alone for laser cooling of a Rb atomic beam can exceed 10^6.

Contributions to increasing the brightness of an atomic beam are not limited to momentum space compression. One must also consider the spatial expansion of the atomic beam in the transverse direction as it crosses the beam of collimating laser light. The minimal transverse expansion Δx for atoms emitted from a point source is of the order of the minimum stopping distance of atoms with transverse velocity v_c. This yields $\Delta x = v_c^2/2a_{max} = \gamma/8k\omega_r \simeq 100\mu m$ for Rb. Since this is typically smaller than the oven hole, this expansion is not of great consequence. In fact, spatial compression using atomic beam focusing as shown in Fig. 7.3, coupled with longitudinal cooling and further collimation, could compress the spatial extent of the beam to $\simeq \sqrt{\gamma/\omega_r}/k$, several μm for Rb. Since the oven hole for typical atomic beams may be a few hundred μm, this may be a factor of 50 in each direction, thus leading to total phase space compression of more than 10^9.

Clearly optical techniques can create atomic beams 10^6 or more times intense than ordinary thermal beams, and also many orders of magnitude brighter. Furthermore, this number could be increased several orders of magnitude if the transverse cooling could produce temperatures below the Doppler temperature (see Chap-

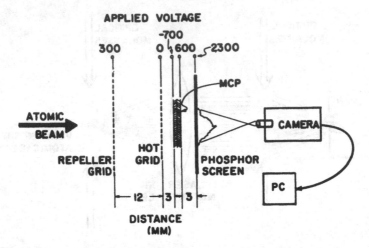

FIGURE 7.4. Schematic diagram of the neutral-atom camera showing the repeller grid, the hot grid, the multichannel plates, and the phosphor screen. Atoms are ionized at the hot grid, directed toward the MCP's by the field between it and the repeller, and accelerated toward the MCP's by the voltage between them and the hot grid. The output electrons excite the phosphor, which is viewed by the TV camera. PC is a personal computer (figure from Ref. 77).

ter 8). For atoms cooled to the recoil temperature of Eq. 5.4, $T_r = \hbar\omega_r/k_B$ where $\Delta p = \hbar k$ and $\Delta x = \lambda/\pi$, the brightness increase could be 10^{17}.

7.3.2   Experiments in One and Two Dimensions

Some of the earliest optical collimation experiments were done in a thermal beam of natural Rb that was produced by an oven at $T \simeq 150°C$ with aperture $\simeq 330~\mu m$ diameter [76, 77]. The beam was mechanically collimated by a defining aperture of diameter $\simeq 330\mu m$ about 24 cm away (see Fig. 7.2). The laser light ($\lambda = 780$ nm) was tuned to select either of the stable Rb isotopes, and experiments were done with each of them. The atomic beam profile was measured with a scanning hot tungsten wire, 25 μm in diameter, 1.3 m away from the region of interaction with the laser beam.

For two-dimensional collimation a single hot wire scan would not provide enough information. Instead of scanning both a vertical and a horizontal hot wire, a new method was devised for observing the spatial distribution of atoms [76, 77]. A heated mesh was placed in the plane perpendicular to the beam (see Fig. 7.4). Ions emitted from the hot grid were accelerated into a pair of multichannel plate electron multipliers whose output electrons were accelerated onto a phosphor-coated screen. The screen was viewed by a standard TV camera whose output was fed to a frame grabber in a computer where the image could be analyzed.

This device has been used for viewing the atomic beam with both one- and two-dimensional collimation. With about half of the molasses laser beam split off to produce a vertically oriented molasses in addition to the horizontal one, full

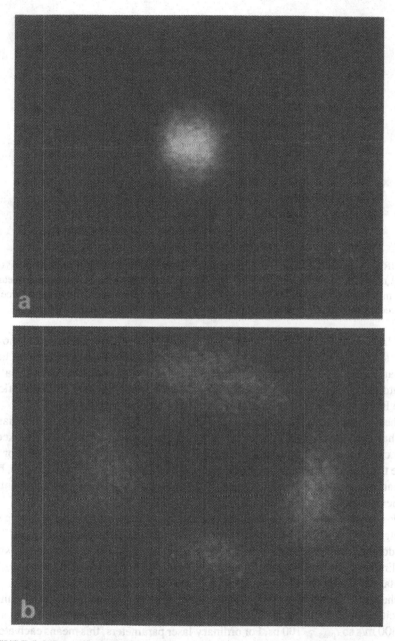

FIGURE 7.5. Image formed by the neutral atom camera of Fig. 7.4 with two-dimensional molasses acting on the atomic beam. The outline of the circular beam spot represents a 6 mm diameter image on the phosphor. The 7 mW molasses laser beam was nearly uniformly intense and rectangular, about 8×20 mm. Its detuning was about -30 MHz for (a) and about $+30$ MHz for (b). Note the collimation for the red detuning and the divergence for the blue detuning. Again the recording time was $1/30$ s and no image averaging was performed (figure from Ref. 76).

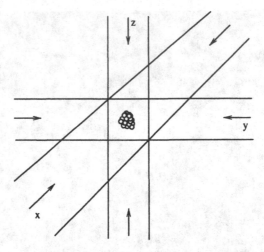

FIGURE 7.6. Schematic diagram of the arrangement of laser beams for 3D optical molasses. Three mutually perpendicular standing waves are formed by reflecting three laser beams from mirrors. Because of the red-detuned laser light, atoms experience a friction force in all directions and are therefore confined in a viscous medium, the optical molasses.

two-dimensional collimation of the atomic beam was achieved. Figure 7.5 shows the results of this experiment. Here the intensity of each beam was only about half of that for the 1D collimation, both light beams were linearly polarized normal to the atomic beam direction, and about 0.4 Gauss was applied along that direction. (The B-field and both polarization vectors were mutually orthogonal.)

The collimated atomic beam spot size shown in Fig. 7.5 of about 1.25 mm diameter has been reduced from its original 6 mm diameter (with the lasers blocked). This corresponds to an increase of *both* brightness *and* intensity by a factor of more than 20. Furthermore, this increase may well be hundreds of times larger, but limitations imposed by the resolution of the imaging system may have prevented its measurement [76].

More recent experiments have imaged the fluorescence of atoms as they pass through a sheet of light to determine the spatial cross section of an atomic beam [78]. It is done by focusing a nearly resonant beam of laser light into a thin sheet using a cylindrical lens (or telescope). The thickness of the sheet is d, determined by the standard Gaussian beam equations and the optics that form it, and the plane of the sheet is oriented perpendicular to the atomic beam. Atoms having longitudinal velocity v pass through it in time $t_{pass} = d/v$. Typical values are $d = 30 \ \mu$m and $v = 300$ m/s so $t_{pass} \sim 100$ ns. For ordinary laser parameters, this means each atom scatters several photons.

The interaction region between the sheet of light and the atomic beam is imaged onto a camera (typically a CCD) using optics with aperture typically around $f/4$ so that the collection efficiency is a few percent, and the quantum efficiency of the CCD results in a total efficiency around 1%. For an atomic beam of flux 10^9 atoms/s through the laser beam, the image information is therefore collected at a rate of 10^7 counts/s, so good images of 10^5 pixels are collected in a few seconds.

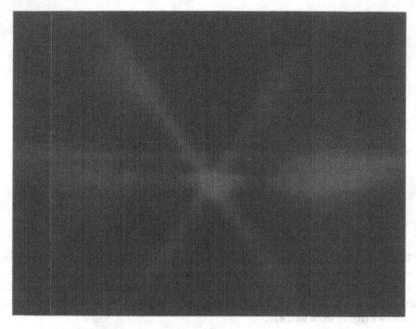

FIGURE 7.7. Photograph of optical molasses in Na taken under ordinary snapshot conditions in the lab at NIST. The upper horizontal streak is from the slowing laser while the three beams that cross at the center are on mutually orthogonal axes viewed from the (111) direction. Atoms in the optical molasses glow brightly at the center (figure from Ref. 81).

7.4 Experiments in Three-Dimensional Optical Molasses

Optical molasses experiments can also work in three dimensions at the intersection of three mutually orthogonal pairs of opposing laser beams (see Ref. 41 and Fig. 7.6). Even though atoms can be collected and cooled in the intersection region, it is important to stress again that this is *not* a trap. That is, atoms that wander away from the center experience no force directing them back. They are allowed to diffuse freely and even escape, as long as there is enough time for their very slow diffusive movement to allow them to reach the edge of the region of the intersection of the laser beams. Because the atomic velocities are randomized during the damping time $1/\omega_r$, atoms execute a random walk with a step size of $v_D/\omega_r = \lambda/2\pi\sqrt{2\varepsilon} \cong$ few μm. To diffuse a distance of 1 cm requires about 10^7 steps or about 30 s [79, 80].

Three-dimensional OM was first observed in 1985 [68]. Preliminary measurements of the average kinetic energy of the atoms were done by blinking off the laser beams for a fixed interval. Comparison of the brightness of the fluorescence before and after the turnoff was used to calculate the fraction of atoms that left the region while it was in the dark. The dependence of this fraction on the duration of the dark interval was used to estimate the velocity distribution and hence the tem-

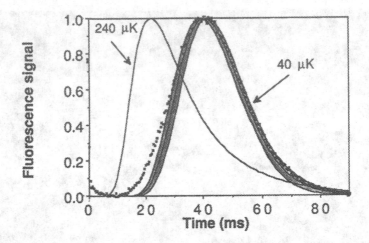

FIGURE 7.8. Data from dropping atoms out of optical molasses into a probe beam about 18 mm below. The calculated time-of-flight spectra are for 240 μK and 40 μK. The shaded area indicates the range of error in the 40 μK calculation from geometric uncertainties. The width of the data is slightly larger than the calculation, presumably because of shot-to-shot instabilities (figure from Ref. 82).

perature. The result was not inconsistent with the two level atom theory described in Sec. 7.2.

Soon other laboratories had produced 3D OM. The photograph in Fig. 7.7 shows OM in Na at the laboratory in the National Bureau of Standards (now NIST) in Gaithersburg. The phenomenon is readily visible to the unaided eye, and the photograph was made under ordinary snapshot conditions. The three mutually perpendicular pairs of laser beams appear as a star because they are viewed along a diagonal.

This NIST group developed a more accurate ballistic method to measure the velocity distribution of atoms in OM [82]. The limitation of the first measurements was determined by the size of the OM region and the unknown spatial distribution of atoms [68]. The new method at NIST used a separate measuring region composed of a 1D OM about 2 cm below the 3D region, thereby reducing the effect of this limitation. When the laser beams forming the 3D OM were shut off, the atoms dropped because of gravity into the 1D region, and the time-of-arrival distribution was measured. This was compared with calculated distributions for T_D and 40 μK as shown in Fig. 7.8. Using a series of plots like Fig. 7.8 it was possible to determine the dependence of temperature on detuning, and that is shown in Fig. 7.9, along with the theoretical calculations for a two-level atom, as given in Sec. 7.2.

It was an enormous surprise to observe that the ballistically measured temperature of the Na atoms was as much as 10 times *lower* than T_D = 240 μK [82], the temperature minimum calculated from the theory. This breaching of the Doppler limit forced the development of an entirely new picture of OM that accounts for the fact that in 3D, a two-level picture of atomic structure is inadequate. The multilevel

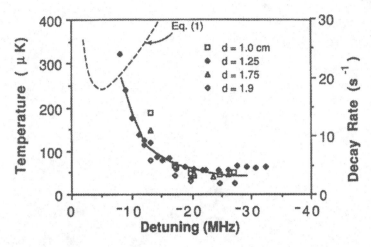

FIGURE 7.9. Temperature *vs.* detuning determined from time-of-flight data for various separations *d* between the optical molasses and the probe laser (data points). The solid curve represents the measured molasses decay rate; it is not a fit to the temperature data points, but its scale (shown at right) was chosen to emphasize its proportionality to the temperature data. The dashed line shows the temperature expected on the basis of the two-level atom theory of Sec. 7.2 (figure from Ref. 82).

structure of atomic states, and optical pumping among these sublevels, must be considered in the description of 3D OM, as discussed in Chapter 8.

These experiments also found that OM was less sensitive to perturbations and more tolerant of alignment errors than was predicted by the 1D, two-level atom theory. For example, if the intensities of the two counterpropagating laser beams forming an OM were unequal, then the force on atoms at rest would not vanish, but the force on atoms with some nonzero drift velocity *would* vanish. This drift velocity can be easily calculated by using Eq. 7.1 with unequal intensities s_{0+} and s_{0-}, and following the derivation of Eq. 7.2. Thus atoms would drift out of an OM, and the calculated rate would be much faster than observed by deliberately unbalancing the beams in the experiments [69].

Chapter 8 describes the startling new view of OM that emerged in the late 1980s as a result of these surprising measurements. The need for a new theoretical description resulting from incontrovertible measurements provides an excellent pedagogical example of how physics is truly an experimental science, depending on the interactions between observations and theory, and always prepared to discard oversimplified descriptions as soon as it is shown that they are inadequate.

8
Cooling Below the Doppler Limit

8.1 Introduction

In response to the surprising measurements of temperatures below T_D, two groups developed a model of laser cooling that could explain the lower temperatures [83, 84]. The key feature of this model that distinguishes it from the earlier picture was the inclusion of the multiplicity of sublevels that make up an atomic state (*e.g.*, Zeeman and hfs). The dynamics of optically pumping atoms among these sublevels provides the new mechanism for producing the ultra-low temperatures [81].

The dominant feature of these models is the non-adiabatic response of moving atoms to the light field. Atoms at rest in a steady state have ground-state orientations caused by optical pumping processes that distribute the populations over the different ground-state sublevels. In the presence of polarization gradients, these orientations reflect the local light field. In the low-light-intensity regime, the orientation of stationary atoms is completely determined by the ground-state distribution: the optical coherences and the excited-state population follow the ground-state distribution adiabatically.

For atoms moving in a light field that varies in space, optical pumping acts to adjust the atomic orientation to the changing conditions of the light field. In a weak pumping process, the orientation of moving atoms always lags behind the orientation that would exist for stationary atoms. It is this phenomenon of non-adiabatic following that is the essential feature of the new cooling process.

Production of spatially dependent optical pumping processes can be achieved in several different ways. As an example consider two counterpropagating laser beams that have orthogonal polarizations, as discussed in Sec. 4.3. The superposition of the two beams results in a light field having a polarization that varies on the wavelength scale along the direction of the laser beams. Laser cooling by such a light field is called polarization gradient cooling. In a three-dimensional optical molasses, the transverse wave character of light requires that the light field always has polarization gradients.

Another way to make optical pumping spatially dependent is to use a standing wave of constant polarization and an additional field such as a uniform dc magnetic field. Since the standing wave light field has nodes and anti-nodes, the rate of the optical pumping compared with the rate of Larmor precession of atoms in the magnetic field changes dramatically over a wavelength. The resulting cooling process is called magnetically induced laser cooling [85] or magnetic orientational cooling [86].

The cooling process that derives from this non-adiabatic following is effective over a limited range of atomic velocities. The force is maximum for atoms that travel a distance $\lambda/4$ during one optical pumping process. If atoms travel at a lower velocity, they will not have reached a very different part of the optical field before a spontaneous emission causes the pumping process to occur; if atoms travel faster, they will already go beyond the largest change in the field before being pumped toward another sublevel [27]. Of course the velocity where this force is effective scales with the characteristic distance over which the optical field changes. Although this is typically $\lambda/4$, it can be much larger for two light waves with oblique $\vec{k}$ vectors.

The nature of this cooling process is fundamentally different from the Doppler laser cooling process discussed in the previous chapter. In that case, the differential absorption from the laser beams was caused by the Doppler shift of the laser frequency, and the process is therefore known as Doppler cooling. In the cooling process described in this chapter, the force is still caused by differential absorption of light from the two laser beams, but the velocity-dependent differential rates, and hence the cooling, relies on the non-adiabaticity of the optical pumping process. Since lower temperatures can usually be obtained with this cooling process, it is called sub-Doppler laser cooling [81, 82, 85].

8.2   Linear ⊥ Linear Polarization Gradient Cooling

One of the most instructive models for discussion of sub-Doppler laser cooling was introduced by Dalibard and Cohen-Tannoudji [83] and their work serves as the basis for this section. They considered the case of orthogonal linear polarization of two counterpropagating laser beams that damps atomic motion in one dimension (see the discussion of polarization gradients in Sec. 4.3). The polarization of this light field varies over half of a wavelength from linear at 45° to the polarization of the two beams, to σ^+, to linear but perpendicular to the first direction, to σ^-, and then it cycles (see Fig. 4.5). To study the effects of this polarization gradient on the cooling process, they considered a $J_g = \frac{1}{2}$ to $J_e = \frac{3}{2}$ transition. This is one of the simplest transitions that shows sub-Doppler cooling.

In the place where the light field is purely σ^+, the pumping process drives the ground-state population to the $M_g = +\frac{1}{2}$ sublevel. This optical pumping occurs because absorption always produces $\Delta M = +1$ transitions, whereas the subsequent spontaneous emission produces $\Delta M = \pm 1, 0$ (see Sec. 4.4). Thus the average $\Delta M \geq 0$ for each scattering event. For σ^--light the population will be pumped toward the $M_g = -\frac{1}{2}$ sublevel. Thus in traveling through a half wavelength in the light field, atoms have to readjust their population completely from $M_g = +\frac{1}{2}$ to $M_g = -\frac{1}{2}$ and back again.

8.2.1 Light Shifts

The interaction between nearly resonant light and atoms not only drives transitions between atomic energy levels, but also shifts their energies as given in Eq. 1.16. These shifts are essentially caused by the Stark effect from the electric field of the light as discussed in Sec. 1.2.1. This light shift of the atomic energy levels plays a crucial role in this scheme of sub-Doppler cooling, and the changing polarization has a strong influence on the light shifts. In the low-intensity limit of two laser beams each of intensity $s_0 I_s$, the light shifts ΔE_g of the ground magnetic substates are given by (see Eq. 1.17a)

$$\Delta E_g = \frac{\hbar \delta s_0 C_{ge}^2}{1 + (2\delta/\gamma)^2}, \tag{8.1}$$

where C_{ge} is the Clebsch-Gordan coefficient that describes the coupling between the atom and the light field (see Sec. 4.5.3). This relation has to be compared with the result obtained in Eq. 1.17a for a two-level atom in a traveling wave. First, the light shift is twice as large, since there are two traveling waves. Second, the coupling has been modified because of the multiplicity of the ground-state, which is expressed by the coefficients C_{eg}^2. Finally, the semiclassical analysis of Sec. 1.2.1 did not take into account the spontaneous emission process, and a more careful analysis [83] leads to the result of Eq. 8.1. The C_{ge}^2's are given in Appendix D for a variety of transition schemes. Since C_{ge} depends on the magnetic quantum numbers and on the polarization of the light field, the light shifts are different for different magnetic sublevels. The ground-state light shift is negative for a laser tuning below resonance ($\delta < 0$) and positive for $\delta > 0$ (see Eq. 1.17a).

In the present case of orthogonal linear polarizations and $J = \frac{1}{2} \rightarrow \frac{3}{2}$, the light shift for the magnetic substate $M_g = \frac{1}{2}$ is three times larger than that of the $M_g = -\frac{1}{2}$ substate when the light field is completely σ^+. On the other hand, when the light field becomes σ^-, the shift of $M_g = -\frac{1}{2}$ is three times larger. So in this case the optical pumping discussed above causes there to be a larger population in the state with the larger light shift. This is generally true for any transition J_g to $J_e = J_g + 1$. A schematic diagram showing the populations and light shifts for this particular case of negative detuning is shown in Fig. 8.1.

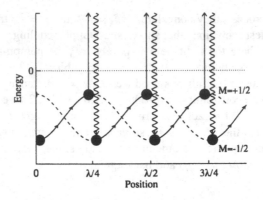

FIGURE 8.1. The spatial dependence of the light shifts of the ground-state sublevels of the $J = \frac{1}{2} \Leftrightarrow \frac{3}{2}$ transition for the case of the lin $\perp$ lin polarization configuration. The arrows show the path followed by atoms being cooled in this arrangement. Atoms starting at $z = 0$ in the $M_g = +\frac{1}{2}$ sublevel must climb the potential hill as they approach the $z = \lambda/4$ point where the light becomes σ^- polarized, and there they are optically pumped to the $M_g = -\frac{1}{2}$ sublevel. Then they must begin climbing another hill toward the $z = \lambda/2$ point where the light is σ^+ polarized and they are optically pumped back to the $M_g = +\frac{1}{2}$ sublevel. The process repeats until the atomic kinetic energy is too small to climb the next hill. Each optical pumping event results in absorption of light at a lower frequency than emission, thus dissipating energy to the radiation field.

8.2.2   Origin of the Damping Force

To discuss the origin of the cooling process in this polarization gradient scheme, consider atoms with a velocity v at a position where the light is σ^+-polarized, as shown at the lower left of Fig. 8.1. The light optically pumps such atoms to the strongly negative light-shifted $M_g = +\frac{1}{2}$ state. In moving through the light field, atoms must increase their potential energy (climb a hill) because the polarization of the light is changing and the state $M_g = \frac{1}{2}$ becomes less strongly coupled to the light field. After traveling a distance $\lambda/4$, atoms arrive at a position where the light field is σ^--polarized, and are optically pumped to $M_g = -\frac{1}{2}$, which is now lower than the $M_g = \frac{1}{2}$ state. Again the moving atoms are at the bottom of a hill and start to climb. In climbing the hills, the kinetic energy is converted to potential energy, and in the optical pumping process, the potential energy is radiated away because the spontaneous emission is at a higher frequency than the absorption (see Fig. 8.1). Thus atoms seem to be always climbing hills and losing energy in the process. This process brings to mind a Greek myth, and is thus called "Sisyphus laser cooling". In this sense it is similar to the Stark cooling discussed in Sec. 6.2.6 [63] and the high intensity "blue cooling" discussed in Sec. 9.3 [87,88].

   The cooling process described above is effective over a limited range of atomic velocities. The damping is maximum for atoms that undergo one optical pumping process while traveling over a distance $\lambda/4$. Slower atoms will not reach the hilltop before the pumping process occurs and faster atoms will already be descending

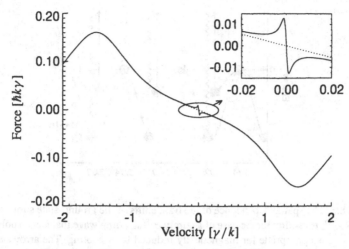

FIGURE 8.2. The force as a function of velocity for an atom in a lin ⊥ lin polarization gradient cooling configuration with $s_0 = 0.5$ and $\delta = 1.5\gamma$. The solid line is the combined force of Doppler and sub-Doppler cooling, whereas the dashed line represents the force for Doppler cooling only. The inset shows an enlargement of the curve around $v = 0$. Note, the strong increase in the damping rate over a very narrow velocity range that arises from the sub-Doppler process.

the hill before being pumped toward the other sublevel. In both cases the energy loss is smaller and therefore the cooling process less efficient.

The damping force $F = -\beta v$ can be estimated from the distance dependence of the energy loss. Denote the optical pumping time by $\tau_p \equiv 1/\gamma_p$ and then the optimum speed is $v_c \cong \gamma_p/k$. The force at this velocity v_c is $F = \Delta W/\Delta z \cong \Delta E\, k \equiv -\beta v_c$. To find the order of magnitude of the friction coefficient β, both the light shift ΔE and the pumping rate γ_p need to be estimated. For a detuning $|\delta| \gg \gamma$, Eq. 2.26 yields the pumping rate $\gamma_p = s_0\gamma^3/4\delta^2$ (the 4 instead of an 8 accounts for the presence of two laser beams), and then choosing $C_{ge}^2 = 1$ in Eq. 8.1 yields $\Delta E = \hbar\gamma^2 s_0/4\delta$. Then the damping rate $\beta/M = \hbar k^2\delta/2M\gamma = \omega_r\delta/\gamma$. The velocity-dependent force has the same order of magnitude as the Doppler force, but its velocity range is γ_p/k. The best result are often obtained with $|\delta| \gg \gamma/2$, so usually this is much smaller than γ/k and therefore β is much larger. Figure 8.2 shows the scale of the forces. It is interesting that β becomes larger when the optical pumping rate γ_p becomes smaller. Although this seems counterintuitive, it happens because v_c becomes smaller when γ_p becomes smaller.

This result is of particular significance because it shows that the friction coefficient for this sub-Doppler process is larger by a factor $(2|\delta|/\gamma)$ than the maximum friction coefficient for Doppler laser cooling. It can be shown that the momentum diffusion coefficient of this process is of the same order of magnitude as that of Doppler cooling, so that the temperature will be smaller than the Doppler temperature by the same factor. Furthermore, it shows that the friction coefficient for this case is independent of intensity, since both ΔE and γ_p are proportional to the intensity.

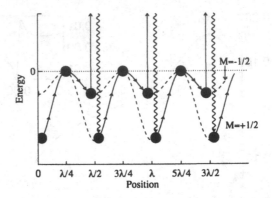

FIGURE 8.3. The spatial dependence of the light shifts of the ground-state sublevels of the $J = 1/2 \leftrightarrow 3/2$ transition for the case of a purely σ^+ standing wave that has no polarization gradient, and is appropriate for magnetically induced laser cooling. The arrows show the path followed by atoms being cooled in this arrangement. Atoms starting at $z = 0$ in the strongly light-shifted $M_g = +1/2$ sublevel must climb the potential hill as they approach the node at $z = \lambda/4$. There they undergo Zeeman mixing in the absence of any light and may emerge in the $M_g = -1/2$ sublevel. They will then gain less energy as they approach the antinode at $z = \lambda/2$ than they lost climbing into the node. Then they are optically pumped back to the $M_g = +1/2$ sublevel in the strong light of the antinode, and the process repeats until the atomic kinetic energy is too small to climb the next hill. Each optical pumping event results in absorption of light at a lower frequency than emission, thus dissipating energy to the radiation field.

8.3   Magnetically Induced Laser Cooling

Although the first models that described sub-Doppler cooling relied on the polarization gradient of the light field as above, it was soon realized that a light field of constant polarization in combination with a magnetic field could also produce sub-Doppler cooling [89]. In this process, the atoms are cooled in a standing wave of circularly polarized light.

There is a simple model using the $J_g = 1/2$ to $J_e = 3/2$ transition to describe this phenomenon [85]. In the absence of a magnetic field, the σ^+ light field drives the population to the $M_g = +1/2$ sublevel. Since the $M_g = +1/2$ sublevel is more strongly coupled to the light field than $M_g = -1/2$, the light shift of this state is larger. Thus atoms traveling through this standing wave will descend and climb the same potential hills corresponding to $M_g = 1/2$ and will experience no average force.

The situation changes if a small transverse magnetic field is applied. Optical pumping processes determine the atomic states in the antinodes of the standing wave light field where the light is strong. But in the nodes, where the intensity of the light field is zero, the small transverse magnetic field precesses the population from $M_g = 1/2$ toward $M_g = -1/2$. Atoms that leave the nodes with $M_g = -1/2$ are returned to $M_g = +1/2$ in the antinodes by optical pumping in the σ^+ light.

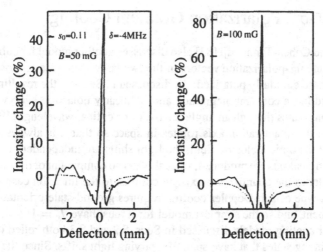

FIGURE 8.4. Typical data of atomic beam collimation using circularly polarized light and a weak magnetic field on a beam of ^{85}Rb atoms (see Fig. 7.2). The scanning hot wire was 1.3 m downstream from the interaction region. The laser parameters are defined as in Eq. 8.1 and Sec. 2.4 (figure from Ref. 90).

This cooling process is depicted in Fig. 8.3 for negative detuning $\delta < 0$. Potential energy is radiated away in the optical pumping process as before, and kinetic energy is converted to potential energy when the atoms climb the hills again into the nodes. The whole process is repeated when the atoms travel through the next node of the light field. Again the cooling process is caused by a "Sisyphus" effect, similar to the case of lin $\perp$ lin. Since this damping force is absent without the magnetic field, it is called magnetically induced laser cooling (MILC).

Efficient cooling by MILC depends critically on the relation between the Zeeman precession frequency ω_Z and the optical pumping rate γ_p in the antinodes. It is clearly necessary that $\gamma_p \gg \omega_Z$ in the antinodes where the light is strong. But as in any cooling process that depends on non-adiabatic processes, there is a limited velocity range where the force is effective. For efficient cooling by MILC, the velocity can not be too small compared to ω_Z/k or atoms will undergo many precession cycles near the nodes and no effective cooling will result. On the other hand, if the velocity is large compared to γ_p/k, then atoms will pass through the antinodes in a time too short to be optically pumped to $M_g = +\frac{1}{2}$ and no cooling will result either. Thus, in addition to the requirement $\delta < 0$, there are two other conditions on the experimental parameters that can be combined to give

$$\omega_Z < kv < \gamma_p. \qquad (8.2)$$

Sub-Doppler cooling has been observed for MILC as shown in Fig. 8.4 for Rb atoms cooled on the $\lambda = 780$ nm transition in one dimension [85]. The width of the velocity distribution near $v = 0$ is as low as 2 cm/s, much lower than the one-dimensional Doppler limit $v_D = \sqrt{7\hbar\gamma/20M} \approx 10$ cm/s for Rb.

8.4 σ^+-σ^- Polarization Gradient Cooling

Dalibard and Cohen-Tannoudji [83] also discussed another model for sub-Doppler cooling where the polarization vectors of the two laser beams are also orthogonal, but in this case circularly polarized. As discussed in Sec. 4.3 the resulting optical electric field has a constant magnitude and is linearly polarized everywhere, but the direction rotates through an angle 2π over one optical wavelength [91]. In the basis where the quantization axis rotates in space so that it is always along the direction of the optical electric field, the light shifts are independent of position, and only π transitions are produced, since there is no component of the optical field perpendicular to the quantization axis. In contrast to the lin $\perp$ lin case discussed earlier, this type of sub-Doppler cooling requires ground-state orientation rather than alignment, and so the simplest model for it must have $J_g = 1$.

The laser cooling models discussed in Secs. 8.2 and 8.3 both relied on optical pumping among states that have spatially varying light shifts. Since the pumping process always preferentially pumps the atoms toward states with larger light shift for a J_g to $J_e = J_g + 1$ transition, a negative detuning will always pump the atoms to states with the lowest energy and therefore cool the atoms by dissipating energy. However, it is clear that the sub-Doppler force in this σ^+-σ^- case cannot rely on a "Sisyphus" effect because the spatially uniform light shifts preclude "hills" and "valleys".

Nevertheless, the non-adiabaticity required for cooling derives from the atomic motion through a region of rotation of the quantization axis. For atoms at rest in the light field, optical pumping tends to redistribute the populations among the magnetic substates according to the local direction of the linearly polarized light, so the $M_g = 0$ sublevel will be populated most strongly and the sublevels with $M_g = \pm 1$ will be populated less. By contrast, moving atoms experience a rotation of the quantization axis, and must be optically pumped in order to follow it. Thus the population of the ground magnetic substates always lags behind the steady state-distribution appropriate to the local field, *i.e.*, the polarization direction.

The authors of Ref. 83 showed that this non-adiabatic following populates the state with $M_g = +1$ more than the state with $M_g = -1$ for atoms traveling toward the laser beam with σ^+ polarization, and vice versa for atoms traveling in the opposite direction. Even a small imbalance in the population can produce a very large damping force. This is because the $M_g = 1$ sublevel scatters the σ^+ light six times more efficiently than the σ^- light because of the different Clebsch-Gordan coefficients (see Appendix D). Since the atoms remain in the $M_g = 1$ sublevel after absorption of σ^+ light followed by spontaneous emission, atoms traveling toward the σ^+ beam scatter much more light from it and experience a large momentum change in the direction opposite to their motion. Atoms traveling toward the σ^- beam are preferentially pumped toward the $M_g = -1$ sublevel from which they strongly scatter light from the σ^--beam and also recoil in the opposite direction.

The atomic motion is clearly damped, and in this case the cooling mechanism also relies on a differential scattering of light from the two laser beams. However, in this case the differential scattering is *not* caused by the difference in Doppler shifts of the two laser beams as in Doppler cooling, but by the imbalance in the populations caused by the time lag in the following of the atomic orientation to the local field.

From the discussion above it is difficult to assess the size of the friction coefficient and the diffusion coefficient for this σ^+-σ^- cooling process. In Ref. 83 it is shown that both coefficients remain smaller by approximately the same factor compared to the lin $\perp$ lin configuration, so comparable final temperatures are to be expected for the two cases [83]. However, the reduction of the friction coefficient can be important in experiments, because cooling times become longer and the effect of perturbations to the cooling process will have a larger impact.

8.5   Theory of Sub-Doppler Laser Cooling

The models discussed in the preceding sections apply to cases where the small J values result in a small number of evolution equations for the density matrix elements. Real atoms have a much richer structure than these simple cases. For example, cooling of the alkalis Na, Rb, and Cs is achieved on the $F = 2 \leftrightarrow 3$, $F = 3 \leftrightarrow 4$, and $F = 4 \leftrightarrow 5$ transitions respectively. It is commonly accepted that the principles of sub-Doppler laser cooling discussed for the simple cases are applicable to these more complicated transitions. An operator description of sub-Doppler laser cooling for any given transition for any given polarization of the laser beams in the presence of external fields has been given in Ref. 92.

The theory considers the case of atoms moving through a monochromatic radiation field of frequency ω_ℓ. In principle the theory is applicable to laser cooling in three dimensions, but only the case of one-dimensional cooling will be described here, where the velocity $\vec{v}$ is in the direction of the laser beams. The electric field is assumed to be classical and is given by

$$\vec{E}(\vec{R}, t) = \vec{E}_+(\vec{R})e^{-i\omega_\ell t} + \vec{E}_-(\vec{R})e^{+i\omega_\ell t}. \qquad (8.3)$$

The atom-field coupling is given by the Rabi operator $\mathcal{R} = \vec{\mu}_{eg} \cdot \vec{E}_+/\hbar$, which has the magnitude of the Rabi frequency. Also, the force operator $\vec{\mathcal{F}}$ in the radiation field can then be written as (see also Sec. 3.1)

$$\vec{\mathcal{F}} = \vec{\nabla}\left(\vec{\mu} \cdot \vec{E}\right) = \hbar\left(\vec{\nabla}\mathcal{R} + \vec{\nabla}\mathcal{R}^\dagger\right). \qquad (8.4)$$

In the low-intensity limit, where the lowest temperatures for sub-Doppler cooling are observed, the optical coherences and the excited state can be eliminated from the evolution equations of the atomic density matrix because they follow the ground-state adiabatically. The force in the steady state can then be written in terms of the ground-state density matrix $\sigma_{gg}(t)$

$$\vec{F}(t) = \text{Tr}\left[\sigma_{gg}(t)\vec{\mathcal{F}}_{eff}(t)\right] \qquad (8.5)$$

with

$$\vec{\mathcal{F}}_{eff}(t) \equiv \frac{-i\hbar}{\gamma/2 + i\delta} \mathcal{R}^\dagger \vec{\nabla} \mathcal{R} + \frac{i\hbar}{\gamma/2 - i\delta} (\vec{\nabla} \mathcal{R}^\dagger) \mathcal{R}. \tag{8.6}$$

After solving the evolution equations for the ground-state density matrix, Eq. 8.5 can then be used to calculate the force.

It is instructive to split the Rabi operator in two parts $\mathcal{R}(R) = \theta(R) \times \Upsilon(R)$ with $\Upsilon(R)$ the Rabi frequency operator containing the position-dependent strength of the electric field and $\theta(R)$ the coupling of the atom to the field. This can also be position dependent because of the changing polarization vector of the local field. Then Eq. 8.6 can be expanded as $\vec{\mathcal{F}}_{eff} = \vec{\mathcal{F}}_1 + \vec{\mathcal{F}}_2 + \vec{\mathcal{F}}_3 + \vec{\mathcal{F}}_4$ with

$$\vec{\mathcal{F}}_1 = i\, F_0\, \frac{\theta^\dagger \theta}{k\gamma^2} \left((\vec{\nabla}\Upsilon^*)\Upsilon - \Upsilon^* \vec{\nabla}\Upsilon \right), \tag{8.7a}$$

$$\vec{\mathcal{F}}_2 = -\frac{2\, F_0\, \delta}{\gamma} \frac{\theta^\dagger \theta}{k\gamma^2} \left((\vec{\nabla}\Upsilon^*)\Upsilon + \Upsilon^* \vec{\nabla}\Upsilon \right), \tag{8.7b}$$

$$\vec{\mathcal{F}}_3 = i\, F_0\, \frac{|\Upsilon|^2}{k\gamma^2} \left((\vec{\nabla}\theta^\dagger)\theta - \theta^\dagger \vec{\nabla}\theta \right), \tag{8.7c}$$

and

$$\vec{\mathcal{F}}_4 = F_0\, \frac{|\Upsilon|^2}{k\gamma^2} \left((\vec{\nabla}\theta^\dagger)\theta + \theta^\dagger \vec{\nabla}\theta \right). \tag{8.7d}$$

Here the factor F_0 in front is given by

$$F_0 = \frac{\hbar k\gamma/2}{1 + (2\delta/\gamma)^2}, \tag{8.8}$$

which is the force on a two-level atom with $s_0 = 1$. The force operator $\vec{\mathcal{F}}_1$ depends on the gradient of the Rabi frequency and is the well-known radiation pressure, proportional to the phase of the field. The operator $\vec{\mathcal{F}}_2$ is the dipole force operator, determined by the amplitude gradient of the field. Both of these are discussed in Sec. 3.2. The force operators $\vec{\mathcal{F}}_3$ and $\vec{\mathcal{F}}_4$ both arise from a gradient in the polarization direction, and they are related to the forces discussed in the preceding sections describing polarization gradient cooling. Note, that $\vec{\mathcal{F}}_3$ depends on the phase gradient of θ, corresponding to a radiative force, whereas $\vec{\mathcal{F}}_4$ depends on the amplitude gradient of θ, corresponding to a redistribution force.

All force operators $\vec{\mathcal{F}}_i$ have a common prefactor given by

$$\vec{\mathcal{F}}_i \propto \frac{\hbar \vec{k} \,|\Omega|^2\, \gamma/2}{(\gamma/2)^2 + \delta^2} = \frac{F_0 |\Omega|^2\, \hat{k}}{(\gamma/2)^2}, \tag{8.9}$$

where the vector $\vec{k}$ derives from the gradient operator $\vec{\nabla}$ on the right-hand side of Eqs. 8.7a–d. This factor is identical to the radiation force in the limit of low

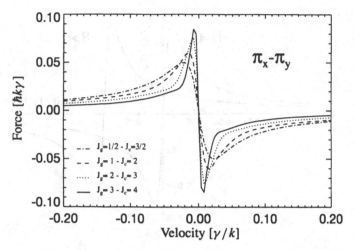

FIGURE 8.5. The calculated force *vs.* velocity curve for the lin ⊥ lin configuration adapted from Ref. 92.

intensity for the case of Doppler laser cooling of a two-level atom in Sec. 3.2. Since the remaining factor is of the order 1, the force in sub-Doppler cooling can never exceed the Doppler force. This is to be expected, since the force still derives from the scattering of light from the two laser beams that form the optical molasses. However, the velocity dependence of the force for the sub-Doppler case, which derives from taking the trace over the density matrix in Eq. 8.5, is much stronger in this case than for the case of Doppler cooling. This leads to an increase in the damping coefficient and, since the momentum diffusion coefficient in the two cases are comparable, leads to lower temperatures as observed in experiments.

After finding the ground-state density matrix by solving the evolution equation, Eq. 8.5 can be used for calculation of the force on an atom. Nienhuis *et al.* [92] studied the case where the atomic velocities are assumed to vary slowly over an optical wavelength. This corresponds to atomic kinetic energies large compared to the potential energy variations caused by the light shift. The periodicity of the problem suggests expansion of the density matrix elements in a Fourier series, and the resulting linear relations for the Fourier coefficients can then be solved numerically. The force is obtained by substitution of the Fourier series in Eq. 8.5, and then averaging over a wavelength.

In Fig. 8.5 the force is plotted for the case of orthogonal linear polarization for transitions from J_g to $J_e = J_g + 1$ for different J_g's. There is an increase of the damping coefficient when J_g is increased that can easily be understood because, for atoms with larger values of J_g, the states with $M_g = \pm J_g$ are less strongly coupled to the light field. For these atoms the optical pumping toward the other states proceeds at a lower rate. For the optimum in the force, the atomic velocity should therefore be lower, whereas the energy loss (the difference in the light shift) remains the same. Since the damping coefficient depends on the ratio of the energy loss to the velocity, it will increase.

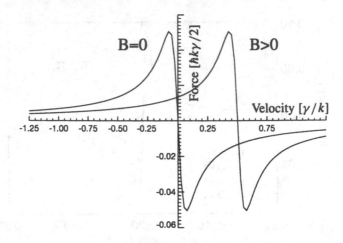

FIGURE 8.6. The calculated force *vs.* velocity curve for the σ^+-σ^- polarization configuration for the case of $B = 0$, and $B \neq 0$ [93]. When $B \neq 0$, the force has a dispersion shape centered at a finite $v_{\text{vsr}} = \omega_Z/k$ where it vanishes. Atoms are cooled to this velocity.

This operator formalism [92] has been extended to the case of laser cooling in stronger magnetic fields by finding operators and calculating forces for the cases described in Secs. 8.9 and 8.10 [94]. An example of the results of such a calculation for the σ^+-σ^- configuration described in Sec. 8.10 is shown in Fig. 8.6. For $B = 0$ there is strong damping to $v = 0$, but for larger B the atoms are damped to a non-zero velocity, in this case ω_Z/k, corresponding to the σ^+-σ^- velocity selective resonance (VSR) illustrated in Fig. 8.12a. Reference 94 shows good agreement between the calculations and the data for all cases, thus putting the VSR view of sub-Doppler laser cooling on firm theoretical grounds.

The momentum diffusion of the atoms can also be obtained using this semiclassical model [92]. It is calculated from the time correlation of the force operator and consists of three terms: (1) the contribution from the random direction of spontaneous emission; (2) the contribution from stimulated processes on a fast time scale caused by the decay of the optical coherences; and (3) the contribution from stimulated processes on a slow time scale caused by the optical pumping among ground states. The first two terms can easily be evaluated at each instant from the local steady-state density matrix. The third term depends on the evolution matrix of the ground-state density matrix and involves an integration over time.

Substitution of both the force and diffusion as a function of velocity in the Fokker-Planck equation (see Sec. 5.4) allows calculation the temporal evolution of the velocity distribution. The results must be cautiously interpreted however, because both the force and the diffusion calculated with the procedure described above assumes that the atoms are in a steady state, *i.e.*, the interaction time is long compared to the optical pumping time.

The operator description can be used to show that lower light intensity results in lower diffusion, but not in a lower damping coefficient. The temperature in the steady state is given by the ratio of the diffusion to the damping coefficient so lower light intensity (or proportionally lower external field) lowers the final temperature. Of course, lowering the intensity also lowers the range for which the semiclassical theory is valid. Furthermore, at very low temperatures, where the recoil of the atom by one photon absorption is comparable to the atomic velocity, the Fokker-Planck equation Eq. 5.20 is no longer valid and a quantum theory is necessary.

8.6   Optical Molasses in Three Dimensions

The theoretical models and experimental results discussed so far in this chapter are all for the case of one dimension. The theoretical models are not easily extended to more dimensions and do not provide the same kind of analytical solutions as does 1D. To enable direct comparison with theory, 1D experiments are required, and these experiments are discussed in Secs. 8.9 and 8.10. One of the limitations of 3D experiments is that they are not able to study cooling schemes without polarization gradients, since the transverse nature of electromagnetic radiation prevents the construction of 3D radiation fields with all polarizations parallel.

One of the outcomes of the models presented in Secs. 8.2 and 8.4 is that the final temperature T_{lim} in polarization gradient cooling scales with the light shift ΔE_g of the ground-states, i.e.,

$$k_B T_{\text{lim}} = b \Delta E_g, \tag{8.10}$$

where ΔE_g is given by Eq. 8.1. The value of the coefficient b depends on the polarization scheme used and is 0.125 for lin $\perp$ lin and 0.097 for σ^+-σ^-. Note that lowering the temperature can easily be achieved by lowering the light shift, either by increasing the detuning δ or decreasing the intensity s_0 (see Eq. 8.1). Since this is a result of the semiclassical theory, the temperature will always be limited by the recoil temperature, as discussed at the end of Sec. 8.5 and in Sec. 8.7.

In the experiments reported by Salomon et al. [95], the temperature was measured in a 3D molasses under various configurations of the polarization. All beams were linearly polarized, but in one configuration the polarization of two counterpropagating beams was chosen to be parallel to one another and in another configuration they were chosen to be perpendicular. Temperatures were measured by a ballistic technique, where the flight time of the released atoms was measured as they fell through a probe a few cm below the molasses region. The sensitivity of the technique was increased by using a specially tailored laser beam to push all but a thin horizontal slice of the atoms out of the molasses just before they were released. In this way the initial vertical position of the remaining atoms was determined more accurately and therefore the fall time was a better measure of their initial vertical velocity.

Results of their measurements are shown in Fig. 8.7a, where the measured temperature is plotted for different detunings as a function of the intensity. For each detuning, the data lie on a straight line through the origin. The lowest temperature obtained is 3 μK, which is a factor 40 below the Doppler temperature and a factor

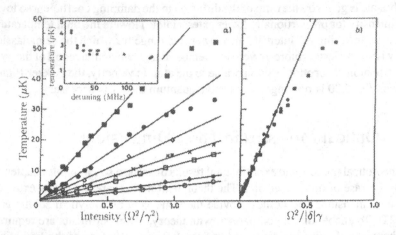

FIGURE 8.7. Temperature as a function of laser intensity and detuning for Cs atoms in an optical molasses from Ref. 95. a) Temperature as a function of the detuning for various intensities. b) Temperature as a function of the light shift. All the data points are on a universal straight line.

15 above the recoil temperature of Cs. If the temperature is plotted as a function of the light shift (see Fig. 8.7b), all the data are on a single universal straight line. The slope of the line is 0.45 for the parallel configuration and 0.35 for the perpendicular configuration. Both slopes are a factor of about 3 higher than the theoretical estimates of 1D and the authors ascribe this discrepancy to the three-fold increase of the number of laser beams.

However, there are a number of differences between the theoretical and experimental situations studied. First, the theory is 1D, whereas the experiments are 3D. Second, the level scheme used in the theory is $J_g = 1/2 \Leftrightarrow J_e = 3/2$ (for lin $\perp$ lin) or $J_g = 1 \Leftrightarrow J_e = 2$ (for σ^+-σ^-), whereas the cooling transition in Cs is a $F_g = 4 \Leftrightarrow F_e = 5$ transition. Third, the polarization gradient in the 1D theory is well-defined, whereas in the 3D experiment atoms see different gradients in different directions and the gradients could change dramatically during the atoms' diffusive movement in the molasses.

In an experiment by Gerz et al. [96] the effect of the angular momentum of the transition on the temperature was studied by exploiting the two isotopes of Rb. In ^{85}Rb the cycling transition is a $F_g = 3 \Leftrightarrow F_e = 4$ transition, whereas in ^{87}Rb the $F_g = 2 \Leftrightarrow F_e = 3$ transition is used. The authors then studied the final temperature as a function of the light shift and found a small effect of the slope on the isotope used. The temperatures for ^{85}Rb are 10% lower under the same laser parameters compared to ^{87}Rb, indicating an increased damping for higher F as predicted by theoretical results [92].

Simulations of the behavior of alkali atoms in an optical molasses were performed by both Molmer [97] and Javanainen [98]. Both authors found that in most cases relation 8.10 holds even in two and three dimensions. However, Javanainen [98] showed that in 3D the temperature is not just given as the ratio of the diffusion averaged over a wavelength and the damping averaged over a wavelength, since the trajectories of the atom are not straight, but severely altered by the cooling process. These aspects have led to the departure from the semiclassical treatments and toward quantum treatments, which are discussed in Chapters 15 and 16.

8.7 The Limits of Laser Cooling

8.7.1 The Recoil Limit

In Sec. 7.2 it was shown that there is a lower limit to Doppler laser cooling that arises from the competition with heating. This heating is caused by the randomness of spontaneous emission, which itself is the irreversible process that is required for laser cooling. It is therefore unavoidable, although its magnitude can be controlled.

This cooling limit is also described there as a random walk in momentum space whose steps are of size $\hbar k$ and whose rate is the scattering rate, $\gamma_p = s_0 \gamma/2$ for zero detuning and $s_0 \ll 1$. As long as the force can be accurately described as a damping force, then the Fokker-Planck equation of Eq. 5.20 is applicable, and the outcome is a lower limit to the temperature of laser cooling given by the Doppler temperature $k_B T_D \equiv \hbar\gamma/2$. A similar argument is used in Sec. 6.4.1 on beam deceleration.

The extension of this kind of thinking to the sub-Doppler processes described in this chapter must be done with some care. In Sec. 8.2.2 it was shown that the damping constant β for polarization gradient cooling is larger than for Doppler laser cooling by the factor $2|\delta|/\gamma$, and a naive application of the consequences of the Fokker-Planck equation would lead to an arbitrarily low final temperature as a result of an arbitrarily large value of β brought about by increasing $|\delta|$. In order to see why this is not true, some careful thinking is needed.

First, the notion that the ratio of D/β gives the final temperature comes from the Fokker-Planck equation, and is only valid for D/β constant, or in particular when D is constant and the force is truly a damping force, that is, $F = -\beta v$. Clearly, this is not true for velocities outside the capture range given by $v_c = \gamma_p/k$ as discussed in Sec. 8.2.2 and shown in Figs. 7.1, 8.5, and 8.6. The difficulty appears because v_c decreases as $\gamma_p \propto 1/\delta^2$ for $\delta \gg \gamma$, and v_c quickly becomes very small, resulting in a very narrow remaining linear region of the force curve (for example, see Fig. 5.3 and its discussion). If the velocity distribution has atoms with $v > v_c$, then these may be still subject to a decelerating force, but not one that is proportional to v [73].

Second, in the derivation of the Fokker-Planck equation in Sec. 5.4, it is explicitly assumed that each scattering event changes the atomic momentum p by an amount that is a small fraction of p as embodied in Eq. 5.10, and this clearly fails when the velocity is reduced to the region of $v_r \equiv \hbar k/M$. Thus the Fokker-Planck equation is not a valid description of the evolution of the velocity distribution in the neighborhood of $v \sim v_r$, and the argument that the final temperature can be arbitrarily reduced by increasing $|\delta|$ is invalid.

This limitation of the minimum steady-state value of the average kinetic energy to a few times $E_r \equiv k_B T_r = (\hbar k)^2/2M$ is intuitively comforting for two reasons. First, one might expect that the last spontaneous emission in a cooling process would leave atoms with a residual momentum of the order of $\hbar k$, since there is no control over its direction. Thus the randomness associated with this would put a lower limit on such cooling of $v_{min} \sim v_r$. Second, the polarization gradient cooling mechanism described above in Sec. 8.2.2 requires that atoms be localizable within the scale of $\sim \lambda/2\pi$ in order to be subject to only a single polarization in the spatially inhomogeneous light field. The uncertainty principle then requires that these atoms have a momentum spread of at least $\hbar k$.

These arguments have been put on firmer ground in a well-studied paper [99]. Its authors provide some analytical and numerical calculations that support these conclusions. They show that the steady-state velocity distribution can be characterized by a single dimensionless parameter U_0/E_r, where U_0 is the light shift of the ground-state. They carefully distinguish the cases where v_c is less than the width of the steady-state distribution at low U_0 resulting, for example, from large δ. This causes the distribution to deviate from Gaussian in the wings so that its width does not correspond to the average value of the kinetic energy (see Fig. 5.3). Figure 8.8 shows their plot of both the average kinetic energy (corresponding to the rms value of momentum) and the width of the velocity distribution at its $1/\sqrt{e}$ height. As shown, the narrowest velocity distribution has a width of a few v_r, corresponding to temperatures $T \sim 10\, T_r$. In this case, the average kinetic energy is much higher because the wings of the velocity distribution contain more fast atoms than would a Gaussian distribution.

8.7.2 Cooling Below the Recoil Limit

The recoil limit discussed in the previous section has been surpassed by two different cooling methods, neither of which can be based in the simple optical notions discussed there. One of these uses optical pumping into a velocity-selective dark state and is described in Chapter 18. The other one uses carefully chosen, counterpropagating laser pulses to induce velocity-selective Raman transitions, and is called Raman cooling [100]. The Raman pulses connect the two hfs sublevels of the ground state of an alkali (in the case of Ref. 100 it was Na), and the two light beams are tuned well away from the atomic resonance to limit spontaneous emission. Since both initial and final states are ground states, their lifetimes are very long and the transitions are correspondingly very narrow, limited only by the adjustable duration of the pulses.

In Ref. 100, Na atoms in a thermal beam are first decelerated using the frequency chirp method of Sec. 6.2.1, then accumulated in a magneto-optical trap as described in Sec. 11.4, and then cooled in a 3D polarization gradient molasses as described in Sec. 8.6 (after the trapping fields were turned off). The cold vapor of atoms was then subjected to a sequence of carefully tailored laser pulses. First there was

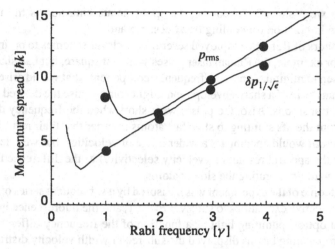

FIGURE 8.8. Plot of the rms momentum (corresponding to the average kinetic energy) and the width at the $1/\sqrt{e}$ height of the steady state momentum distribution vs. Ω/γ for Cs atoms cooled with light detuned by $\delta = -15\gamma$. These curves would coincide for Gaussian distributions. The heavy dots show the result of the Monte Carlo simulations, whereas the lines are from numerical integration of the optical Bloch Eqs. (figure adapted from Ref. 99).

a pulse of a single frequency that excited all atoms within a velocity range γ/k out of the hfs sublevel $F_g = 1$, and therefore populated $F_g = 2$ by spontaneous emission. Since the atoms had been cooled well below the Doppler limit, almost all the atoms were optically pumped this way.

Then there was a short pulse of counterpropagating beams containing frequencies tuned far from atomic resonance, but whose difference $\Delta\omega$ was only slightly less than hfs splitting ω_{hfs}. Therefore atoms travelling toward the higher frequency laser were Doppler-shifted into resonance for the Raman transition from $F_g = 2$ to $F_g = 1$. The resonance condition for the Raman transition is $2kv = \omega_{hfs} - \Delta\omega$, and for atoms cooled to a few times the recoil velocity v_r, it would be necessary to have $\Delta\omega \sim 250$ kHz below ω_{hfs}. Atoms therefore underwent a momentum change of $2\hbar k$ opposite to their motion, and were transferred to sublevel $F_g = 1$. Then another optical pumping pulse returned them to $F_g = 2$, and they suffered a random momentum change of magnitude $\hbar k$. Atoms travelling in the opposite direction were unaffected because the Raman transition was Doppler-shifted further from resonance. This process was repeated several times, and each time the frequency difference between the beams was brought closer to the hfs splitting so that atoms were swept toward zero velocity. Then the entire sequence was repeated, but now with the counterpropagating beams' directions reversed, so that atoms with opposite velocity were also swept toward zero velocity.

In each sequence, the slowing was compromised by the randomness of the recoil from spontaneous emission, but the well-defined momentum change of $2\hbar k$ opposite to the atomic velocity from the Raman pulses dominated, and on average the atoms were slowed. Of course, the spontaneous emission is the necessary dissipative interaction for cooling (see Secs. 5.5 and 8.7).

The authors of Ref. 100 employed several very clever schemes to optimize their results. For example, the Raman laser pulses were not square, but had shapes carefully chosen to minimize spurious frequency components that would arise from the Fourier transform of a sharp envelope and might compromise the desired sequence of atomic transitions. Also, the pulses were short when the frequency difference was far from the hfs splitting to slow fast atoms in order that their resulting broad spectral width would encompass a wider range of velocities, but were lengthened to narrow the spectral resolution (velocity selectivity) as the difference frequency shifted toward decelerating the slower atoms.

The outcome of the experiment was measured by a subsequent series of similarly velocity-selective excitations to sublevel $F_g = 1$, and the fluorescence induced by additional optical pumping light as a function of the frequency difference of the counterpropagating beams displayed the sub-recoil width velocity distribution.

8.8   Sisyphus Cooling

The reader may note that there are several apparently different cooling schemes described in Secs. 6.2.6, 8.2.2, 8.3, and 9.3 that all share a similar energy loss mechanism. This section summarizes the similarities of these schemes, and discusses the underlying physical principles.

The motion of atoms in a spatially varying potential causes an exchange between kinetic and potential energy. If the potential is periodic in space, so is this exchange. The simplest and most dramatic example is for the light shift of a two-level atom in a standing wave detuned from atomic resonance. In this case, the sinusoidal potential for the ground-state is exactly out of phase with that of the excited state because their light shifts have opposite signs as shown in Fig. 9.2. For multilevel atoms, different sublevels may be subject to different potentials as shown in Figs. 8.1 and 8.3.

The presence of multiple potentials enables a velocity-dependent energy loss mechanism when the populations of the states of moving atoms can be manipulated in some appropriate way. In general, if atoms can be in states that give up large amounts of kinetic energy as they move *up* the potential, and then be switched to states that gain back a smaller amount of kinetic energy as they move *down* the potential, the net effect of multiple cycles is to extract energy from the atoms. In some cases when the multiple potentials are out of phase with one another, the switching can be arranged in such a way that atoms are always moving up a potential hill. Of course, that energy must be accounted for, and it is easily seen that if the switching is done by optical pumping, then the fluoresced light is bluer than the absorbed light, and the energy is radiated away. Thus there is a cyclic refrigeration process that converts kinetic energy into potential energy, and eventually radiates away the potential energy.

Any cooling process must be both dissipative and irreversible in order to satisfy thermodynamics. In the examples discussed above the dissipative aspect is apparent in the velocity dependence of the force. Atoms moving through a light field are driven to adjust their internal state to the changing conditions. However, this adjustment cannot be instantaneous and a certain time lag in their internal state arises. Of course, this non-adiabatic character depends on the atomic velocity. For low velocities the time lag is small and small energy losses result. For higher velocities the time lag increases and higher losses occur, until the velocity is too high for atoms to have a significant response to the light field changes. Thus there is a range of velocities where atoms experience a damping force that is proportional but opposite to their velocities.

The irreversible aspect is the optical pumping process. As long as there are spontaneous emission events, light is radiated into the unoccupied modes of the radiation field and lost from the atom-laser system. Since this radiation is generally of higher frequency than the laser light, energy is taken out of the atom-laser system and the atoms are cooled. Since the phase of the spontaneously emitted light is random, information from the atomic system is lost and the process is therefore irreversible. By contrast, stimulated emission puts light back into the radiation field that is driving the atoms, and leaves the atomic wavefunction with a fixed phase relation to the optical field. This process is responsible for the light shift that produces the spatially varying potential in many of the examples discussed below.

The major differences between the cooling schemes of the four sections cited above are in the way the optical pumping is tailored to obtain the desired effect. In the case of atoms excited into Rydberg states described in Sec. 6.2.6, excitation occurs in the small region of strong electric field where the laser beams are focused between a pair of small electrodes. Moving atoms gain potential energy climbing out of the field region, and when they undergo spontaneous decay outside this region where the atomic potential energy is higher, they radiatively dissipate the potential energy they gained. Slow atoms decay before they travel very far, so they lose little kinetic energy, but fast ones go further uphill and thus lose more energy. By contrast, ground-state atoms moving into the field have much smaller Stark shifts, and undergo negligible kinetic energy gains. Thus the atoms move along a level potential until they are excited in the field, then they climb a hill until they decay, which may not be until they reach the flat region at the top of the hill, and then they move along a level potential and repeat the process.

In the second case, described in Sec. 8.2.2, a given pair of atomic ground-state magnetic sublevels experiences a spatially varying light shift as atoms move through a polarization gradient. Atoms are optically pumped between sublevels as they move through the regions of varying polarization in just the right way to keep them always moving uphill. The third case is described in Sec. 8.3 where there is a low intensity standing wave and a weak $\vec{B}$ field perpendicular to its $\vec{k}$ vector. Atoms are shifted among their ground-state sublevels by Larmor precession at the nodes, and by optical pumping at the antinodes, in just the right way to produce a cyclic energy loss as they move through the light field. The Larmor precession is not irreversible, but the optical pumping is, thus satisfying thermodynamic criteria (see Sec. 5.1).

In the fourth case, discussed in Sec. 9.2, the atoms literally lose their mechanical energy to the Earth's gravity, just like Sisyphus. They fall into the evanescent wave field in a hfs state that undergoes a large light shift so that their kinetic energy is stored as optical potential energy. Then they are optically pumped into a hfs state with a much smaller light shift so the stored potential energy is sharply reduced and they cannot bounce back to their original height when they are ejected upward by the field. At their upper turning point they are optically pumped back to their original hfs state with little exchange of mechanical energy. In the fifth case, the high-intensity standing wave of Sec. 9.3, atoms can only decay from a point where the excited-state component of their wavefunctions is significant, and this is at the top of potential hills for blue-tuned light. Thus they also dissipate their potential energy into the radiation field.

The physical notions described in these five examples are often categorized as "Sisyphus" cooling because of the obvious connection with an ancient Greek myth [81, 83]. Furthermore, the concept can be extended to include magneto-optical cooling effects to a finite velocity as described in Secs. 8.9 and 8.10. Velocity selective resonances are also closely related to Sisyphus cooling in a moving reference frame.

8.9   Cooling in a Strong Magnetic Field

The sub-Doppler cooling model called MILC discussed in Sec. 8.3 was restricted to the condition of Eq. 8.2 where $\omega_Z < \gamma_p$ in the antinodes. Thus the perturbation of the transverse magnetic field on the magnetic substates could be neglected and the precession causes only a redistribution of the population at the nodes of the light field. At larger magnetic fields this picture breaks down and there are new phenomena in sub-Doppler laser cooling.

With $\omega_Z > \gamma_p$ it is convenient to choose the quantization axis along the magnetic field direction and describe the optical pumping as a small perturbation to the Zeeman effect. The optical Bloch equations (OBE) for the atomic density matrix given in Sec. 2.3 can then be solved by transforming to a frame rotating with a frequency ω_Z around the magnetic field direction [93]. Neglecting terms that oscillate at twice the rotation frequency, an approximate solution to the OBE can be found that shows resonances for the particular velocity v_{vsr} defined by $2\vec{k} \cdot \vec{v}_{vsr} \equiv \pm \omega_Z$ [93]. The nature of these velocity selective resonances (VSR) can be inferred from Fig. 8.9. By choosing the quantization axis along the magnetic field direction, the ground-states are split by an amount $\hbar\omega_Z$ and the light field can now induce both σ and π transitions. The Raman transitions between two ground-states become resonant when the opposite Doppler shifts of the two laser beams combine to match the difference in ground-state energies. The condition for this resonance is $(\vec{k}_1 - \vec{k}_2) \cdot \vec{v}_{vsr} = \omega_Z$.

The force for this case can be written as

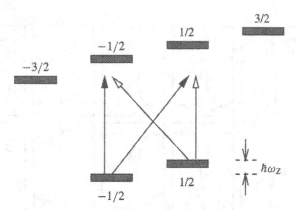

FIGURE 8.9. Optical excitation for a $J = {}^1\!/_2 \Leftrightarrow {}^3\!/_2$ transition for VSR in σ^+ polarized light with a strong $\vec{B}$-field perpendicular to the $\vec{k}$ vectors. In this case $\vec{B}$ is the appropriate choice for the quantization axis. The energy levels are split by the Zeeman interaction. The dark arrows indicate absorption from one laser beam, light arrows from the other. Stimulated emission processes follow the same arrows downward, but are not shown. The resonance condition is satisfied for both sets of arrows (each set has one dark and one light), but for one set π absorption is from one laser beam and σ stimulated emission is into the other, and vice versa for the other set.

$$F = \frac{-\beta(v - v_{\text{vsr}})}{1 + ((v - v_{\text{vsr}})/v_c)^2}, \tag{8.11}$$

where both the friction coefficient β and capture velocity v_c are comparable to the corresponding parameters for other sub-Doppler processes [83]. However, the cooling no longer drives the atoms toward $v = 0$, but to $v = v_{\text{vsr}} = \omega_Z/2k$, and this resonance velocity depends only on the magnetic field strength. Since the cooling process relies on the velocity of the atoms to shift the Raman transition into resonance, the model is called velocity selective resonance (VSR). Note that the laser frequency can be detuned far from atomic resonance and thus the Doppler shift of the laser light does not produce any appreciable difference in the absorption rate of the two beams. However, the Doppler shift is important for the Raman resonance of the transition between the two ground-states, which can be much narrower than the natural width γ [100].

Measurements in a one-dimensional molasses of Rb atoms have verified this model [93]. Figure 8.10 shows the result when the magnetic field is increased from small values up to 1 Gauss. There is cooling toward zero velocity at small magnetic field, as discussed in Sec. 8.3. However, at larger magnetic field the central peak starts to split into two peaks, symmetric around the center of the profile. The splitting of the two peaks Δv_p is plotted in Fig. 8.11 as a function of the magnetic field for various values of the detuning and the laser intensity for the two isotopes of Rb. The straight lines are given by the condition $\Delta v_p = 2v_{\text{vsr}} = \omega_Z/k$ with the appropriate g_F-factor for each isotope (see Eq. 4.4b). There is good agreement between the data and the VSR model.

Figure 8.10 also shows the results for blue detuning of the laser in the lower traces. Since the force vs. velocity curve is reversed when the detuning is changed

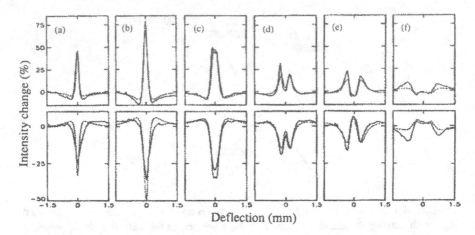

FIGURE 8.10. The spatial distribution of an atomic-beam of ^{85}Rb 1.3 m downstream from the molasses as measured by scanning a hot wire for negative (top) and positive (bottom) detuning (see Fig. 7.2). The laser parameters are $s_0 = 0.25$ and $\delta = \pm0.67\gamma$ and the magnetic field is (a) 0.057 G, (b) 0.114 G, (c) 0.23 G, (d) 0.40 G, (e) 0.57 G, and (f) 1.14 G. The solid lines are experimental data and the dashed lines are the results from the model (figure from Ref. 93).

from negative to positive, there is heating for positive detuning at small B field. However, at large B field there is also clearly sub-Doppler cooling at zero velocity. This is related to the cooling by blue-tuned optical molasses discussed in Sec. 9.3, where a coherence is established between the ground and excited states of a two-level atom [87, 88, 101, 102]. In the present case, however, the coherence is established between two magnetic sublevels by the strong magnetic field, and the laser intensity is always very low.

8.10   VSR and Polarization Gradients

The VSR picture can be extended to include the effects of polarization gradient cooling processes [103], and Fig. 8.12 shows some of the most interesting cases. To establish the Raman resonance condition, the possible two-photon couplings between different states need to be examined carefully. In the case of a strong magnetic field these rely solely on the direction of the magnetic field and of the polarization vectors of the two laser beams. From the energy difference in the splitting of the coupled ground-states the resonance velocities can then be obtained. Since the light can drive only transitions with $\Delta M = 0, \pm1$, the total number of resonances in each case is limited to five, namely, $\Delta M_g = 0, \pm1$, and ±2.

One of the simplest examples of the VSR picture is the case of one-dimensional laser cooling with the σ^+-σ^- polarization gradient scheme in a strong magnetic field pointing along the σ^- laser beam direction. The resonance condition can then only be fulfilled between two ground-state levels with $\Delta M_g = +2$ (see

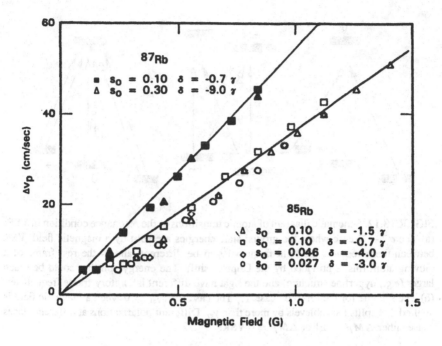

FIGURE 8.11. The separation between the peaks for many data sets, including those of Fig. 8.10, *vs.* magnetic field strength for the $F = 3 \Leftrightarrow 4$ transition in ^{85}Rb and the $F = 2 \Leftrightarrow 3$ transition in ^{87}Rb. Symbols denote experimental points for various intensities ($0.25 \leq s_0 \leq 10$) and detunings ($1 \leq |\delta| \leq 10\gamma$), where the average longitudinal velocity ($v \sim 350$ m/s) was used to convert the deflection angle into a transverse velocity. The solid lines are for the resonance condition. The laser parameters are defined as in Eq. 8.1 and Sec. 2.4 (figure from Ref. 93).

Fig. 8.12a). Since one laser beam has a σ^+-polarization and the other beam has a σ^--polarization, there is no cooling toward the opposite resonance velocity $v_{\text{vsr}} = -\omega_Z/k$ because the Raman transition is Doppler tuned out of resonance at this velocity. Cooling in this scheme will be toward $v_{\text{vsr}} = +\omega_Z/k$, which is twice the resonance velocity in the case of MILC (see Sec. 8.3). The experimental results clearly show cooling to one velocity $v_{\text{vsr}} = \omega_Z/k$. The atoms are not simply deflected, but also cooled toward this resonance velocity, and the width of the peaks in all cases is below the Doppler limit.

Another interesting example of VSR is the case of orthogonal linear polarizations, where the magnetic field is directed along the polarization vector of one of the laser beams. This beam therefore induces $\Delta M = 0$ transitions, whereas the other beam induces $\Delta M = \pm 1$ transitions (Fig. 8.12b). Thus the selection rules can only be satisfied for $\Delta M_g = \pm 1$, or equivalently $v_{\text{vsr}} = \pm\omega_Z/2k$. Since the problem is not symmetric with respect to reversing the atomic velocity, cooling toward $v_{\text{vsr}} = +\omega_Z/2k$ is different from cooling toward $v_{\text{vsr}} = -\omega_Z/2k$. Experiments corroborate this model in great detail.

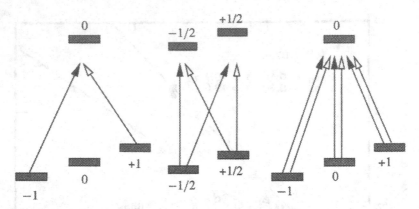

FIGURE 8.12. Schematic diagram of atomic transitions at the resonance condition in VSR. (a) The σ^+-σ^- case where the ground-state energies are split by a magnetic field. VSR between them requires the light frequencies to be different, and in the rest frame of a moving atom this is provided by the Doppler shift. The energy splitting could be much larger (e.g., hyperfine structure) and the light have different laboratory-frame frequencies. (b) The scheme for the lin $\perp$ lin case. (c) The case for $v_{vsr} = 0$ when a magnetic field is applied that splits the sublevels by more than γ_p. Different polarizations at different places cause either $\Delta M_F = \pm 1$ or $\Delta M_F = 0$ VSR.

There is one more case where the atoms are cooled in the lin $\perp$ lin polarization gradient configuration, but now with the magnetic field direction at an angle of 45° with respect to both polarization vectors. Each laser beam can induce transitions $\Delta M = 0, \pm 1$ in a quantization frame along $\vec{B}$, so all five resonance conditions can be fulfilled. The resonance condition for $v_{vsr} = 0$ is particularly interesting (see Fig. 8.12c). The force in this case derives from a redistribution process that transfers photons from one laser beam to the other in stimulated processes that returns atoms to their original ground-state. In all three of these VSR cases, the measurements of Refs. 93 and 103 confirm these models in considerable detail.

9

The Dipole Force

9.1 Introduction

Atoms in an inhomogeneous light field, such as a standing wave, experience a force that derives from the spatial gradient of their light shifts called the dipole force (see p. 10). The simple case of the dipole force on atoms in a standing wave can also be viewed as absorption from one beam followed by stimulated emission into the other of the two counterpropagating beams that constitute the standing wave. The ordering of these sequential events determines the direction of the force, and is itself determined by the relative phase of the counterpropagating beams at the position of the atom. This relative phase, of course, determines the slope of the envelope of the standing wave. These forces can be very much larger than the maximum value of the dissipative force $F_{max} = \hbar k \gamma / 2$ (see Eq. 3.14) because the dipole force is not limited by the requirement for spontaneous decay from the excited state. Since the slope of the potential associated with the light shift increases with light intensity without limit, the force can be arbitrarily large.

A simple appeal to symmetry shows that the dipole force on atoms in a monochromatic standing wave vanishes when averaged over a wavelength. Various asymmetric processes can interfere with such averaging, resulting in a net average force that can either be conservative or not. For velocity-dependent forces, the symmetry breaking caused by spontaneous emission also provides the irreversibility, and the most well-studied case is optical molasses at high intensity (see Sec. 9.3).

For other inhomogeneous light fields, such as the optical trap at the focus of a Gaussian beam discussed in Sec. 11.2.1, absorption followed by stimulated emission from beams with different $\vec{k}$-vectors causes the net force. Similar ideas apply, but are not so easily visualized.

9.2   Evanescent Waves

One particularly important example of the dipole force occurs near the boundary between two dielectric media. For the particular case of glass and vacuum, a light beam that is totally internally reflected at the interface leaves an evanescent wave penetrating into the vacuum. The evanescent intensity decays exponentially, perpendicular to the surface, with characteristic length $\lambda/2\pi$. The sharp intensity gradient can produce a large force [104] that can reflect atoms incident on the surface, and constitutes a very appealing mirror for atoms. A very thorough review of this topic has been given in an excellent article [105].

The maximum atomic kinetic energy that can be reflected in such a scheme is comparable to the light shift at the surface of the dielectric, and this is largest for $\Omega \gg \delta$, given by $\hbar\Omega/2$ (see p. 8). Thus atoms incident on a surface with velocity v and a component normal to the surface of $v \sin\theta$ can be reflected by a light shift of $\hbar\Omega/2 = Mv^2 \sin^2\theta/2$. Because this scales as θ^2 for small angles, and the light intensity scales as Ω^2, the maximum incident angle that can be reflected scales with the fourth root of the light intensity. Thus evanescent wave reflection is indeed a very weak process for thermal velocity atoms.

Even with much more powerful lasers, the domain where $\Omega \gg \delta$ is subject to spontaneous emission at the maximum possible rate $\gamma/2$, and since the atomic reflection time is very many times $1/\gamma$, there are many scattering events. These lead to diffuse reflection through random heating, as well as atomic decoherence, both quite undesirable for an atomic mirror. Therefore evanescent wave mirrors must work in the region where $\Omega \ll \delta$ and the light shifts are considerably smaller, given by $\hbar\Omega^2/4\delta$, in order to ensure elastic reflection.

Grazing incidence reflection from such evanescent waves was demonstrated as early as 1987 [106,107], and falling atoms were even bounced from such an atomic trampoline [108]. However, the use of such a scheme for a general atomic mirror is severely limited by the power requirements.

The reflection of two-level atoms from an evanescent wave is purely conservative because the force derives purely from the light shift. Even spontaneous emission cannot change the nature of the force on a two-level atom. However, a multilevel atom can undergo optical pumping among different ground-states, and a scheme can be arranged to cause energy loss of bouncing atoms. The simplest example comprises the two ground hfs states of an alkali atom, along with one of the excited states as shown in Fig. 9.1. The evanescent wave must be tuned blue of both transitions, and the detuning from each of them differs by the hfs splitting. If the detuning from the higher-frequency transition δ_1 that excites the lower hfs state is chosen to be a small fraction of the hfs splitting, then the detuning δ_2 from

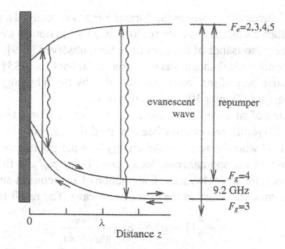

FIGURE 9.1. The gravitational Sisyphus cooling mechanism for Cs atoms bouncing on an evanescent wave. The left side shows the increasing energy of both ground hfs states as atoms approach the blue-detuned evanescent field. If they approach in the lower hfs sublevel and bounce back in the upper one as a result of optical pumping near their turning point (transitions shown nearest $z = 0$), they gain internal energy but lose mechanical energy because they don't bounce back to their original height. They are returned to the lower hfs sublevel by the weak repumping laser (transitions shown far from $z = 0$) during their long, slow upward flight and thus radiate away the internal energy gained close to $z = 0$. The cooling cycle does not happen in the reverse direction because the large light shifts caused by the evanescent wave tune the repumper too far from resonance. The right side shows the relative laser frequencies.

the lower-energy transition that excites the upper hfs state is $\delta_2 = \delta_1 + \omega_{hfs}$, and this could be considerably larger than δ_1 as shown in Fig. 9.1.

Atoms bouncing from a horizontal evanescent wave mirror spend most of their time where their velocity is minimum near their turning points. If they are dropped in their lower hfs state, but are excited by the evanescent light because of its relatively small detuning, and undergo spontaneous emission to the upper hfs state, they will experience a much smaller light shift on the way back up and, consequently, will not bounce back to their original height. Hence they lose mechanical energy because the light shift has raised their ground-state energy considerably as shown in Fig. 9.1. The presence of a weak repumping beam, directed downward, not only returns the rising atoms to the lower hfs state near their upper turning point, but also further reduces their energy. Note that this repumping beam will not be effective when the atoms are near their lower turning point in the evanescent wave because of the large light shift there.

Thus there is a cyclic process by which gravitational energy is converted into optical potential energy via the light shift in the strong evanescent wave, and then the change of state leaves the atoms in the upper hfs state at their next turning point. Optical pumping then radiates away an energy $\hbar\omega_{hfs}$, which is larger than the energy separation at the lower turning point (see Fig. 9.1), and the refrigeration cycle is closed. In some sense, this gravitational Sisyphus cooling is the closest of all to the original Greek myth, where the former King was really always working against gravity. Such a scheme has been realized in a hybrid optical trap discussed in Sec. 11.2.3 where thousands of bounces have been observed [109]. Note, that a nearly identical cooling mechanism was proposed as far back as 1983 [110] except that the conservative part of the force was provided by the inhomogeneous field of a magnetic trap (see Chapter 10) instead of gravity.

The limiting case of such cooling is the loss of energy ΔE in a single bounce. This elementary Sisyphus process has been studied theoretically by several authors [105, 111, 112] who showed that the energy loss per bounce can be readily calculated in terms of the spontaneous decay probability p_{sp} and the branching ratio q for decay into the lower hfs state. Calculation of p_{sp} requires an integration of γ_p along the atomic trajectory in the evanescent wave. The result is

$$\Delta E = -\frac{2}{3}(1-q)p_{sp}E_{kin}\frac{\omega_{hfs}}{\omega_{hfs}+\delta_1}, \qquad (9.1)$$

where E_{kin} is the incoming kinetic energy. Such a single bounce energy loss was first observed in 1995 by grazing incidence reflection of an atomic beam of Na from an evanescent wave mirror with appropriate detuning [113]. The authors report a loss of $\simeq 50\%$ of the kinetic energy associated with the component of velocity perpendicular the evanescent wave mirror.

9.3 Dipole Force in a Standing Wave: Optical Molasses at High Intensity

The low-intensity description of optical molasses given in Sec. 7.2 is appropriate when excited atoms return to the ground state preferentially by spontaneous emission because the rate of stimulated emission is much smaller than γ. When this is not the case, there is a fundamental change in the nature of the optical force.

On p. 10 the eigenvalues of the total Hamiltonian $\mathcal{H}$ were shown to oscillate spatially in a standing wave, and the corresponding spatial dependence of the atomic energy was interpreted as an oscillatory force [87]. Moving atoms experience this potential, and exchange kinetic with potential energy while they move through the optical field as if moving up and down hills. The possibility for a velocity-dependent force arises because the atoms can undergo spontaneous decays from the oscillatory potential of one pair of the dressed states to one of another pair, and those potential curves may not necessarily be the same ones.

Consider an atom moving as shown in Fig. 9.2. For $\delta > 0$, the state $|\phi_1\rangle$ of Eq. 1.19 is a pure ground state in the node where $\Omega = 0$ so that $\Omega' = \delta$ and its energy is lowest. An atom cannot undergo spontaneous emission at such a place because it is a pure ground state, but must decay where there is some excited state mixed in by the atom-field interaction, as shown in Fig. 9.2. If the decay is to another state $|\phi_1\rangle$ in a different pair, the atomic motion in the potential is unaffected, but if the decay is to a state $|\phi_2\rangle$ in a different pair, the phase of the hill climbing and descending is reversed.

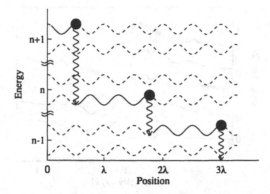

FIGURE 9.2. The spatial dependence of the energy levels of Fig. 1.4 in a standing wave. Two-level atoms moving in such a light field can decay as shown, and the most probable decay point is from the tops of the hills. Since atoms may fall to the bottoms of hills, energy is lost on the average (figure adapted from Ref. 87).

By contrast, an atom beginning in state $|\phi_2\rangle$ is more likely to undergo a spontaneous decay near a node where it is a pure excited state than at an antinode where its wavefunction has an admixture of ground state. If the decay is to a state $|\phi_1\rangle$, again the phase of motion in the oscillatory potential is reversed. The result of such a sequence of decays is that atoms climb more hills than they descend on the average, radiating light of average frequency higher than ω_ℓ from the tops of the hills. Thus they convert kinetic energy into potential energy, and then radiate away the potential energy. Their motion is thereby damped by the light field, and the atoms are cooled, just as in other Sisyphus cooling mechanisms described in Sec. 8.8. Needless to say, the mechanism described here depends on atomic motion and so energy loss is quite small for low velocities and vanishes for $v = 0$. This cooling scheme works only for the case of $\delta > 0$, which is exactly opposite of the detuning needed for laser cooling in low-intensity light as discussed in Sec. 7.2 for optical molasses.

The force that actually slows the atoms derives from the light shift which is the reversible exchange of momentum between the atoms and the light field via absorption followed by stimulated emission. Atoms are excited by light from one beam and stimulated to emit into the other, thus exchanging $2\hbar k$ with the field for each cycle. The rate of these processes is not limited by γ, but increases with the light intensity. This force can therefore be very much larger than the $\hbar k\gamma/2$ limit associated with spontaneous emission processes (see Eq. 6.2). Spontaneous emission and the Doppler shift are not involved in producing the force.

Such stimulated processes produce optical coherences between the ground and excited states, manifest by the mixing of the atomic states. However, in the absence of the spontaneous emission that causes the velocity-dependent damping force, atoms that move through the light field experience no average force because these stimulated processes of momentum exchange between atom and field can occur in either direction with equal likelihood. Movement along the oscillatory potential is

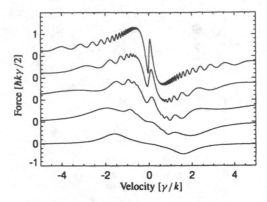

FIGURE 9.3. The velocity dependence of the force on atoms moving in counterpropagating beams with various intensities but with fixed detuning of $\delta = -1.5\gamma$. The bottom curves (lowest intensity) show ordinary optical molasses similar to Fig. 7.1. At higher intensity, near the top, some small wiggles appear that develop into a sign reversal near $v = 0$ and into the Doppleron resonances at v far from 0. The saturation intensity of the curves shown are $s_0 = 1, 5, 25, 100$ and 1000.

conservative and therefore does not produce any cooling. Only spontaneous decay that causes transitions between the manifolds is both irreversible and dissipative, and so can cause the compression of phase space volume.

This "blue cooling" at high intensity has been demonstrated for several atoms [45, 88, 114, 115]. More quantitative calculations show how the force reverses sign for a fixed δ as the light intensity increases from low to high [101], as well as how heating for $\delta > 0$ at low intensity becomes cooling at high intensity (see Fig. 9.3). There are also velocity-resonant phenomena that are often explained in terms of virtual bosons called Dopplerons [102]. The cooling mechanism is reminiscent of the Stark cooling of Rydberg states described above in Section 6.2.6 [63], where atoms move uphill more than downhill because of spontaneous emission between different states. Thus it falls into the category of Sisyphus cooling as discussed in Section 8.8.

9.4   Atomic Motion Controlled by Two Frequencies

9.4.1   Introduction

In the early days of laser cooling, the view of two-level atoms moving in a monochromatic laser beam provided the fundamental picture. The topics that could be described this way included atomic beam slowing and cooling, optical molasses, optical dipole traps, lattices and band structure effects, and a host of others. Within a few years it became clear that this simple view was inadequate, and that the multiple level structure of atoms was necessary to explain some experiments. Perhaps the most dramatic impact came from the discovery of cooling below the Doppler temperature, that could only be explained by polarization gradient cooling in atoms with multiple ground-state levels as described in Chapter 8. Other examples requiring such multiple atomic levels are the MOT (multiple excited state levels, see Chapter 11), along with velocity selective resonances and velocity selective coherent population trapping (both requiring multiple ground-state levels as discussed in Chapters 8 and 18 respectively).

Thus the extension from two-level to multilevel atoms gave an astounding richness to the topic of atomic motion in single-frequency optical fields. It might be expected that a comparable multitude of new phenomena is to be found for the motion of two-level atoms in multifrequency fields, however, this subject has not received as much attention.

Laser cooling requires an irreversible process to provide the dissipation needed to compress the volume of phase space, and this is achieved by spontaneous emission. In the two-level atom case (Doppler cooling), both the force *and* the dissipation are provided by the incoherent process of absorption followed by spontaneous emission. By contrast, in most cases of laser cooling that depend on the multiple levels of real atoms, such as Sisyphus cooling in a polarization gradient, the force results from the frequent coherent sequence of absorption followed rapidly by *stimulated* emission (the dipole force). Comparatively infrequent spontaneous emission, that results in optical pumping among the multiple ground-state levels, provides the required irreversible process.

9.4.2 Rectification of the Dipole Force

In 1987 Kazantzev made some of the earliest proposals for two-frequency light fields where the dipole force does not vanish when averaged over a wavelength [116]. The first example to be considered here of an optical force derived from only stimulated processes that survives such spatial averaging is the case of a light field composed of two standing waves of different frequencies. The resulting phenomenon is called rectification of the dipole force, and can be applied to a two-level atom. The parameters of the two light fields are chosen so that the light shift from second field ΔE_2 is enough to cause a spatial modulation of the detuning of the first field δ_1, while the force of field 2 is still very much smaller than that from field 1 [116, 117].

On p. 10 there is a discussion about why the spatial dependence of the light shift in a single-frequency standing wave is not sinusoidal, even though the intensity distribution is. Such a spatial dependence of the light shift ΔE_1 is plotted for a standing wave field in the solid curve of part (a) in Fig. 9.4, directly from Eq. 1.16. Field 1 has $\delta_1 = 3\gamma$ and $\Omega_1 = 40\gamma$. Curve (b) of Fig. 9.4 shows the dipole force, which is the gradient of the light shift of the solid curve in (a). Its spatial average clearly vanishes, as expected, because of the symmetry about $F = 0$. Of course, this represents the energy and force for only one of the two dressed states. The energy and force of the other partner of the pair has a complementary spatial dependence but their populations are different as well as spatially varying. Note that even with the small detuning, the spontaneous emission that occurs at the

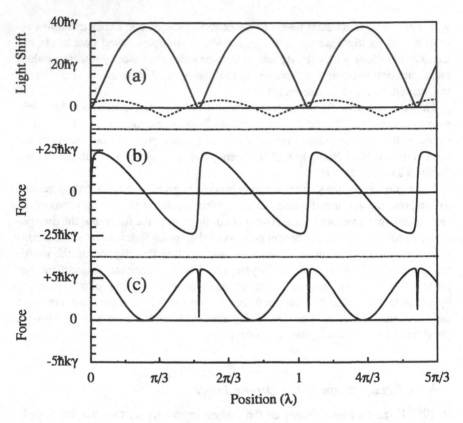

FIGURE 9.4. (a) The solid curve shows the light shift of the ground state of a two level atom in the standing wave of field 1, $\Omega_1 = 40\gamma$ tuned close to resonance, $\delta_1 = 3\gamma$. Note that it is *not* sinusoidal, as discussed on p. 10. The dotted curve shows the spatial variation of the detuning caused by the smaller light shift of the more detuned field 2, which has $\Omega_2 = 50\gamma$ and $\delta_2 = -200\gamma$. (b) The gradient of curve (a), corresponding to the ground state force from field 1. (c) The total force on the atoms, calculated from Eq. 3.16. Here the gradient of a suitably modified curve (a) is not appropriate because the atom spends considerable time in the excited state whose light shift is opposite, and an average must be taken. This is accomplished using the density matrix calculation that generates Eq. 3.16. Because the sign of the light shift, and hence of the dipole force, depends on the detuning, choosing the relative spatial phase of these standing waves carefully results in a rectification of the force as a result of the spatial variation of the detuning of the strong field 1 caused by the small light shift of field 2.

maximum rate $\gamma_p \sim \gamma/2$, is still about 80 times slower than the rate of stimulated processes at Ω_1.

The presence of a second standing wave with a relatively small maximum light shift ΔE_2 can strongly affect the spatial dependence of the energy levels connected by field 1, and thus the force. The dotted curve in part (a) of Fig. 9.4 shows the detuning of field 1 produced by the light shift ΔE_2 of these energy levels. The standing wave of field 2 is displaced to the right by the choice of the phase, and its dotted curve is displaced downward by the 3γ detuning of field 1. Field 2 has $\delta_2 = -200\gamma$ and $\Omega_2 = 25\gamma$ for each counterpropagating field that makes the standing wave, and so δ_2 dominates the radical $\sqrt{\Omega^2 + \delta^2}$ everywhere. Its light shift is only $\Delta E_2 \simeq -\hbar\Omega_2^2/4\delta_2 \approx 3\hbar\gamma$ compared with the $\simeq 40\hbar\gamma$ of field 1, and the force from it is correspondingly much smaller.

Because ΔE_2 is small, it can be simply treated as a spatially varying shift of the atomic energy levels relative to the frequency of field 1. (A complete solution of the optical Bloch equations 2.21 shows that over a range of appropriate parameter values these light shifts can be considered separately [117].) With this shift, the detuning of field 1 from atomic resonance varies between δ_1 and $\delta_1 - 2\Delta E_2/\hbar$, approximately $\pm3\gamma$ as shown in the dotted curve of Fig. 9.4a. (The factor of 2 arises because both ground and excited states of the transition have light shifts, and they're opposite.)

Since both the light shift and the dipole force reverse sign with the detuning as shown in Sec. 1.2.1, the corresponding spatial dependence of the force is changed from the curve of Fig. 9.4b to the curve of Fig. 9.4c. It is mostly positive, its spatial average does not vanish, and atoms are thus always subject to an average positive force. The oscillatory dipole force of field 1 has been "rectified" by the light shift caused by field 2 with its well-chosen spatial phase. Making the force negative, instead of positive as shown, would require that the second standing wave be shifted spatially by $\lambda/2$ so that the sign of its relative spatial phase with the first standing wave is reversed.

The reader should be cautioned that this rectified force is *not* suitable for cooling atoms. It is completely conservative as a result of the absence of spontaneous emission, and so it is not surprising that the rectified force cannot compress the phase space volume of a sample of atoms. Furthermore, it depends sensitively on the detunings of both beams, and the Doppler shifts of a collection of atoms with different speeds would render this scheme inoperable.

9.4.3 The Bichromatic Force

In contrast to the rectification described above, choosing somewhat different parameters for light of two frequencies can indeed produce a force appropriate for slowing and cooling a thermal sample. Reference 118 describes a dramatic demonstration that exploits the enormous dipole force on a two-level atom in a two-frequency light field for slowing a thermal beam in a very short distance. As above, the dipole force derives from the frequent coherent sequence of absorption followed rapidly by stimulated emission. Spontaneous emission among the dressed state levels interrupts this coherent process and provides a non-vanishing force over large distances and a wide velocity range, as well as the required irreversible process for cooling. The resulting force is no longer limited to $\hbar k\gamma/2$. This doesn't happen in ordinary single-frequency Sisyphus cooling because the spatial average of the dipole force vanishes, and with a single frequency, there is no "judicious choice" of a spatial phase shift to rectify the force.

The physical principles of bichromatic slowing have been laid out quite clearly in Refs. 118–120.[1] Consider atomic motion along the axis of counterpropagating light beams containing two frequencies. Each beam contains both frequencies, and they are detuned from atomic resonance by $\pm\delta$ (difference frequency $= 2\delta$). This is in strong contrast to the rectification case discussed above where the detunings are quite different. Each beam can be thought of as an amplitude-modulated single carrier frequency at the atomic resonance, and the modulation period is π/δ.

The equal intensities of each beam are chosen so that the envelope of one "pulse" of the beats satisfies the condition of a π-pulse for the atoms: ground-state atoms are coherently driven to the excited state and vice versa. This π-pulse condition is found by adding the electric fields of the light of each frequency, or correspondingly the associated Rabi frequencies $\Omega_1 = \Omega_2 \equiv \Omega$ to find

$$\Omega_{total} = 2\Omega \cos \delta t, \tag{9.2}$$

because the difference frequency between the two beams is 2δ. Then the π-pulse condition is found from

$$\int_{-\pi/2\delta}^{\pi/2\delta} 2\Omega \cos \delta t \, dt = \pi. \tag{9.3}$$

The result is $\Omega = \pi\delta/4$.

Since atoms are subject to these π-pulses alternately from one beam direction and then the other, they are coherently driven between the ground and excited states, and the force on them can become very large. This is because the first π-pulse causes excitation along with momentum transfer in one direction, and the second π-pulse causes stimulated emission along with momentum transfer *along the same direction*. The magnitude of the momentum transfers in each full cycle is $2\hbar k$ and the repetition rate of these processes is δ/π, so that the optimum total force is on the order of $2\hbar k\delta/\pi$. This is *very* much larger than the usual maximum radiative force $\hbar k\gamma/2$ given by Eq. 6.2, principally because it is a coherently controlled rapid momentum exchange whose rate is limited only by laser power through the π-pulse condition.

The mechanism described above requires two additional features to be applicable to atomic laser cooling: (1) there must be some directional asymmetry so that its spatial average doesn't vanish, as does the usual dipole force in a standing wave as described above, and (2) it must be velocity dependent so that it can compress the phase space volume occupied by the atomic sample. The first condition is satisfied by a careful choice of the relative intervals between the counterpropagating "pulses" of light, and exploitation of the random nature of spontaneous emission. If the oppositely traveling pulses are unevenly spaced, spontaneous emission is more likely to occur in the longer interval than in the shorter one as long as $\tau \gg \pi/\delta$. Then the force will have an average in the direction of the pulse following the longer interval. Atoms that get out of phase with this preferred order by suffering

---

[1] Warning, the definition of γ in Refs. 118 and 119 is $\gamma \equiv 1/2\tau$.

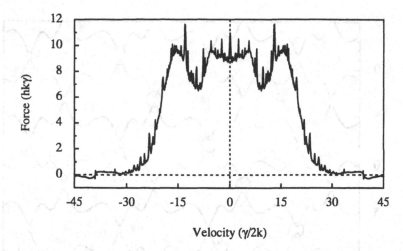

FIGURE 9.5. The velocity dependence of the bichromatic force for $\Omega = 43\gamma$ and $\delta = 39\gamma$ where $\gamma \equiv 1/\tau$. The average value of the force is only about 3/4 of that expected from the π-pulse model of $\hbar k\delta/\pi$. The calculation is done using the program of Ref. 118. The small wiggles near $v = \pm40\gamma/k$ are ordinary optical molasses.

spontaneous emission during the short interval are more likely to undergo correction than to remain out of phase. The optimum duration for the shorter interval is $\pi/4\delta$ so that atoms typically spend $1/4$ of their time in the wrong phase. Then another interval of $\pi/4\delta$ is needed to negate the resulting undesired force, leaving half the time, $\pi/2\delta$ for the desired force. Thus the average force is $\hbar k\delta/\pi$, half the optimum estimated above, and this is borne out by numerical calculations shown in Fig. 9.5 [118,119].

The velocity dependence arises from more subtle effects that derive from the combination of light shifts and Doppler shifts. A large force requires a large value of δ and a correspondingly large value of Ω to satisfy the π-pulse condition, suggesting that the dressed atom picture is appropriate for description of the system. For velocities near $\pm\delta/2k$ the combination of light shifts and Doppler shifts optimize the transfer of momentum between the counterpropagating light beams.

Consider each beam as a carrier frequency at ω_ℓ that is 100% amplitude modulated with period π/δ. This allows a description of the problem as having two frequencies instead of four, thereby providing a much simpler picture. Atoms moving at velocity v see these counterpropagating modulated beams at the Doppler shifted carrier frequencies $\omega_\ell \pm kv$. The dressed state manifolds of atoms exposed to a single optical frequency contain two states separated by Ω' (see Eq. 1.16 and Fig. 1.2). But in the presence of two frequencies equally detuned above and below resonance by $\pm kv$, each pair of states in a manifold becomes an infinite ladder separated by kv [121].

This is relatively easy to see by enumerating the dressed states of energies going upward from that of $|g, m, n\rangle$, where g represents the ground state, m represents the photon number of the red-detuned field mode, and n represents the photon number of the blue-detuned mode. The energy of $|e, m-1, n\rangle$ is $\hbar kv$ higher than

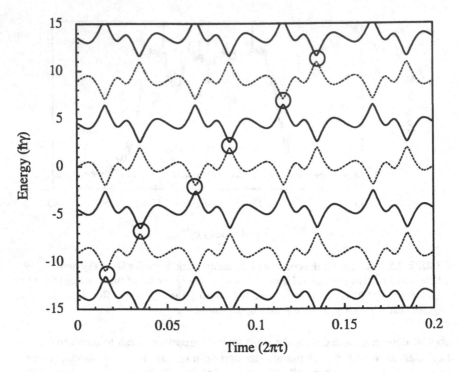

FIGURE 9.6. The time dependence of the dressed state energy levels of an atom moving at velocity $v = 4.5\,\gamma/k$ in a bichromatic field of detuning $\delta = 10\,\gamma$. When the fields of both frequencies have the same amplitude, the splitting at $t = 0$ is $\pm kv$ because the light shift vanishes, as discussed in the text. An atom beginning at the lower left evolves toward the upper right by Landau-Zener transitions at the circles, so the laser field gains energy $\sim 7\hbar kv$ in time $0.3\,\pi\tau$. This model force of $2\hbar k\delta/\pi$ needs to be reduced by a factor of 2 resulting from spontaneous emissions as explained in the text.

that of $|g, m, n\rangle$, and the energy of $|g, m - 1, n + 1\rangle$ is higher by another $\hbar kv$. This ladder extends infinitely upward through $|e, m - 2, n + 1\rangle, |g, m - 2, n + 2\rangle,$ $|e, m - 3, n + 2\rangle$, etc., and similarly downward, as shown in Fig. 9.6.

   If the two light beams were constant and uniform, the interaction that produces the light shift could shift these levels neither upward nor downward because they would always be repelled by a neighboring state. However, the light fields are both modulated and out of phase with one another, so the energy shifts are not zero, but are time dependent. The levels are shifted apart by whichever beam is strongest at any particular instant, and these alternate.

   With proper choice of Ω these energy levels undulate up and down in time by amounts that first "touch" the neighboring state below and then the neighboring state above, so the atoms can make Landau-Zener transitions through the infinite ladder of states. If the phases are chosen so that the atoms' potential energy increases through the array of states $|g, m, n\rangle$ enumerated above, their kinetic energy decreases and they are slowed. The rate of energy loss is $\hbar kv$ divided by π/δ, and setting this equal to the force × velocity gives once again the huge deceleration force $\hbar k\delta/\pi$.

For velocities smaller than $\simeq \delta/2k$ there are Doppleron-like higher order processes that also contribute to the force, and for velocities larger than this the force drops rapidly to zero (see Fig 9.5). The huge net force $F = \hbar k\delta/\pi$ has a nearly constant magnitude over a velocity range of $v = \pm\delta/2k$, as shown in Fig. 9.5 [118]. Shifting this force curve along the velocity axis to make it suitable for beam slowing is accomplished simply by opposite shifts of the two carrier frequencies by $\pm\delta/2$ so that the picture given above is correct in a frame moving at $v = \delta/2k$. References 118 and 119 provide a good description of how the bichromatic force depends on velocity, and how to exploit it for beam slowing.

9.4.4 Beam Collimation and Slowing

The very large bichromatic force can provide a dramatic improvement in atomic beam collimation. This happens because the time required to stop atoms from an initial transverse velocity v_x is $Mv_x/2F = \pi/2\omega_r \simeq 25$ μs for a typical $\omega_r \simeq 2\pi \times 10$ kHz, where $\omega_r = \hbar k^2/2M$ is the recoil frequency. This *very* short time is independent of v_x and δ because the force $F = \hbar k\delta/\pi$ is proportional to $\delta = kv_x$. For transverse collimation of a thermal atomic beam, only a few mm length along the beam is required for an interaction time of $\simeq 10 - 50$ μs. For a collimation angle $\theta \simeq 0.1$, the beam would expand by only a small fraction of 1 mm, and the corresponding laser parameters are quite modest: a laser beam would need only a few mW of power for such collimation.

Of course, the bichromatic force has a fixed sign, and can only operate on atoms diverging in one direction. However, the direction of the force can be reversed by shifting the relative phase of the two frequencies by π, and this can be done by reflection from a mirror $\pi c/2\delta \simeq 1-2$ m away, a distance easily accommodated on an optical table. This returning beam could collimate atoms diverging in the opposite direction a few mm downstream. It is also possible to exploit the Doppler shift by tilting the laser beams with respect to the forward direction, thereby choosing $-v_x = 0$. The extension of this collimation scheme to two dimensions is straightforward. Thus the bichromatic force can be used to capture the atoms emitted from a thermal atomic beam in a cone of angle about 0.1 radian.

It is quite appealing to consider the application of this bichromatic force to both the collimation and the slowing of a beam of metastable helium in the 2^3S state (He*). There are several examples of sources of He* that operate at the temperature of liquid nitrogen (77 K) where the atoms have a mean velocity $\simeq 800$ m/s [122,123]. The required detuning δ and laser power for effective slowing can be found for the He* transition at $\lambda = 1.083$ μm, for which $\gamma/k = 1.7$ m/s. Slowing atoms to 100 m/s requires $\delta/k \simeq 700$ m/s so that $\delta = 400\gamma$ (the frequencies would be shifted so that the velocity range of the force is from $v = 100$ to 800 m/s, centered at 450 m/s). Thus the force $\hbar k\delta/\pi$ is $\simeq 250$ times larger than the radiative force, $\hbar k\gamma/2$, and the slowing distance is correspondingly reduced from 1.8 m to less than 1 cm! The slowing time $\pi/2\omega_r$ is only $\simeq 6\mu$s. One of the most important advantages of this very short slowing length and time is the reduction of atom loss by several orders of magnitude that accompanies the usual 1–2 m slowing length associated with the radiative force. Needless to say, the broad velocity range covered by the bichromatic force removes the necessity of Doppler compensation, typically a multi-kilowatt Zeeman magnet.

Even though the π-pulse model suggests $\Omega = \pi\delta/4$, as given by Eq. 9.3, numerical calculations have shown that $\Omega = 1.1\,\delta$ gives better results for the bichromatic force [109, 119]. Then $\delta = 400\gamma$ leads to a required saturation parameter of $s = 2\Omega^2/\gamma^2 \sim 4 \times 10^5$. Since the saturation intensity for this transition is 0.16 mW/cm^2, the required intensity is only $\sim 6 \times 10^4$ mW/cm^2, corresponding to 600 mW in a 1 mm diameter beam. Since the slowing distance is so short, a 1 mm diameter beam is practical. With the advent of fiber laser amplifiers having high gain and capable of few W output at $\lambda = 1083$ nm [124], such an experiment seems quite feasible.

10
Magnetic Trapping of Neutral Atoms

10.1 Introduction

Magnetic trapping of neutral atoms has the potential for use in very many areas, including high-resolution precision spectroscopy, collision studies, Bose-Einstein condensation, and atom optics. Although ion trapping, laser cooling of trapped ions, and trapped ion spectroscopy were known for many years [125], it was only in 1985 that neutral atoms were first trapped [126]. Such experiments offer the capability of the spectroscopic ideal of an isolated atom at rest, in the dark, available for interaction with electromagnetic field probes.

Confinement of neutral atoms depends on the interaction between an inhomogeneous electromagnetic field and an atomic multipole moment. Although Earnshaw's theorem prohibits an electrostatic field from stably trapping a charged particle (similarly for magnetic fields and monopoles), dipoles may be trapped by a local field minimum (local maxima are forbidden [127]). Unperturbed atoms do not have electric dipole moments because of their inversion symmetry, and therefore electric (*e.g.*, optical) traps require induced dipole moments that must be produced by mixing states of opposite parity. This is often done with nearly resonant optical fields, thus producing the optical traps discussed in Chapter 11. On the other hand, many atoms have ground- or metastable-state magnetic dipole moments that may be used for trapping them magnetically.

In order to confine any object, it is necessary to exchange kinetic for potential energy in the trapping field, and in neutral atom traps, the potential energy must be stored as internal atomic energy. There are two immediate and extremely important consequences of this requirement. First, the atomic energy levels will necessarily shift as the atoms move in the trap, and these shifts will affect the precision of spectroscopic measurements, perhaps severely. Since one of the potential applications of trapped atoms is in high-resolution spectroscopy, such inevitable shifts must be carefully considered. For example, the quantum states of motion of trapped atoms must be well characterized in order to interpret spectroscopic measurements [128]. Furthermore, the spatial distribution of a magnetically confined Bose condensate that constitutes the trap's ground state must be well understood for a wide variety of studies.

Second, practical traps for ground-state neutral atoms are necessarily very shallow compared with thermal energy because the energy level shifts that result from convenient size fields are typically considerably smaller than $k_B T$ for $T = 1$ K. Neutral atom trapping therefore depends on substantial cooling of a thermal atomic sample, and is often connected with the cooling process. In most practical cases, atoms are loaded from magneto-optical traps, where they have been efficiently accumulated and cooled to mK temperatures (see Sec. 11.4), or from optical molasses, where they have been optically cooled to μK temperatures (see Sec. 7.2).

The small depth of neutral atom traps also dictates stringent vacuum requirements, because an atom cannot remain trapped after a collision with a thermal energy background gas molecule. Since these atoms are vulnerable targets for thermal energy background gas, the mean free time between collisions *must* exceed the desired trapping time. The cross section for destructive collisions is quite large because even a gentle collision (*i.e.*, large impact parameter) can impart enough energy to eject an atom from a trap. At pressure P sufficiently low to be of practical interest, the trapping time is $\sim (10^{-8}/P)$ s, where P is in Torr.

10.2 Magnetic Traps

The Stern-Gerlach experiment in 1924 first demonstrated the mechanical action of inhomogeneous magnetic fields on neutral atoms having magnetic moments, and the basic phenomenon was subsequently developed and refined, for example, into the use of magnetic hexapole lenses for focusing and state selecting atoms in beams in the 1950s [129, 130]. An atom with a magnetic moment $\vec{\mu}$ can be confined by an inhomogeneous magnetic field because of an interaction between the moment and the field. This produces a force given by

$$\vec{F} = \vec{\nabla}(\vec{\mu} \cdot \vec{B}). \tag{10.1}$$

Several different magnetic traps with varying geometries that exploit the force of Eq. 10.1 have been studied in some detail in the literature. The general features of the magnetic fields of a large class of possible traps has been presented [131]. These include designs with coaxial coils (quadrupole and hexapole) as well as others, most notably the Ioffe trap discussed below.

W. Paul originally suggested a quadrupole trap comprised of two identical coils carrying opposite currents (see Fig. 10.1). This trap clearly has a single center

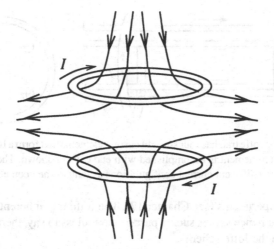

FIGURE 10.1. Schematic diagram of the coil configuration used in the quadrupole trap and the resultant magnetic field lines. Because the currents in the two coils are in opposite directions, there is a $|\vec{B}| = 0$ point at the center.

where the field is zero, and is the simplest of all possible magnetic traps. When the coils are separated by 1.25 times their radius, such a trap has equal depth in the radial (x-y plane) and longitudinal (z-axis) directions [131]. Its experimental simplicity makes it most attractive, both because of ease of construction and of optical access to the interior. Such a trap was used in the first neutral atom trapping experiments at NIST on laser-cooled Na atoms for times exceeding 1 s, and that time was limited only by background gas pressure [126].

The magnitude of the field is zero at the center of this trap, and increases in all directions as

$$B = A\sqrt{\rho^2 + 4z^2}, \tag{10.2}$$

where $\rho^2 \equiv x^2 + y^2$, and the field gradient A is constant (see Ref. 131). The field gradient is fixed along any line through the origin, but has different values in different polar directions because of the 4 in Eq. 10.2. Therefore the force of Eq. 10.1 that confines the atoms in the trap is neither harmonic nor central, and angular momentum is not conserved. Later in this chapter there is some discussion about classical motion and quantum states in this potential.

The requisite field for the quadrupole trap can also be provided in two dimensions by four straight currents as indicated in Fig. 10.2. The field is translationally invariant along the direction parallel to the currents, so a trap cannot be made this way without additional fields. These are provided by end coils that close the trap, as shown (see Ref. 131 for calculations and plots of the field of this Ioffe trap).

Although there are very many different kinds of magnetic traps for neutral particles, this particular one has played a special role. As described in Secs. 10.4.1 and 10.4.2, there are certain conditions required for trapped atoms not to be ejected in a region of zero field such as occurs at the center of a quadrupole trap. This problem is not easily cured, so the Ioffe trap has been used in some of the Bose

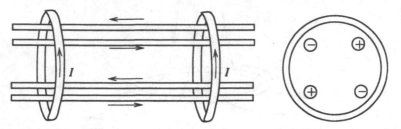

FIGURE 10.2. The Ioffe trap has four straight current elements that form a linear quadrupole field. The axial confinement is accomplished with end coils as shown. These fields can be achieved with many different current configurations as long as the geometry is preserved.

condensation experiments (see Chapter 17). The Ioffe trap inherently has $|\vec{B}| \neq 0$ everywhere, and hence serves such a purpose. Needless to say, there are a number of variations on the Ioffe scheme.

10.3   Classical Motion of Atoms in a Magnetic Quadrupole Trap

There are several motivations for studying the motion of atoms in a magnetic trap. Knowing their positions may be important for trapped atom spectroscopy [132, 133]. Optical cooling of atoms once they are trapped may depend upon knowing both their positions and velocities so that laser beams of proper polarization and direction can be applied. Simply studying the motion for its own sake has turned out to be an interesting problem because the distorted conical potential of the quadrupole trap does not have analytic solutions, and its bound states are not well known.

Because of the dependence of the trapping force on the angle between the field and the atomic moment (see Eq. 10.1), the orientation of the magnetic moment with respect to the field must be preserved as the atoms move about in the trap. Otherwise the atoms may be ejected instead of confined by the fields of the trap. This requires velocities low enough to ensure that the interaction between the atomic moment $\vec{\mu}$ and the field $\vec{B}$ is adiabatic, especially when the atom's path passes through a region where the field magnitude is small and therefore the energy separation between the trapping and non-trapping states is small. This is especially critical at the low temperatures of the Bose condensation experiments. Therefore energy considerations that focus only on the trap depth are not sufficient to determine the stability of a neutral atom trap: orbit and/or quantum state calculations and their consequences must also be considered.

10.3.1   Simple Picture of Classical Motion in a Trap

For the two-coil quadrupole magnetic trap of Fig. 10.1, stable circular orbits of radius ρ in the $z = 0$ plane can be found classically by setting $\mu \nabla B = M v^2 / \rho$, so $v = \sqrt{\rho a}$, where $a \equiv \mu \nabla B / M$ is the centripetal acceleration supplied by the field gradient (cylindrical coordinates are appropriate). Such orbits have an angular frequency of $\omega_T = \sqrt{a/\rho}$. For traps of few cm size and a few hundred Gauss depth, $a \sim 250$ m/s^2, and the fastest trappable atoms in circular orbits have $v_{max} \sim 1$ m/s so $\omega_T / 2\pi \sim 20$ Hz. Because of the anharmonicity of the potential, the orbital frequencies depend on the orbit size, but in general, atoms in lower energy orbits have higher frequencies.

In order for the quadrupole trap to work, the atomic magnetic moments must be oriented so that they are repelled from regions of strong field. This orientation may be produced by the optical pumping process that occurs during Zeeman compensated laser deceleration and cooling of atoms (see Sec. 6.2.2), but it must be preserved while the atoms move around in the trap even though the trap fields change directions in a very complicated way. The condition for adiabatic motion can be written as $\omega_Z \gg |dB/dt|/B$, where $\omega_Z = \mu B/\hbar$ is the Larmor precession rate in the field. The orbital frequency for circular motion is $\omega_T = v/\rho$, and since $v/\rho = |dB/dt|/B$ for a uniform field gradient, the adiabaticity condition is

$$\omega_Z \gg \omega_T. \qquad (10.3)$$

More general orbits must satisfy a similar condition.

For the two-coil quadrupole trap, the adiabaticity condition can be easily calculated. Using $v = \sqrt{\rho a}$ for circular orbits in the $z = 0$ plane, the adiabatic condition for a practical trap ($A \sim 1$ T/m) requires $\rho \gg (\hbar^2/M^2 a)^{1/3} \sim 1$ μm as well as $v \gg (\hbar a/M)^{1/3} \sim 1$ cm/s. Note that violation of these conditions (i.e., $v \sim 1$ cm/s in a trap with $A \sim 1$ T/m) results in the onset of quantum dynamics for the motion (deBroglie wavelength $\approx$ orbit size). This is precisely the domain of the quantum motion studied in Chapter 15 and the Bose condensation experiments discussed in Chapter 17.

Since the non-adiabatic region of the trap is so small (less than 10^{-18} m^3 compared with typical sizes of ~ 2 cm corresponding to 10^{-5} m^3), nearly all the orbits of most atoms are restricted to regions where they are adiabatic. Therefore most of such laser-cooled atoms stay trapped for many thousands of orbits corresponding to several minutes. At laboratory vacuum chamber pressures of typically 10^{-10} Torr, the mean free time between collisions that can eject trapped atoms is ~ 2 min, so the transitions caused by non-adiabatic motion are not likely to be observable in atoms that are optically cooled. However, evaporative cooling (see Chapter 12) reduces the average total energy of a trapped sample sufficiently that the orbits are confined to regions near the origin so such losses dominate, and several schemes have been developed to prevent such losses from non-adiabatic transitions (see Sec. 10.4.3).

10.3.2 Numerical Calculations of the Orbits

Atoms that are near the center of a two-coil quadrupole trap see the potential $U = \mu A \sqrt{\rho^2 + 4z^2}$ [131] that is appropriate as long as an atom stays in a fixed

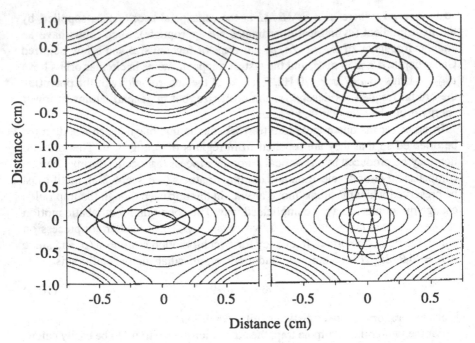

FIGURE 10.3. Orbits of atoms in a magnetic trap found from solving the classical equations of motion. These closed trajectories are shown superposed on the potential contours of the trap. The orbit on the upper left clearly has enough energy to escape, but is confined by bouncing off the field close to the coils, which are located outside the corners of each plot. Very small changes in the initial conditions lead to open orbits.

Zeeman sublevel so its interaction with the field is determined. In any direction, U rises linearly with distance from the origin, but the force is not central anywhere except along the lines $z = 0$ and $\rho = 0$, and therefore orbital angular momentum is not conserved. This asymmetric anharmonic potential does not yield analytical solutions, but numerical integration of Hamilton's classical equations of motion have been reported [134]. In addition to the circular orbits discussed above, there are other closed orbits in the planes containing the symmetry axis. Some of these are shown, along with the potential contours, in Fig. 10.3.

In addition to these closed orbits, there are unclosed, bounded orbits that cannot be adequately described by plots such as those shown in Fig. 10.3. Instead, Poincaré sections were made for each crossing of the $z = 0$ plane and some of these are shown in Fig. 10.4 [134]. These show relatively smooth island chains that break into successively smaller islands at higher energy. As the energy of the trapped particle is increased toward the depth of the trapping potential, the motion becomes more and more erratic, and the islands of stability begin to break up and proliferate.

When the atomic kinetic energy approaches the trap depth, the motion clearly becomes chaotic, as shown in Fig. 10.4. However, there are regions where the potential is higher than in the saddle points at the trap threshold, and there are

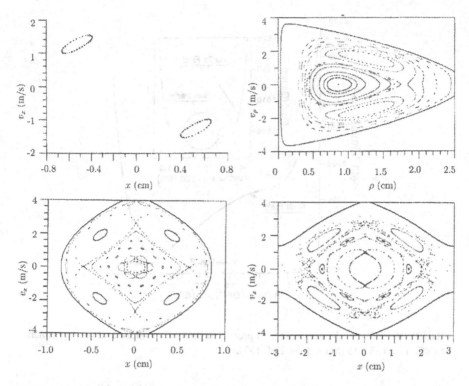

FIGURE 10.4. Poincaré plots of open orbits for magnetically trapped atoms. The same calculations that led to the plots of Fig. 10.3 are used for these. In the upper left plot, an orbit similar to the upper left of Fig. 10.3 is started with slightly different parameters, and an energy low enough to stay trapped. The orbit is quasi-regular. In contrast to the other three, the plot on the upper right is for non-zero angular momentum. The two lower plots are for different energies, showing the breakup of the island chains.

orbits where atoms bounce off these high points as shown in Fig. 10.3. Therefore the surface of section plots show islands even for energies above the trap depth.

10.3.3 Early Experiments with Classical Motion

There has been a large number of successful neutral atom magnetic trapping experiments. The first one used an atomic beam slowed by the Zeeman compensation technique described in Sec. 6.2.2 [126]. The atoms were allowed to drift out of the solenoid and into the center of a two-coil trap. Then a short pulse of nearly resonant light brought the drift velocity to zero, and the field coils were quickly turned on. Atoms could be trapped for 1 s, but this value was limited only by the vacuum of 10^{-8} Torr. The density of trapped atoms was estimated to be $10^3/\text{cm}^3$, several orders of magnitude below the density of the background gas. This first demonstration showed that ideas of neutral atom trapping could indeed be put into practice.

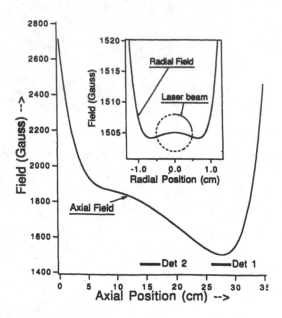

FIGURE 10.5. Plot of the magnetic field profile for the Ioffe trap used in the experiments of Refs. 132, 133, 135 (figure from Ref. 133).

Later, similar experiments in a much better vacuum achieved 2 min. trapping time [50]. In these experiments the authors used a multisection superconducting solenoid magnet whose first section had a decreasing field to serve as a Zeeman-compensated atom slower and whose second section formed a carefully designed magnetic trap [133]. This Ioffe trap, which has been discussed in Sec. 10.2 and by Pritchard [110] and others [131], has a bias field to discriminate against transitions that can eject atoms from it, as discussed in Secs. 10.3.1 and 10.4.2. It is constructed from four straight parallel currents arranged to form a quadrupole, and two current loops that form the ends. A schematic diagram is shown in Fig. 10.2.

This trap has been used for the first successful experiments in both optical and rf spectroscopy as well as laser cooling of trapped neutral atoms. The optical absorption spectrum of Na atoms in this trap is shown in Fig. 10.6 [133]. A low-intensity probe beam is directed through the trap on axis, and the transmission through the trapped atoms is recorded. The Zeeman splitting caused by the dc bias field isolates the transition between the sublevels with $M_F = 2$ of the ground state and $M_F = 3$ of the excited state, which tunes linearly with magnetic field. Thus the absorption spectrum reflects the number of atoms subject to a particular value of B.

The spectrum can be used to establish the spatial distribution of the trapped atoms, since the field profile is known. As the spectrum shows, there is no absorption below the bottom of the trap, and the highest absorption appears where the trap's field gradient is smallest. The relative strength of the two absorption peaks,

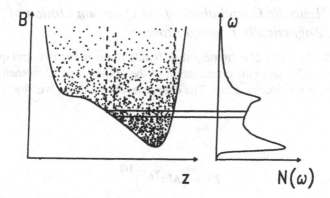

FIGURE 10.6. Distribution of atoms in the magnetic trap of Fig. 10.5 (shaded area). The corresponding optical absorption spectrum is shown on the right (figure from Ref. 133).

as well as the shape of the curve between them, can be used to determine that most of the atoms lie in the deeper portion of the trap (see Fig. 10.6). From this one can estimate the average energy of the trapped atoms, and extract a temperature. The initial experiments indicated that all energies below the trap depth were equally likely to be populated indicating that the temperature is higher than the trap depth. Later experiments by the same authors demonstrated laser cooling of these magnetically trapped atoms, and the use of optical absorption spectroscopy to measure the lowered temperature [135].

The ground-state hyperfine structure of Na allows for rf spectroscopy on magnetically trapped atoms The inhomogeneous magnetic field results in a position-sensitive resonance condition for the rf transition, and therefore enables further study of the spatial distribution of the trapped atoms, much like NMR imaging in medical diagnosis. This, too, can be used to extract the energy distribution of the atoms, and has also been used to observe laser cooling of magnetically trapped atoms [135, 136].

10.4   Quantum Motion in a Trap

Modern techniques of laser and evaporative cooling have the capability to cool atoms to energies where their deBroglie wavelengths are on the micron scale. Such cold atoms may be readily confined to micron size regions in magnetic traps with easily achievable field gradients, and in such cases, the notion of classical orbits is inappropriate. The motional dynamics must be described in terms of quantum mechanical variables and suitable wavefunctions. Furthermore, the distribution of atoms confined in various quantum states of motion in quadrupole as well as other magnetic traps is critical for interpreting the measurements on Bose condensates (see Chapter 17).

10.4.1   Heuristic Calculations of the Quantum Motion of Magnetically Trapped Atoms

Studying the behavior of extremely slow (cold) atoms in the two-coil quadrupole trap begins with a heuristic quantization of the orbital angular momentum using $Mr^2\omega_T = n\hbar$ for circular orbits. The energy levels are then given by

$$E_n = \frac{3}{2}E_1 n^{2/3},$$ (10.4a)

where

$$E_1 = \left(Ma^2\hbar^2\right)^{1/3},$$ (10.4b)

which corresponds to about 5 kHz. For velocities of optically cooled atoms of a few cm/s, $n \sim 10\text{--}100$. By contrast, evaporative cooling (see Chapter 12) can produce velocities ~ 1 mm/s resulting in $n \sim 1$. It is readily found that $\omega_Z = n\omega_T$, so that the adiabatic condition of Eq. 10.3 is satisfied only for $n \gg 1$. The lower-lying (small-n) bound states, whose orbits are confined to a region near the origin where the field is small, are strongly coupled to unbound states as a result of the motion (dynamic coupling), and these are rapidly ejected from the trap. On the other hand, the large-n bound states are less coupled because they spend most of their time in a stronger field, and thus satisfy the condition of adiabaticity of the orbital motion relative to the Larmor precession. In this case the separation of the rapid precession from the slower orbital motion is reminiscent of the Born-Oppenheimer approximation for molecules.

10.4.2   Three-Dimensional Quantum Calculations

The quantum mechanical description of atomic motion in a two-coil quadrupole trap begins with the Schrödinger equation $\mathcal{H}\Psi = E\Psi$. The Hamiltonian is $\mathcal{H} = p^2/2M + V$, where $V = -\vec{\mu} \cdot \vec{B}$, and it has off diagonal elements that arise from the inhomogeneity of the $\vec{B}$ field. Then Ψ can be expanded in a basis of products of spatial coordinates in the trap and internal angular momentum variables: $\Psi = \sum c_{ik}\psi_i(r, \theta, \phi)|J, M_J\rangle_k$. It is possible to find a transformation Λ that diagonalizes V with respect to the magnetic quantum number M_J so that $\Lambda V \Lambda^{-1} = V_d$. Thus the Schrödinger equation becomes

$$(\Lambda p^2/2M + V_d\Lambda - E\Lambda)\Psi = 0.$$ (10.5a)

Defining $\Phi \equiv \Lambda\Psi$ leads to

$$\frac{1}{2M}[\Lambda, p^2]\Lambda^{-1}\Phi + \frac{p^2}{2M}\Phi + (V_d - E)\Phi = 0.$$ (10.5b)

Eq. 10.5b is an eigenvalue equation for Φ that can be solved for its eigenfunctions by leaving out the first term with the commutator $[\Lambda, p^2]$ and then treating it later as a perturbation.

The spatial part of Φ in the remaining unperturbed equation is found by exploiting the cylindrical symmetry of the trap so that its radial and angular parts can be separated. These angular solutions for the external motion are combinations of spherical harmonics. The remaining radial equation has a centrifugal repulsion term proportional to $1/r^2$ that helps reduce the wave function at the origin for all but the lowest orbital angular momenta. The numerical solutions for the unperturbed equation have been discussed in some detail for the case of atoms with two magnetic states, $J = 1/2$ [128].

The perturbation that is the commutator of the first term in Eq. 10.5b is nonzero because the spatially dependent terms of V, and hence of Λ, do not commute with p. Its diagonal matrix elements represent energy shifts and its off-diagonal elements drive transitions between the different M_J eigenfunctions of the unperturbed equation. Such transitions couple bound and continuum (*i.e.*, trapped and free) unperturbed states, and therefore correspond to the Majorana transitions that can cause atoms to be ejected. Thus it becomes clear how the dynamics of the atomic orbits cause atoms to escape from the trap. This first term has a complicated functional dependence upon the angular and spin variables. Its effect on the widths and energies of the bound and unbound quantum states of atoms in the trap field has been carefully described by numerical calculations [128].

10.4.3   Experiments in the Quantum Domain

The quadrupole trap was chosen for extended discussion in this chapter because it was the first to succeed and is the easiest to analyze. However, its applicability is limited because very low energy atoms can escape by non-adiabatic processes as embodied in Eq. 10.3 and described in Sec. 10.4.1. Several different ways to avoid this loss have been considered, and the two that have been most successful up to now are discussed below.

The naive idea of adding a uniform dc magnetic field to the quadrupole trap will not eliminate the region of zero field, but simply shift it to a different location. However, using two orthogonal sets of coils driven with out-of-phase ac currents to add a slowly rotating uniform field moves the point of zero field continuously. If the rotation frequency is faster than the atomic orbital frequency, atoms may seek the hole at the bottom of the trap but not attain it. The authors of Ref. 137 were able to compare the loss rate of atoms from this Time-Orbiting Potential (TOP) trap with that from an ordinary magnetic quadrupole trap with similar parameters, and showed that it indeed allowed tight magnetic confinement with $\simeq 100$ times smaller losses. Such a trap was used to produce the first reported Bose condensation where atoms must be confined to the lowest trapped state [137, 138].

Needless to say, the orbits and/or quantum states of the TOP trap are not the same as those of the pure quadrupole trap discussed above. Instead, the effective trap potential has a rounded bottom shape but no zero field point. Up to now, there have been no direct calculations of the orbits or quantum states in a TOP trap.

The Ioffe trap discussed in Secs. 10.2 and 10.3.3 offers another solution to the loss problem. In this trap the field never vanishes, and its minimum value of $|\vec{B}|$ is adjustable by varying the straight and end-coil currents. The cost of this flexibility is a shallower trap with an elongated trapping volume as shown in Figs. 10.2 and 10.5 [131]. This type of trap can also confine atoms in its lowest quantum state, and has been used in two variations to produce Bose condensation.

One of these variations is the replacement of wires and currents by permanent magnets [139]. Such a trap is compact and reliable, but suffers from the inability to change or turn off the fields. The second variation is the replacement of each of the four straight currents by two parallel, separated loops, making eight additional coils along with the end coils, for a total of ten [140]. This "butterfly" or "cloverleaf" trap offers more complete optical access to its interior.

11

Optical Traps for Neutral Atoms

11.1 Introduction

The force on atoms confined in the magnetic traps described in Chapter 10 arises from the permanent magnetic dipole moments of the atoms in the inhomogeneous field of the trap. By contrast, the inversion symmetry of atomic wave functions prevents them from having permanent electric dipole moments, so optical trapping of neutral atoms by electrical interaction must proceed by inducing a dipole moment. This can be accomplished either by electrostatic fields or by nearly resonant optical frequency fields. Inducing appropriate dipole moments with dc fields can be accomplished in atoms that have a sufficiently close-lying energy states of opposite parity (this excludes most atomic ground states but favors Rydberg states). By contrast, there are several types of optical traps that employ various configurations of laser beams [40, 41]. These produce not only the mixing of atomic states of opposite parity needed to provide dipole moments for interaction with the field, but also the strong field gradients appropriately arranged for such trapping.

Chapter 3 describes two kinds of optical forces, labeled radiative and dipole, that are each discussed in some detail in Chapters 6 and 9 respectively. Both of these forces play important roles in the purely optical traps described in the first part of this chapter. Optical traps for two level atoms that depend purely on the radiative force can not work because of the optical Earnshaw theorem discussed in Sec. 11.3. However, optical pumping between the sublevels of complicated atoms can indeed produce traps that depend purely on the radiative force, because the force is not simply proportional to the intensity but also depends on the internal state of the atoms. Thus the premise of the optical Earnshaw theorem doesn't hold.

FIGURE 11.1. A single focused laser beam produces the simplest type of optical trap.

Furthermore, hybrid traps in which both forces play a role have been demonstrated. Moreover, the force in those types of optical traps that depend purely on the dipole force are proportional to ∇I, not I itself, and so they, too, are not restricted by the optical Earnshaw theorem.

For such dipole optical traps, the oscillating electric field of a laser induces an oscillating atomic electric dipole moment that interacts with the laser field. If the laser field is spatially inhomogeneous, the interaction and associated energy level shift of the atoms (ac Stark shift or light shift, see Eqs. 1.17) varies in space and therefore produces a potential, just as in the sub-Doppler cooling schemes described in Chapter 8. The force from this potential is called the dipole force (see Sec. 3.2 and Chapter 9). When the laser frequency is tuned below atomic resonance ($\delta < 0$), the sign of the interaction is such that atoms are attracted to the maximum of laser field intensity, whereas if $\delta > 0$, the attraction is to the minimum of field intensity. Note that these traps may require additional cooling to offset the concomitant radiative heating. Atoms may be captured near the nodes or antinodes of optical standing waves, even in 3D, thereby making an array of microscopic optical traps called an optical lattice and discussed in Chapter 16.

The second, and larger part of this chapter describes the most widely used, and therefore perhaps the most important of all optical traps. This uses an inhomogeneous magnetic field to Zeeman tune the atomic transition frequencies so that the radiative force on the atoms varies with position. The optical field is relatively weak so that the dipole force on the atoms is negligible, and the magnetic field gradient is sufficently small that the magnetic force is dominated by the radiative force. This magneto-optical trap is relatively simple to build, captures atoms easily, and is quite robust against realistic experimental conditions such as alignment errors, laser frequency instabilities, magnetic field imperfections, and a host of others. For these and other reasons, it has become the workhorse of cold atom physics, and has also appeared in dozens of undergraduate laboratories.

11.2   Dipole Force Optical Traps

11.2.1   Single-Beam Optical Traps for Two-Level Atoms

The simplest imaginable trap consists of a single, strongly focused Gaussian laser beam (see Fig. 11.1) [141, 142] whose intensity at the focus varies transversely with r as

$$I(r) = I_0 e^{-r^2/w_0^2}, \qquad (11.1)$$

where w_0 is the beam waist size. Such a trap has a well-studied and important macroscopic classical analog in a phenomenon called optical tweezers [143–145].

With the laser light tuned below resonance ($\delta < 0$), the ground-state light shift is everywhere negative, but largest at the center of the Gaussian beam waist. Ground-state atoms therefore experience a force attracting them toward this center given by the gradient of the light shift which is found from Eq. 1.17a, and for $\delta \gg \Omega$ and $\delta \gg \gamma$ is found to be

$$F \simeq -\frac{\hbar}{4\delta}\nabla(\Omega(r)^2) = -\frac{\hbar\gamma^2}{8\delta I_s}\nabla I(r), \qquad (11.2)$$

since $\Omega^2 = \gamma^2 I/2I_s$. For the Gaussian beam, this transverse force at the waist is harmonic and is given by

$$F \simeq \frac{\hbar\gamma^2}{4\delta}\frac{I_0}{I_s}\frac{r}{w_0^2}e^{-r^2/w_0^2}. \qquad (11.3)$$

In the longitudinal direction there is also an attractive force, but it is a bit more complicated and depends on the details of the focusing. Thus this trap produces an attractive force on atoms in three dimensions.

Although it may appear that the trap does not confine atoms longitudinally because of the radiation pressure along the laser beam direction, careful choice of the laser parameters can indeed produce trapping in 3D. This can be accomplished because the radiation pressure force decreases as $1/\delta^2$ (see Eq. 3.14), but by contrast, the dipole force only decreases as $1/\delta$ for $\delta \gg \Omega$ (see Eq. 3.16). If $|\delta|$ is chosen to be sufficiently large, atoms spend very little time in the untrapped (actually repelled) excited state because its population is proportional to $1/\delta^2$. Thus a sufficiently large value of $|\delta|$ both produces longitudinal confinement and maintains the atomic population primarily in the trapped ground state. A given laser power can produce a maximum intensity, and a corresponding light shift and trap depth, that is inversely proportional to the area of the beam spot, πw_0^2. Thus a large numerical aperture is required for focusing such a beam.

The first optical trap was demonstrated in Na with light detuned below the D-lines [142]. With 220 mW of dye laser light tuned about 650 GHz below the Na transition and focused to a $\sim 10~\mu$m waist, the trap depth was about $15\hbar\gamma$ corresponding to 7 mK. Single-beam dipole force traps can be made with the light detuned by a significant fraction of its frequency from the atomic transition. Such a far-off-resonance trap (FORT) has been developed for Rb atoms using light detuned by nearly 10% to the red of the D_1 transition at $\lambda = 795$ nm [146]. Between 0.5 and 1 W of power was focused to a spot about 10 μm in size, resulting in a trap 6 mK deep where the light scattering rate was only a few hundred/s. The trap lifetime was more than half a second.

There is a qualitative difference when the trapping light is detuned by a large fraction of the optical frequency. In one such case, Nd:YAG light at $\lambda = 1064$ nm was used to trap Na whose nearest transition is at $\lambda = 596$ nm [147]. In a more extreme case, a trap using $\lambda = 10.6~\mu$m light from a CO_2 laser has been used to trap

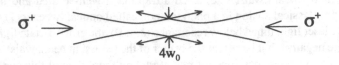

FIGURE 11.2. Focused laser beams of the simple light trap discussed in the text.

Cs whose optical transition is at a frequency $\sim$ 12 times higher ($\lambda = 852$ nm) [148]. For such large values of $|\delta|$, calculations of the trapping force can not exploit the rotating wave approximation as was done for Eq. 1.17a, and the atomic behavior is similar to that in a dc field. It is important to remember, that for an electrostatic trap Earnshaw's theorem precludes a field maximum, but that in this case there is indeed a local 3D intensity maximum of the focused light.

11.2.2 Hybrid Dipole Radiative Trap

A variation of this trap combines the dipole force with the radiation pressure force (see Eq. 3.14). In this design, shown schematically in Fig. 11.2, two less tightly focused laser beams with Gaussian transverse intensity profiles are directed coaxially and oppositely, with their foci slightly separated [149]. Again the frequency is below resonance, so the dipole force produces transverse confinement. The scattering force produces axial confinement because atoms axially displaced from the equilibrium point midway between the two foci experience increased intensity in one beam and decreased intensity in the other. The unbalance results in a net scattering force that pushes them back to the equilibrium point.

Such a trap both cools and heats the atoms. Although Doppler cooling reduces the kinetic energy of the trapped atoms, two associated heating mechanisms necessarily destabilize such laser traps. One is the heating or momentum diffusion arising from the random direction of both absorption and spontaneous emission of light (fluctuations in the scattering force). More important at high intensity is the heating associated with fluctuations in the dipole force that are best discussed in the dressed atom picture described in Sec. 9.3. Fluorescent decay from an excited state may land atoms in either of the two types of states shown in Fig. 9.2. Since the optical forces in these states have opposite signs, atoms decaying spontaneously down the ladder of dressed states (see Fig. 9.2) experience a fluctuating force that has no correlation with their motion, and are therefore heated. The fluctuations of the force do not saturate with intensity, and hence cannot be compensated by making a deeper trap using high intensity light. The result is that the steady-state kinetic energy of atoms in such a trap, resulting from balance between the heating and cooling mechanism, is always about equal to the trap depth. Atoms are thus continuously boiled out of the trap.

The characteristics of such optical dipole force traps have been studied by Gordon and Ashkin [27]. To obtain a trap that is $\sim$ 100 mK deep corresponding to $v \sim \gamma/k$, the saturation parameter s_0 should be as high as $\sim 10^8$ and the detuning

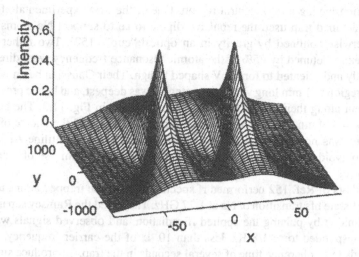

FIGURE 11.3. The light intensity experienced by an atom located in a plane 30 μm above the beam waists of two quasi-focused sheets of light traveling parallel and arranged to form a V-shaped trough. The x and y dimensions are in μm (figure from Ref. 152).

as large as $\sim 10^6 \gamma$ ($\sim 10^{13}$ Hz). The damping rate of the atomic kinetic energy by Doppler cooling is then about 100 times smaller than the heating rate by diffusion in momentum space. Such a trap is inherently unstable, but since the escape time of the atoms can be as large as several seconds, such dipole traps can work provided they are accompanied by effective cooling.

Variations of this trapping scheme have been discussed that include damping from auxiliary light beams [150], alternating light beams rapidly to avoid standing waves and thus large heating from dipole force fluctuations [144], and optical molasses [142]. The first reported optical trap used an alternation on the μs time scale between trapping fields that both confined and heated the atoms, and optical molasses that cooled them before they could escape very far [142]. The trapping light had the configuration of a single tightly focused laser beam as shown in Fig. 11.1.

11.2.3 Blue Detuned Optical Traps

One of the principle disadvantages of the optical traps discussed in Secs. 11.2.1 and 11.2.2 is that the negative detuning attracts atoms to the region of highest light intensity. This results in significant spontaneous emission unless the detuning is a large fraction of the optical frequency such as the Nd:YAG laser trap [147] or the CO_2 laser trap [148]. More important in some cases is that the trap relies on Stark shifting of the atomic energy levels by an amount equal to the trap depth, and this severely compromises the capabilities for precision spectroscopy in a trap [151].

Attracting atoms to the region of *lowest* intensity would ameliorate both of these concerns, but such a trap requires positive detuning (blue), and an optical configuration having a dark central region. One of the first experimental efforts at a blue detuned trap used the repulsive dipole force to support Na atoms that were otherwise confined by gravity in an optical "cup" [152]. Two rather flat, parallel beams detuned by 25% of the atomic resonance frequency were directed horizontally and oriented to form a V-shaped trough. Their Gaussian beam waists formed a region $\simeq 1$ mm long where the potential was deepest, and hence provided confinement along their propagation direction as shown in Fig. 11.3. The beams were the $\lambda = 514$ nm and $\lambda = 488$ nm from an argon laser, and the choice of two frequencies was not simply to exploit the full power of the multiline Ar laser, but also to avoid the spatial interference that would result from use of a single frequency.

The authors of Ref. 152 performed rf spectroscopy on the trapped atoms using the ground-state hfs transition of Na at 1.77 GHz. They used the Ramsey separated fields technique by pulsing the applied rf radiation, and observed signals whose width corresponded to $\simeq 1/8$ Hz, less than 10^{-10} of the carrier frequency. This corresponds to a coherence time of several seconds in the trap. To produce similar coherence times in an atomic fountain (see Sec. 13.7.2) would require a fountain 20 m high, which seems quite impractical. In spite of the residual Stark shifts that limit the use of such a trap for a clock, however, the authors point out that it would be valuable for precision spectroscopy of relative quantities, for example accurate comparison of the hfs frequency in electric fields for an electric dipole moment search.

Obviously a hollow laser beam would also satisfy the requirement for a blue-detuned trap, but conventional textbook wisdom shows that such a beam is not an eigenmode of a laser resonator [153]. Some lasers can make hollow beams, but these are illusions because they consist of rapid oscillations between the TEM_{01} and TEM_{10} modes of the cavity. Nevertheless, Maxwell's equations permit the propagation of such beams, and in the recent past there have been several studies of the LaGuerre-Gaussian modes that constitute them, most notably by L. Allen [154–156], M. Padgett [157, 158] and the thesis of M. Beijersbergen [159]. The several ways of generating such hollow beams have been tried by many experimental groups and include phase and amplitude holograms, hollow waveguides, axicons or related cylindrical prisms, stressing fibers, and simply mixing the TEM_{01} and TEM_{10} modes with appropriate cylindrical lenses.

An interesting experiment has been performed using the ideas of Sisyphus cooling with evanescent waves as discussed in Secs. 8.7 and 9.2 combined with a hollow beam formed with an axicon [109]. In the previously reported experiments with atoms bouncing under gravity from an evanescent wave field [108, 160], they were usually lost to horizontal motion for several reasons, including slight tilting of the surface, surface roughness, horizontal motion associated with their residual motion, and horizontal ejection by the Gaussian profile of the evanescent wave laser beam. The authors of Ref. 109 simply confined their atoms in the horizontal direction by surrounding them with a wall of blue detuned light in the form of a vertical hollow beam. Their gravito-optical surface trap cooled Cs atoms to $\simeq 3$ μK at a density of $\simeq 3 \times 10^{10}$/cm^3 in a sample whose $1/e$ height in the gravitational field was only 19 μm. Simple ballistics gives a frequency of 450 bounces/s, and the $\simeq$ 6 s lifetime (limited only by background gas collisions) corresponds to several thousand bounces. However, at such low energies the deBroglie wavelength of the atoms is $\simeq$ 1/4 μm, and the atomic motion is no longer accurately described classically, but requires the deBroglie wave methods of Chapter 15.

11.2.4 Microscopic Optical Traps

In a standing wave the light intensity varies from zero at a node to a maximum at an antinode in a distance of $\lambda/4$. Since the light shift, and thus the optical potential, vary on this same scale, it is possible to confine atoms in wavelength-size regions of space.

Of course, such tiny traps are usually very shallow, so loading them requires cooling to the μK regime. The momentum of such cold atoms is then so small that their deBroglie wavelengths are comparable to the optical wavelength, and hence to the trap size. In fact, the deBroglie wavelength equals the size of the optical traps ($\lambda/2$) when the momentum is $2\hbar k$, corresponding to a kinetic energy of a few μK. Thus the atomic motion in the trapping volume is not classical, but must be described quantum mechanically. Even atoms whose energy exceeds the trap depth must be described as quantum mechanical particles moving in a periodic potential that display energy band structure [161]. Such effects have been observed in very careful experiments as described in Secs. 16.3 and 16.4.

Atoms trapped in wavelength-sized spaces occupy vibrational levels similar to those of molecules. The optical spectrum can show Raman-like sidebands that result from transitions among the quantized vibrational levels [162, 163] as shown in Fig. 16.6. These quantum states of atomic motion can also be observed by stimulated emission [162, 164] and by direct rf spectroscopy [165, 166]. Considerably more detail about atoms in such optical lattices is to be found in Chapter 16.

There is one very special case of atoms trapped in a wavelength-size region where such quantum effects are not important. The magnetic dipole transition between the two ground hfs states of H atoms has been used to trap them in a microwave cavity [167, 168]. Since the transition strength is very much weaker than that of an optical transition, the strength of the trapping field had to be correspondingly stronger. In the optical case such a strong field could cause undesired spontaneous emission, but that is absent for the microwave transition, so there needn't be a large detuning. The physical principles discussed above still apply completely.

Apart from this case of trapping with microwaves, progress in optical trapping of laser cooled atoms has evolved toward quantization of the atomic center of mass motion or external coordinates. The classical description of atomic motion that assumes that atoms have arbitrary position and momentum has become outmoded. In this new quantum picture of atomic motion, atomic position and momentum need to be considered as quantum mechanical variables, as discussed in Chapter 15.

11.3   Radiation Pressure Traps

One of the basic limitations of dipole traps comes from the large saturation parameters needed for confinement. To overcome this problem traps have been proposed that rely on the scattering force to cool and trap atoms [169]. These designs include either four or six Gaussian beams that converge on a small volume where atoms are trapped. However, such traps cannot be stable as long as the trapping force is proportional to light intensity [169]. This can be simply understood by considering that the flow of optical energy cannot be directed inwards everywhere on the surface of the trapping volume, and thus the force cannot be directed inwards everywhere. Since this is similar to the Earnshaw's theorem for electrostatics, it is called the optical Earnshaw theorem. However, for atoms that have multiple ground states whose absorption probabilities are not all the same, various configurations of laser beams can be used to make stable optical traps [170].

One such example is the trap relying on optical pumping demonstrated for Cs atoms [171]. Here an arrangement of six diverging beams of modest power with various circular polarizations directed toward the center was able to confine over 10^7 atoms in a sub-mm size volume. The special feature of this trap was the absence of any magnetic field, thereby enabling extremely rapid switching of the trapping force. Furthermore, trapped atoms experience no Zeeman shifts that could complicate precision spectroscopy.

11.4   Magneto-Optical Traps

11.4.1   Introduction

The most widely used trap for neutral atoms is a hybrid, employing both optical and magnetic fields, to make a magneto-optical trap (MOT) first demonstrated in 1987 [172]. The operation of a MOT depends on both inhomogeneous magnetic fields and radiative selection rules to exploit both optical pumping and the strong radiative force [172, 173]. The radiative interaction provides cooling that helps in loading the trap, and enables very easy operation. The MOT is a very robust trap that does not depend on precise balancing of the counterpropagating laser beams or on a very high degree of polarization. The magnetic field gradients are modest and can readily be achieved with simple, air-cooled coils. The trap is easy to construct because it can be operated with a room-temperature cell where alkali atoms are captured from the vapor. Furthermore, low-cost diode lasers can be used to produce the light appropriate for all the alkalis except Na, so the MOT has become one of the least expensive ways to produce atomic samples with temperatures below 1 mK.

Trapping in a MOT works by optical pumping of slowly moving atoms in a linearly inhomogeneous magnetic field $B = B(z) \equiv Az$, such as that formed by a magnetic quadrupole field as discussed in Sec. 10.2. Atomic transitions with the simple scheme of $J_g = 0 \rightarrow J_e = 1$ have three Zeeman components in a magnetic field, excited by each of three polarizations, whose frequencies tune with field (and

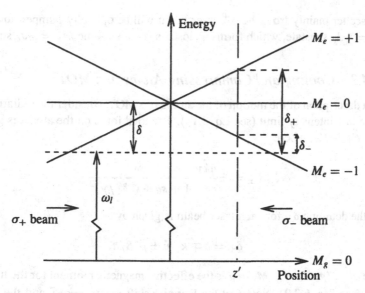

FIGURE 11.4. Arrangement for a MOT in 1D. The horizontal dashed line represents the laser frequency seen by an atom at rest in the center of the trap. Because of the Zeeman shifts of the atomic transition frequencies in the inhomogeneous magnetic field, atoms at $z = z'$ are closer to resonance with the σ^- laser beam than with the σ^+ beam, and are therefore driven toward the center of the trap.

therefore with position) as shown in Fig. 11.4 for 1D. Two counterpropagating laser beams of opposite circular polarization, each detuned below the zero field atomic resonance by δ, are incident as shown.

Because of the Zeeman shift, the excited state $M_e = +1$ is shifted up for $B > 0$, whereas the state with $M_e = -1$ is shifted down. At position z' in Fig. 11.4 the magnetic field therefore tunes the $\Delta M = -1$ transition closer to resonance and the $\Delta M = +1$ transition further out of resonance. If the polarization of the laser beam incident from the right is chosen to be σ^- and correspondingly σ^+ for the other beam, then more light is scattered from the σ^- beam than from the σ^+ beam. Thus the atoms are driven toward the center of the trap where the magnetic field is zero. On the other side of the center of the trap, the roles of the $M_e = \pm 1$ states are reversed and now more light is scattered from the σ^+ beam, again driving the atoms towards the center.

The situation is analogous to the velocity damping in an optical molasses from the Doppler effect as discussed in Sec. 7.2, but here the effect operates in position space, whereas for molasses it operates in velocity space. Since the laser light is detuned below the atomic resonance in both cases, compression and cooling of the atoms is obtained simultaneously in a MOT.

So far the discussion has been limited to the motion of atoms in 1D. However, the MOT scheme can easily be extended to 3D by using six instead of two laser beams. Furthermore, even though very few atomic species have transitions as simple as $J_g = 0 \rightarrow J_e = 1$, the scheme works for any $J_g \rightarrow J_e = J_g + 1$ transition. Atoms that scatter mainly from the σ^+ laser beam will be optically pumped toward the $M_g = +J_g$ substate, which forms a closed system with the $M_e = +J_e$ substate.

11.4.2   Cooling and Compressing Atoms in a MOT

For a description of the motion of the atoms in a MOT, consider the radiative force in the low intensity limit (see Eq. 3.14). The total force on the atoms is given by $\vec{F} = \vec{F}_+ + \vec{F}_-$, where

$$\vec{F}_\pm = \pm \frac{\hbar \vec{k} \gamma}{2} \frac{s_0}{1 + s_0 + (2\delta_\pm/\gamma)^2} \tag{11.4a}$$

and the detuning $\delta_\pm$ for each laser beam is given by

$$\delta_\pm = \delta \mp \vec{k} \cdot \vec{v} \pm \mu' B/\hbar. \tag{11.4b}$$

Here $\mu' \equiv (g_e M_e - g_g M_g)\mu_B$ is the effective magnetic moment for the transition used (see Sec. 6.2.2). Note that the Doppler shift $\omega_D \equiv -\vec{k} \cdot \vec{v}$ and the Zeeman shift $\omega_Z = \mu' B/\hbar$ both have opposite signs for opposite beams.

When both the Doppler and Zeeman shifts are small compared to the detuning δ, the denominator of the force can be expanded as in Sec. 7.2 and the result becomes

$$\vec{F} = -\beta \vec{v} - \kappa \vec{r}, \tag{11.5}$$

where the damping coefficient β is defined in Eq. 7.2. The spring constant κ arises from the similar dependence of $\vec{F}$ on the Doppler and Zeeman shifts, and is given by

$$\kappa = \frac{\mu' A}{\hbar k} \beta. \tag{11.6}$$

The force of Eq. 11.5 leads to damped harmonic motion of the atoms, where the damping rate is given by $\Gamma_{MOT} = \beta/M$ and the oscillation frequency $\omega_{MOT} = \sqrt{\kappa/M}$. For magnetic field gradients $A \approx 10$ G/cm, the oscillation frequency is typically a few kHz, and this is much smaller than the damping rate that is typically a few hundred kHz. Thus the motion is overdamped, with a characteristic restoring time to the center of the trap of $2\Gamma_{MOT}/\omega_{MOT}^2 \approx$ several ms for typical values of the detuning and intensity of the lasers. Note that this restoring force is larger than the purely magnetic force of Chapter 10 by a factor $\approx kz$, so it dominates when atoms are more than a few wavelengths from the center of the trap.

The steady-state temperature of atoms in a MOT is expected to be comparable to the temperature for optical molasses. Since the polarizations of the counterprop-agating laser beams are opposite, it seems that sub-Doppler temperatures could be achieved in a MOT. Sub-Doppler processes in 1D rely on a detailed balance between optical pumping and the local polarization, and in 3D such a balance is disturbed by the laser beams in the other directions. In the 3D light fields of the MOT there are always polarization gradients and the light shifts are spatially dependent, leading to Sisyphus cooling. Detailed studies of polarization gradient processes in a MOT [174–176] show that for sufficiently low intensities the temperature of the MOT is indeed below the Doppler limit and proportional to the light shift (see Eq. 8.10). The proportionality constant b depends on the atomic transition and the polarization gradient.

Since the MOT constants β and κ are proportional, the size of the atomic cloud can easily be deduced from the temperature of the sample. The equipartition of the energy of the system over the degrees of freedom requires that the velocity spread and the position spread are related by

$$k_B T = m v_{\text{rms}}^2 = \kappa z_{\text{rms}}^2. \tag{11.7}$$

For a temperature in the range of the Doppler temperature, the size of the MOT should be of the order of a few tenths of a mm, which is generally the case in experiments.

11.4.3   Capturing Atoms in a MOT

Although the approximations that lead to Eq. 11.5 for the force hold for slow atoms near the origin, they do not apply for the capture of fast atoms far from the origin. In the capture process, the Doppler and Zeeman shifts are no longer small compared to the detuning, so the effects of the position and velocity can no longer be disentangled. However, the full expression of Eq. 11.4 for the force still applies and the trajectories of the atoms can be calculated by numerical integration of the equation of motion [177].

Simulations of the motion can exploit the different time scales of the problem. The shortest one is the spontaneous lifetime $\tau = 1/\gamma$, which is of the order of 20 ns. Since this is much smaller than the damping time, there is a large number of spontaneous emission cycles during the slowing so it can be assumed that there is a continuous force acting on the atoms. The second time scale is the damping time that is of the order of several ms. In this time interval the atoms are slowed and captured in the MOT. The slowest time scale is the lifetime of the atoms in the MOT, which is of the order of 1 s under good vacuum conditions.

Figure 11.5a shows the results of a simulation of the trajectories of Na-like atoms that enter the MOT with a certain velocity. The simulation is carried out in 1D and the laser beams are assumed to interact with the atoms over a range comparable to the diameter of the laser beams. For sufficiently low velocities the atoms are immediately slowed down by the Doppler cooling process when they enter the MOT region. After this short deceleration period their velocities are within the range of the overdamped motion of Eq. 11.5, and the atoms are compressed to the center of the trap with the same rate, as shown by the straight line in the (z, v)-plane. Atoms entering at higher velocities are slowed down by the tail of the Lorentz profile, and if they are completely stopped before the end of the MOT region, they can be captured. The capture velocity v_c of the MOT is thus given by the incoming velocity for which atoms are completely stopped when they reach the opposite edge of the MOT region. In this simulation, v_c is approximately 55 m/s.

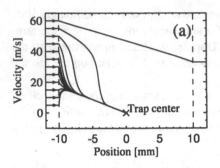 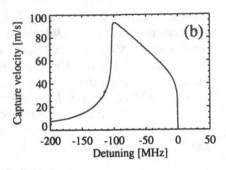

FIGURE 11.5. Numerical simulation of the capture process in 1D of the MOT for a $J_g = 0 \rightarrow J_e = 1$ transition. (a) Trajectories for Na atoms with such a simplified structure entering the MOT region with different initial velocities, which is increased between different trajectories by 5 m/s. Here $s_0=10$ and $\delta = -30$ MHz. For low enough velocities the atoms are collected in the center of the trap and remain trapped. (b) Dependence of the capture velocity v_c on the detuning of the laser from resonance. The largest v_c is obtained for a detuning of ≈ -100 MHz $\approx -10\gamma$.

Figure 11.5b shows v_c as a function of detuning of the laser light from resonance. For small negative detunings v_c increases with increasing detuning, and reaches a maximum near $\delta = -100$ MHz $\approx -10\gamma$. If the detuning is increased further, the atoms can only be cooled in the tail of the Lorentz profile, which is not sufficient to completely slow them to rest.

The situation becomes more complicated when real atoms are considered. In the alkalis the ground state has $L = 0$ so $J = S$ and $F = I + J$. Cooling and trapping is achieved by using the $F_g = I + S \rightarrow F_e = F_g + 1$ cycling transition. This system is closed, i.e., spontaneous emission to the ground state is always to the same F_g-state because of the selection rule $\Delta F = 0, \pm 1$. However, another excited hfs state $F_e = F_g$ is close by, and only a small excitation rate to that state leads to a loss of atoms caused by spontaneous emission to the $F_g' = I - S$ ground state. Since the hyperfine splitting in the ground state is very large, atoms are confined to this state and are no longer cooled and trapped. In order to prevent this, a second laser beam, called a repumper, has to be used and this is tuned to the $F_g' = I - S \rightarrow F_e = F_g' + 1$ transition. The $F_e = F_g' + 1$ state can then decay to the original $F_g = I + S$ state.

The hyperfine structure in the excited state changes the detuning dependence of the MOT characteristics considerably [177]. For example, if the laser is red detuned from the cycling transition by more than half the splitting between adjacent hyperfine states, the frequency is closer to resonance with the adjacent hyperfine state, and furthermore, is detuned to the blue. Then the cooling becomes heating and the atoms can no longer be trapped.

On the other hand, the hyperfine structure also allows other cycling transitions to be used for cooling and trapping. For Na it was found [172, 177] that the transition $F_g = 1 \rightarrow F_e = 0$ can be used to trap atoms, where the $F_g = 1 \rightarrow F_e = 1$ transition is used to cool the atoms. Repumping is achieved by tuning the repumper

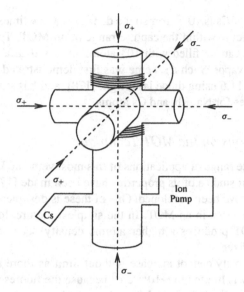

FIGURE 11.6. The schematic diagram of a MOT shows the coils and the directions of polarization of the six light beams. It has an axial symmetry and various rotational symmetries, so some exchanges would still result in a trap that works, but not all configurations are possible. Atoms are trapped from the background vapor of Cs that arises from a piece of solid Cs in one of the arms of the setup.

to the $F_g = 2 \rightarrow F_e = 1$ transition. This so-called type II trap is much weaker than the more common type I trap described earlier, and leads to a much larger trap volume. However, since the density of the MOT is limited by collision processes, the larger volume allows for the storage of more atoms and the type II MOT usually appears much brighter than the type I MOT.

The capture velocity of a MOT is serendipitously enhanced because atoms traveling across it experience a decreasing magnetic field just as in beam deceleration described in Sec. 6.2.2 [173]. This enables resonance over an extended distance and velocity range because the changing Doppler shift of decelerating atoms can be compensated by the changing Zeeman shift as atoms move in the inhomogeneous magnetic field. Of course, it will only work this way if the field gradient A does not demand an acceleration larger than the maximum acceleration a_{max} (see Sec. 6.2). Thus atoms are subject to the optical force over a distance that can be as long as the trap size, and can therefore be slowed considerably.

The very large velocity capture range v_c of a MOT can be estimated by using $F_{max} = \hbar k \gamma / 2$ and choosing a maximum size of a few cm for the beam diameters. Thus the energy change can be as large as a few K, corresponding to $v_c \sim 100$ m/s [173], as in Fig. 11.5b. The number of atoms in a vapor with velocities below v_c in the Boltzmann distribution scales as v_c^4 (see Sec. 5.2), and there are enough slow atoms to fall within the large MOT capture range even at room-temperature, because a few K includes 10^{-4} of the atoms. A more conservative estimate of the capture range might cost another factor of 10, but this is still a very large number of atoms for most room-temperature vapors. For example, at a temperature of 300 K, the vapor pressure of Cs is 10^{-5} Torr so the density is a few times 10^{11} atoms/cm^3 leaving 10^7 atoms/cm^3 within the capture range of an MOT. Thus a Cs MOT in a modest size cell can be filled with 10^9 atoms in less than a second from the room-temperature vapor. Such a scheme was first demonstrated in 1990 with the trap shown in Fig. 11.6 using diode laser light [178], and has since been repeated in many laboratories for Na, Rb, and Cs atoms.

11.4.4   Variations on the MOT Technique

Because of the wide range of applications of this most versatile kind of atom trap, a number of careful studies of its properties have been made [173, 179–186], and several variations have been developed. One of these is designed to overcome the density limits achievable in an MOT. In the simplest picture, loading additional atoms into an MOT produces a higher atomic density because the size of the trapped sample is fixed.

However, the density cannot increase without limit as more atoms are added. The atomic density is limited to $\sim 10^{11}$/cm^3 because the fluorescent light emitted by some trapped atoms is absorbed by others as discussed on p. 27, and this diffusion of radiation presents a repulsive force between the atoms [183, 184]. Another limitation lies in the collisions between the atoms, and as discussed in Chapter 14, the collision rate for excited atoms is much larger than for ground-state atoms. Adding atoms to a MOT thus increases the density up to some point, but adding more atoms then expands the volume of the trapped sample. In some cases the radiation pressure may cause the sample to break up into a central cloud surrounded by an orbiting ring [183, 184] driven by asymmetries in the magnetic field or laser beam profiles. Photographs of some of these atomic clouds are shown in Fig. 11.7. In addition, certain kinds of collisions among the trapped atoms may also play a role in limiting the density to a similar value.

One way to overcome this limit is to have much less light in the center of the MOT than at the sides. Simply lowering the laser power is not effective in reducing the fluorescence because it will also reduce the capture rate and trap depth. But those advantageous properties can be preserved while reducing fluorescence from atoms at the center if the light intensity is low only in the center.

The repumping process for the alkali atoms provides an ideal way of implementing this idea [187]. If the repumping light is tailored to have zero intensity at the center, then atoms trapped near the center of the MOT are optically pumped into the "wrong" hfs state and stop fluorescing. They drift freely in the "dark" at low speed through the center of the MOT until they emerge on the other side into the region where light of both frequencies is present and they begin absorbing again. Then they feel the trapping force and are driven back into the "dark" center of the trap. Such a MOT has been operated at MIT [187] with densities close to 10^{12}/cm^3, and the limitations are now from collisions in the ground state rather than from multiple light scattering and excited state collisions (see Chapter 14).

Another variation of the MOT is designed to produce spin-polarized atoms. In a usual MOT, the orientation of the atomic spins varies throughout the trap volume because of the varying direction of the quadrupole magnetic field and the

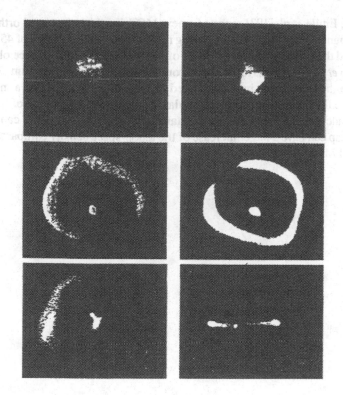

FIGURE 11.7. Spatial distribution of atoms trapped in a MOT whose beams are slightly misaligned. When there are less than 10^8 atoms, they form a central clump, but with more than that there is an orbiting group of atoms as well. (b) and (d) show time exposures of this, but (c) shows the clump distinctly when the camera is strobed at 110 Hz. (e) and (f) show a full ring from the top and side (figure from Ref. 184).

different optical polarizations. However, a different trap has been built where two of the three pairs of laser beams are misaligned in the "racetrack" arrangement, and more coils have been added to change the field symmetry [188, 189]. In this case the trap can work adequately even when the two beams in the third pair have the same polarization and one pair of coils produces a uniform field. Atoms are therefore subject to a strong optical pumping toward a particular alignment, and the total sample has a 75% spin alignment [189].

In a third variation, the number of laser beams has been reduced from six to four and arranged in tetrahedral symmetry similar to Fig. 16.3b [190, 191]. There are advantages to this arrangement apart from the simplicity of fewer laser beams. First, capturing atoms from a slowed atomic beam is enormously simplified because there is no laser light copropagating with the atoms. Second, the restrictions on polarization purity may be relaxed. Of course, it is a bit more difficult to produce such a configuration of laser beams, but for certain applications, it is certainly advantageous.

Finally, Emile *et al.* [192] reported a new MOT, in which they used orthogonal pairs of counterpropagating beams having relative polarization angles of 45°. They interpreted the trapping as being a result of a new magneto-optical force observed by Grimm *et al.* [193]. This force arises from a redistribution of light from one laser beam into the other beam by a stimulated process in the presence of a magnetic field. Since this force arises from a stimulated process, the magnitude of the force can be made much larger than the spontaneous force. Therefore one can expect that this trap can have a larger increase of the phase-space density compared to the traditional MOT.

12
Evaporative Cooling

12.1 Introduction

Laser cooling leads to the production of samples of atoms with low temperature and high density. In the 1920s Bose and Einstein predicted that for sufficiently low temperature and high density, a gas of atoms undergoes a phase transition that is now called Bose-Einstein condensation (BEC—see Chapter 17). This phase transition is predicted to occur at a phase space density $\rho \equiv n\lambda_{dB}^3 \cong 2.612$, where n is the density of the gas and $\lambda_{dB} = h/M\bar{v} = h/\sqrt{3Mk_BT}$ is the deBroglie wavelength of the atoms. For ordinary gases at room temperature and pressure, $\rho \sim 10^{-6}$, but in a practical atomic beam oven, $\rho \sim 3 \times 10^{-10}$.

Achieving BEC has been one of the holy grails in physics for many years, and from the beginning of laser cooling it was clear that this could be one of the possible routes for achieving it. With laser cooling one can obtain μK temperatures with small loss of atoms, so that the phase space density can be increased. However, in the mid 1990s it became clear that the increase in phase space density by laser cooling of alkali atoms had reached its limit. If the density of the sample becomes too large, light scattered by one atom is reabsorbed by others, causing a repulsion between them. For resonant light, the optical thickness of a sample of atoms that has been laser cooled to the recoil limit and compressed to $\rho \sim 1$ is only one optical wavelength, so light can neither enter nor escape a reasonably sized sample.

The increase of density also leads to an increase in the collision rate. The collision rate between atoms with one in the excited state (S+P collisions) is also much larger at low temperatures than the rate for such collisions with both atoms in the ground state (S+S collisions). Since S+P collisions are generally inelastic, and since the inelastic energy exchange generally leads to a heating of the atoms, increasing the density increases the loss of cold atoms. To achieve BEC, resonant light should therefore be avoided, and thus laser cooling alone is not the most likely route for achieving BEC.

A more promising route to BEC is the technique of evaporative cooling. This method is based on the preferential removal of those atoms from a confined sample with an energy higher than the average energy, followed by a rethermalization of the remaining gas by elastic collisions. Although evaporation is a process that occurs in nature, it was applied to atom cooling for the first time in 1988 [194]. One way to think about evaporative cooling is to consider cooling of a cup of coffee. Since the most energetic molecules evaporate from the coffee and leave the cup, the remaining atoms obtain a lower temperature and are cooled. Furthermore, it requires the evaporation of only a small fraction of the coffee to cool it by a considerable amount. Thus even though the method results in the removal of some of the atoms in a trap, those that remain have much lower average energy (temperature) and so they occupy a smaller volume near the bottom of the trap, thereby increasing their density. Since both the temperature and the volume decrease, the phase space density increases.

This chapter describes a model of evaporative cooling. Since such cooling is not achieved for single atoms but for the whole ensemble, an atomic description of the cooling process must be replaced by thermodynamic methods. These methods are completely different from the rest of the material in the book, and will therefore remain rather elementary.

12.2   Basic Assumptions

Evaporative cooling works by the preferential removal of atoms having an energy higher than the average energy, as suggested schematically in Fig. 12.1. If the atoms are trapped, it can be achieved by lowering the depth of the trap, thereby allowing the atoms with energies higher than the trap depth to escape, as discussed first by Hess [195]. Elastic collisions in the trap then lead to a rethermalization of the gas. To sustain the cooling process the trap depth can be lowered continuously, achieving a continuous decrease of the temperature. Such a process is called forced evaporation. Although more refined techniques have been developed, this technique was first employed for evaporative cooling of hydrogen [136, 194, 196, 197].

Several models have been developed for this process, but the simplest one was developed by Davis et al. [198], and is mainly of pedagogical value [199]. In this model the trap depth is lowered in one single step and the effect on the thermodynamic quantities, such as temperature, density and volume, is calculated. Although the process can be repeated and the effects of multiple steps added up cumulatively, forced evaporative cooling is a continuous process and should be described by other models. However, the results of the simple model provide considerable insight to the process without resorting to tedious calculations.

In many models of evaporative cooling the following assumptions are made:

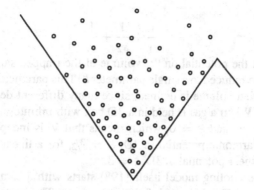

FIGURE 12.1. Principle of the evaporation technique. Once the trap depth is lowered, atoms with energy above the trap depth can escape and the remaining atoms reach a lower temperature.

1. The gas behaves sufficiently ergodically, *i.e.*, the distribution of atoms in phase space (both position and momentum) depends only on the energy of the atoms and the nature of the trap.

2. The gas is described by classical statistics and is assumed to be far from the transition point to the BEC phase ($\rho \ll 1$).

3. The quantum mechanical scattering is pure s-wave, *i.e.*, the temperature is sufficiently low that all higher partial waves do not contribute to the cross section (see Sec. 14.2). Furthermore, the cross section for elastic scattering is energy-independent and is given by $\sigma = 8\pi a^2$, where a is the scattering length. Also, it is assumed that the ratio of elastic to inelastic collision rates is sufficiently large that the elastic collisions dominate.

4. Evaporation preserves the thermal nature of the distribution, *i.e.*, the thermalization is much faster than the rate of cooling.

5. Atoms that escape from the trap neither collide with the remaining atoms nor exchange energy with them. This is called full evaporation.

The simple model uses all of these assumptions, and their implications will be discussed later in the chapter.

12.3 The Simple Model

The first step in applying this simple model is to characterize the trap by calculating how the volume of a trapped sample of atoms changes with temperature T. Consider a trapping potential that can be expressed as a power law given by

$$U(x, y, z) = \epsilon_1 \left| \frac{x}{a_1} \right|^{s_1} + \epsilon_2 \left| \frac{y}{a_2} \right|^{s_2} + \epsilon_3 \left| \frac{z}{a_3} \right|^{s_3}, \qquad (12.1)$$

where a_j is a characteristic length and s_j the power for a certain direction j. Then one can prove [200] that the volume occupied by trapped atoms scales as $V \propto T^\xi$, where

$$\xi \equiv \frac{1}{s_1} + \frac{1}{s_2} + \frac{1}{s_3}. \tag{12.2}$$

Thus the effect of the potential on the volume of the trapped sample for a given temperature can be reduced to a single parameter ξ. This parameter is independent of how the occupied volume is defined, since many different definitions lead to the same scaling. When a gas is held in a 3D box with infinitely high walls, then $s_1 = s_2 = s_3 = \infty$ and $\xi = 0$, which means that V is independent of T, as expected. For a harmonic potential in 3D, $\xi = 3/2$, for a linear potential in 2D $\xi = 2$, and for a linear potential in 3D, $\xi = 3$.

The evaporative cooling model itself [198] starts with a sample of N atoms having a temperature T held in an infinitely deep trap. The strategy for using the model is to choose a finite quantity η, and then (1) lower the trap depth to a value $\eta k_B T$, (2) allow for a thermalization of the sample by collisions, and (3) determine the change in phase space density ρ.

Only two parameters are needed to completely determine all the thermodynamic quantities for this process (the values after the process are denoted by a prime). One of these is $v \equiv N'/N$, the fraction of atoms remaining in the trap after the cooling. The other[1] is γ, a measure of the decrease in temperature caused by the release of hot atoms and subsequent cooling, modified by v, and defined as

$$\gamma \equiv \frac{\log(T'/T)}{\log(N'/N)} = \frac{\log(T'/T)}{\log v}. \tag{12.3}$$

This yields a power-law dependence for the decrease of the temperature caused by the loss of the evaporated particles, namely, $T' = T v^\gamma$. The dependence of the other thermodynamic quantities on the parameters v and γ can then be calculated.

The scaling of $N' = Nv$, $T' = T v^\gamma$, and $V' = V v^{\gamma\xi}$ can provide the scaling of all the other thermodynamic quantities of interest by using the definitions for the density $n = N/V$, the phase space density $\rho = n\lambda_{dB}^3 \propto nT^{-3/2}$, and the elastic collision rate $k_{el} \equiv n\sigma v \propto nT^{1/2}$. The results are given in Table 12.1. For a given value of η, the scaling of all quantities depends only on γ. Note that for successive steps j, v has to be replaced with v^j.

In order to determine the change of the temperature in the cooling process, it is necessary to consider in detail the distribution of the atoms in the trap. The density of states for an ideal gas in free space is given by [201]

$$D(E) = \frac{2\pi(2M)^{3/2}VE^{1/2}}{h^3}. \tag{12.4}$$

However, for atoms in a trap the density of states is affected by the trapping potential $U(x, y, z)$, and becomes [200]

$$D(E) = \frac{2\pi(2M)^{3/2}}{h^3} \int_V \sqrt{E - U(x, y, z)} \, d^3r. \tag{12.5}$$

---

[1] This γ is not to be confused with the natural width of the excited state.

thermodynamic variable	symbol	exponent q
Number of atoms	N	1
Temperature	T	γ
Volume	V	$\gamma\xi$
Density	n	$1 - \gamma\xi$
Phase space density	ρ	$1 - \gamma(\xi + 3/2)$
Collision rate	k	$1 - \gamma(\xi - 1/2)$

TABLE 12.1. Exponent q for the scaling of the thermodynamic quantities $X' = X\nu^q$ with the reduction ν of the number of atoms in the trap.

The fraction of atoms remaining in the trap after decreasing the trap depth to $\eta k_B T$, becomes

$$\nu = \frac{1}{N} \int_0^{\eta k_B T} D(E) e^{-(E-\mu)/k_B T} dE, \tag{12.6}$$

where the exponential factor stems from the Maxwell-Boltzmann distribution of the atoms (see Sec. 5.2), and μ is the chemical potential. For $\eta = \infty$, $\nu = 1$ and this determines the chemical potential μ for N atoms [200]. Substituting this relation for μ into Eq. 12.6 yields

$$\nu = \int_0^{\eta} \Delta(\epsilon) e^{-\epsilon} d\epsilon, \tag{12.7}$$

where the reduced energy is defined as $\epsilon \equiv E/k_B T$. Furthermore, the reduced density of states $\Delta(\epsilon)$ is given by

$$\Delta(\epsilon) \equiv \frac{\epsilon^{\xi + 1/2}}{\Gamma(\xi + 3/2)}, \tag{12.8}$$

with $\Gamma(x)$ the complete gamma function. Figure 12.2 shows the reduced density of states as a function of $\tilde{\epsilon} = \epsilon/(\xi + 3/2)$ for various values of ξ. The scaling of ϵ is performed so that the reduced density of states is nearly independent of ξ. The results for different potentials can therefore be compared directly.

The integral in Eq. 12.7 can be written in terms of the incomplete gamma function Γ_{inc} to give

$$\nu = \frac{\Gamma_{inc}(\xi + 3/2, \eta)}{\Gamma(\xi + 3/2)}. \tag{12.9}$$

Note that the fraction of atoms remaining is fully determined by the final trap depth η for given potential characterized by the trap parameter ξ.

The averaged reduced energy $\bar{\epsilon}$ of the atoms before truncation is given by

$$\bar{\epsilon} = \frac{\int_0^{\infty} \epsilon \Delta(\epsilon) e^{-\epsilon} d\epsilon}{\int_0^{\infty} \Delta(\epsilon) e^{-\epsilon} d\epsilon} = \frac{\Gamma(\xi + 5/2)}{\Gamma(\xi + 3/2)} = \xi + 3/2. \tag{12.10}$$

The average energy $\bar{\epsilon}'$ after truncation is given by the same expression, when the upper boundary is changed from ∞ to η. The average energy is thus

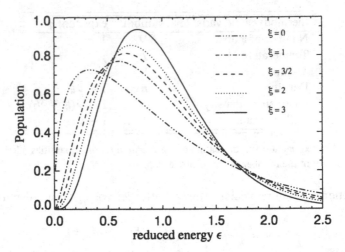

FIGURE 12.2. Reduced density of states $\Delta(\epsilon)$ as a function of the scaled energy $\bar{\epsilon} = \epsilon/(\xi + 3/2)$ for various trapping potentials, indicated by their parameter ξ.

$$\bar{\epsilon}' = \frac{\Gamma_{inc}(\xi + 5/2, \eta)}{\Gamma_{inc}(\xi + 3/2, \eta)}. \tag{12.11}$$

Since the average energy is directly proportional to the temperature, the ratio T'/T is given by

$$\frac{T'}{T} = \frac{\bar{\epsilon}'}{\bar{\epsilon}} = \nu^{\gamma}, \tag{12.12}$$

or

$$\gamma = \frac{\log(T'/T)}{\log(N'/N)} = \frac{\log(\bar{\epsilon}'/\bar{\epsilon})}{\log \nu}. \tag{12.13}$$

For each evaporated atom the energy carried away ϵ_{out} is given by

$$\epsilon_{out} = \frac{\bar{\epsilon} - \bar{\epsilon}'}{1 - \nu} = (\xi + 3/2)\frac{1 - \nu^{\gamma+1}}{1 - \nu}. \tag{12.14}$$

For large η, the value of ν approaches 1 so the denominator $(1 - \nu)$ can be treated as small. Then

$$\gamma = \frac{\epsilon_{out}}{\xi + 3/2} - 1. \tag{12.15}$$

so in that case, γ is just the excess energy above the average energy, which is carried away by the evaporated atoms.

The results of the model are given in Fig. 12.3. Apart from the 3D box potential ($\xi = 0$) the results for the number of atoms and the temperature are nearly identical for the different potentials. However, for a stronger potential (larger ξ) the decrease in the volume with decreasing temperature is much larger and therefore the increase in density n is much larger. Not only does this lead to a larger increase in phase space density ρ, but this is also important for the rethermalization of the atoms. As the results show, the elastic collision rate also increases strongly for a stronger potential. This way the rethermalization speeds up considerably and the cooling process can be accelerated. In the case of a weak potential (ξ between 0 and 1) the collision rate decreases for all values of η and therefore the cooling process eventually stops. Thus the model indicates that BEC cannot be obtained in such potentials.

12.4   Speed and Limits of Evaporative Cooling

12.4.1   Boltzmann Equation

Although this simple model shows many aspects of the evaporation process, it does not provide information about its time scale. Experimental results show that ~ 2.7 elastic collisions are necessary to rethermalize the gas [202]. In order to model the rethermalization process, Luiten et al. [203] have discussed a model based on the Boltzmann equation. In their model, the evolution of the phase space density $\rho(\vec{r}, \vec{p})$ is calculated. This evolution is not only caused by the trapping potential, but also by collisions between the particles. Only elastic collisions, whose cross section is given by $\sigma = 8\pi a^2$, with a the scattering length, are considered. This leads to the Boltzmann equation [201].

The Boltzmann equation is solved numerically by dividing phase space density into a large number of bins and calculating the flow of particles from one bin to another at each instant. As an example of their method, they solved the Boltzmann equation for a flat initial distribution, corresponding to an infinite temperature. Then at $t = 0$ the trap depth was lowered and the change in phase space density was calculated after various collision times. The resulting phase space density, normalized to the initial distribution, is shown in Fig. 12.4. The effect of the thermalization is clearly evident from the figure. After 64 collisions, the number of slow atoms has increased by a factor 60. In the same figure, the authors indicate with a dashed line a Maxwell-Boltzmann (MB) distribution. Clearly the "real" distribution is always very close to a MB distribution, apart from a very small region in energy close to the top of the trap. However, the authors argue that there is a difference between evaporation and thermalization. Although the distribution remains mainly MB, the restoration of the high-energy tail by collisions takes much more than four collisions.

12.4.2   Speed of Evaporation

So far the speed of the evaporative cooling process has not been considered. As an extreme example, consider the case of an extremely large value of η where one just has to wait for a single event where one particle has all the energy of the system. Evaporation of that single particle then cools the whole system to zero temperature [199]. More realistically one can consider the following two cases. If the trap depth is ramped down too quickly, the thermalization process does not have time to run its course and the process becomes less efficient. On the other

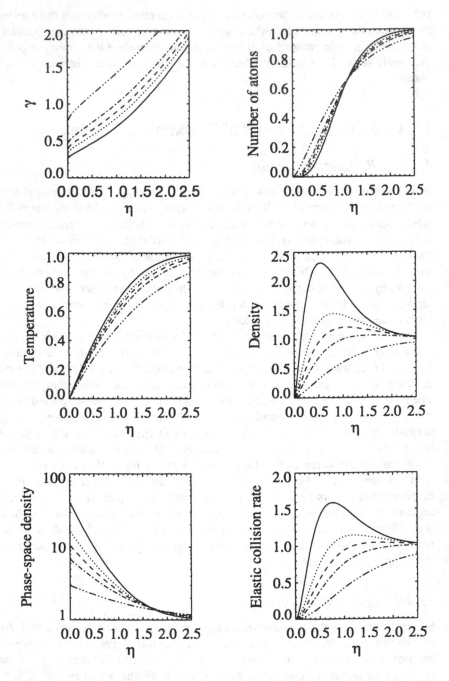

FIGURE 12.3. Result of the model for evaporation for different values of ξ (see Fig. 12.2) for the thermodynamic quantities: (1) γ, (2) Number of atoms, (3) Temperature, (4) Density, (5) Phase-space density, and (6) Elastic collision rate (figure adapted from Ref. 198).

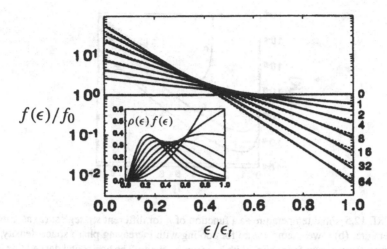

FIGURE 12.4. Evolution of the phase space density ρ as a function of the normalized energy ϵ/ϵ_t after a number of elastic collisions, where ϵ_t denotes the final trap depth. At $t = 0$, ρ_0 is assumed to be flat, but because of evaporative cooling ρ increases for small ϵ. The axis on the right-hand side is the number of elastic collisions (from Ref. 203).

hand, if the trap depth is ramped down too slowly, the loss of particles by inelastic collisions becomes important, thereby making the evaporation inefficient.

The speed of evaporation can be found from the principle of detailed balance [199]. It states that elastic collisions produce atoms with energy larger than $\eta k_B T$ at a rate that is given by the number of atoms with energy larger than this divided by their collision time. The velocity of atoms with this energy is given by $v = \sqrt{2\eta k_B T/M} = \bar{v}\sqrt{3\eta/2}$, where $\bar{v}$ is the average velocity for given temperature (see Eq. 5.9). The fraction of atoms in the MB-distribution with $\epsilon > \eta$ for large η is given by

$$f(\epsilon > \eta) = e^{-\eta}\sqrt{3\eta/2}. \tag{12.16}$$

The elastic collision rate is given by $k_{el} = n\sigma v$. The rate of evaporated atoms dN/dt becomes

$$\frac{dN}{dt} = -Nf(\epsilon > \eta)k_{el} = -n\sigma\bar{v}\eta e^{-\eta}N \equiv -\Gamma_{ev}N. \tag{12.17}$$

The average elastic scattering rate depends on the relative velocity and not on the average velocity of the atom (see Table 5.1). Thus the average of k_{el} is $\bar{k}_{el} = 4n\sigma\bar{v}/\sqrt{3\pi}$. The ratio of the evaporation time and the elastic collision time then becomes

$$\frac{\tau_{ev}}{\tau_{el}} = \frac{\sqrt{2}e^{\eta}}{\eta}. \tag{12.18}$$

Note that this ratio increases exponentially with η.

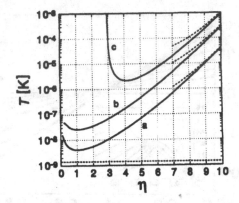

FIGURE 12.5. Final temperature as a function of η for different strategies: (a) asymptotic temperature, (b) lowest temperature for cooling with increasing phase space density, and (c) lowest temperature for cooling with increasing density. The horizontal dashed line is the limit for evaporative cooling T_e (figure from Ref. 203).

12.4.3 Limiting Temperature

In the models discussed so far, only elastic collisions have been considered, *i.e.*, during such collisions kinetic energy is only redistributed between the partners. However, if part of the internal energy of the colliding partners is exchanged with their kinetic energy in the collision, then it is inelastic. The inelasticity of the collision can cause problems for two reasons: (1) the internal energy released can cause the atoms to heat up, and (2) the atoms can change their internal states, and the new states may no longer be trapped. In each case, inelastic collisions can lead to trap loss and are therefore not desirable. For evaporative cooling, elastic collisions are referred to as good collisions and inelastic collisions as bad collisions. It is important that the ratio of the number of good to bad collisions is large.

Apart from collisions with the background gas and three-body recombination, there are two inelastic processes that are important for evaporative cooling of alkali atoms: dipolar relaxation and spin relaxation. Since both of these are inelastic processes, the collision rate $n k_{\mathrm{dip}}$ for them at low energies becomes constant [1]. Here k_{dip} is the velocity-independent inelastic collision rate. Since the elastic collision rate is given by $k_{el} = n \sigma v_{\mathrm{rel}}$, the ratio of good (= elastic) to bad (= relaxation) collisions goes down when the temperature does. This limits the temperature to a value T_e near that where the ratio between good and bad collisions becomes unity, and T_e is given by

$$k_B T_e = \frac{\pi M k_{\mathrm{dip}}^2}{16 \sigma^2}. \qquad (12.19)$$

The limiting temperature for the alkalis is of the order of 1 nK, depending on the values for σ and k_{dip}. In practice, however, this ratio has to be considerably larger than unity, and so the practical limit for evaporative cooling occurs when the ratio is $\sim 10^3$ [199].

Group	Atom	N (10^6)	n $(10^{12}\mathrm{cm}^{-3})$	T $(\mu\mathrm{K})$	ρ (10^{-6})	η_{tot}
Rice	^{7}Li	200	0.07	200	7	1.7
		0.1	1.4	0.4		
MIT	^{23}Na	1000	0.1	200	2	1.9
		0.7	150	2		
JILA	^{87}Rb	4	0.04	90	0.3	3.0
		0.02	3	0.17		

TABLE 12.2. Results obtained with evaporative cooling for the achievement of BEC [199]. The first line in each case represents the starting point and the second line represents the end point.

In the model of Ref. 203, the authors discuss different strategies for evaporative cooling, and a summary of their results is given in Fig. 12.5. Note that even for the strategy of the lowest temperature, the final temperature is higher than T_e.

12.5   Experimental Results

In all the earliest experiments that achieved BEC, the evaporative cooling was "forced" by inducing rf transitions to magnetic sublevels that are not bound in the magnetic trap (see Sec's. 10.3 and 10.4). The experiments described in Sec. 10.3.3 laid the groundwork for this technique. Atoms with the highest energies can access regions of the trap where the magnetic field is stronger, and thus their Zeeman shifts would be larger. A correspondingly high-frequency rf field would cause only these most energetic atoms to undergo transitions to states that are not trapped, and in so doing, the departing atoms carry away more than the average energy. Thus a slow sweep of the rf frequency from high to low would continuously shave off the high-energy tail of the energy distribution, and thereby continuously drive the temperature lower and the phase space density higher.

In Table 12.2 the results of evaporative cooling from the first three groups that have obtained BEC is given. The success of evaporative cooling using this rf shaving technique demonstrates that it is much easier to select high energy atoms and waste them than it is to cool them.

Part III

Applications

13
Newtonian Atom Optics and its Applications

13.1 Introduction

Atom optics is a new field that has emerged as a result of the capabilities of laser cooling. Devices depending on both material components and carefully arranged electromagnetic fields have been demonstrated. However, neutral atoms do not penetrate matter, so the only material devices that can be used for atom optics must function as masks, gratings, zone plates, and slits. Apart from simple masking, the principal effect of these intensity modulators is deBroglie wave diffraction, and so their discussion is left to Chapter 15. By contrast, atoms traveling in inhomogeneous electromagnetic fields, for example an optical standing wave, can experience a dipole force as discussed in Chapter 9. Thus the trajectories of atoms can be altered by the fields so that it becomes possible to control the motion of atoms using devices analogous to those in optics, including mirrors, lenses, beam splitters, retardation plates, *etc.*.

Newtonian atom optics refers to the domain of atomic motion that is classical in the sense that atoms are considered as point particles whose motion can be described by Newton's laws. In this domain, atoms can be localized, and their position and momentum can be known simultaneously. The analogies are readily made to geometrical optics, where light is considered to be described as rays that are lines drawn perpendicular to the optical wavefronts. However, there are at least two distinct examples where particle atom optics has no analogy in classical optics. These are the effect of gravity that arises from the atomic mass, and the dissipative processes that allow laser cooling and other forms of phase space compression. On the other hand, the domain called wave optics where diffraction and interference must be considered, is more akin to the quantum states of motion of atoms. Such motion is discussed in several places in this book, including Chapters 15 through 18.

One device that could be appropriate to this chapter is a beam splitter. However, just as in optics, the atom optical beam splitter results in two (or more) beams that have a high degree of relative coherence. Since this property is appropriate for deBroglie wave optics, beam splitters are discussed in Chapter 15. For similar reasons, diffractive optics such as gratings and zone plates are also discussed in Chapter 15.

13.2   Atom Mirrors

One of the first proposals for an atomic mirror was made by Cook and Hill [104], who suggested reflecting atoms from the evanescent wave of laser radiation leaking into the vacuum when light is totally internally reflected at a vacuum-dielectric interface (see Sec. 9.2). If the light is detuned blue from resonance, the atoms are repelled by the intensity gradient because they are attracted to the weak field region, and thus are reflected back into the vacuum. This technique was demonstrated by Balykin et al. [204], who specularly reflected Na atoms off an internally illuminated quartz plate. The laser light was nearly resonant with one of the hyperfine components of the D_2-line of Na, so only atoms impinging on the plate in the $F = 2$ ground state were reflected, whereas atoms in the $F = 1$ ground state were unaffected. In this way they could achieve a quantum-state selectivity of around 100. Balykin and Letokhov [205] suggested that a pair of concave mirrors based on this principle would be the ideal arrangement for building an atomic cavity. One of the basic limitations of such a cavity, however, would be gravity. Since the atomic trajectories would always be perturbed by the gravitational force, the lifetime of atoms in the cavity would be limited.

This problem can be overcome by dropping cold atoms from an optical molasses held a few mm above a concave surface. Atoms released from the molasses fall down and are then reflected by the mirror. Although several bounces are possible on this "atomic trampoline", early experiments [108,206] reported only one or two bounces. In an improved version of their earlier experiment, Aminoff et al. [160] showed that atoms can bounce as many as eight times before they are lost from the cavity. The losses were attributed to light scattering during reflection, collisions with background atoms, and scattering of stray light. Going to higher detunings reduced the losses caused by scattering, but the signal-to-noise ratio decreased as well. More recently, a hybrid gravito-optical trap was demonstrated that used a hollow laser beam for lateral confinement and showed thousands of bounces [109].

Another kind of atomic mirror has been proposed that depends on the large static dipole moment of Rydberg atoms. The force derives from the presence of such atoms in inhomogeneous dc fields (see Sec. 6.2.6). Imagine that an atom undergoes optical excitation to a Rydberg state in an inhomogeneous field. The laser detuning and beam position are chosen to excite a particular sublevel as a result of the spatially varying Stark shifts. If the energy of the chosen sublevel increases strongly with field, then the atom would be attracted toward a weaker field region. Atoms in such states would be repelled by a strongly concentrated field such as that found near the edge of capacitor plates.

Suppose atoms were incident in their ground state from a field-free region into a region with a strong dc electric field gradient. Since the Stark shifts of the ground state are small the atoms would travel freely. If they traversed a thin sheet of light tuned to excite them to a Rydberg state, they would then experience a strong force deriving from the field gradient. The resulting force can be arranged to deflect or focus the atoms [207] and could also be arranged to reflect atoms back along their paths. Carefully tailored fields and well-chosen Rydberg states could combine to produce a very effective atomic mirror.

Atoms can also be reflected by a strongly inhomogeneous magnetic field, even in their ground states because the ground-state magnetic moments can be large. The first suggestion for such magnetic mirrors was made in Ref. 208, but it was based on a large scale field and was not very practical. Later people experimented with arrangements of miniature permanent magnets, on the mm scale. A much more clever approach was described in 1998 [209]. The authors used the strong field gradients near the surface of recorded magnetic media such as floppy disks or video tape. In their most recent experiments they were able to demonstrate about a dozen bounces.

One problem of this type of mirror can arise from the possible reorientation of the magnetic moments of different atoms resulting from their passage through different field regions. Because the magnetization of the material is essentially random, it is quite likely that there will be strongly varying or near-zero field regions in the neighborhood of the surface of the magnetized material. Atoms traversing these regions might undergo non-adiabatic transitions, i.e., spin flips, as a result of their motion (see Sec. 10.3.1). This results in a kind of decoherence that can affect deBroglie wave reflection, as discussed in Sec. 15.5.

13.3   Atom Lenses

13.3.1   Magnetic Lenses

The first lens for neutral atoms was devised in 1951 and used the inhomogeneous field of a hexapole array of magnets [129]. The principal idea is to produce a magnetic field that varies quadratically with distance from the axis so that the field gradient, and thus the force, is harmonic (see Sec. 10.2).

It is straightforward to expand a solution to Laplace's equation for the magnetic potential $\Phi(r, \theta)$ with boundary conditions of six-fold symmetry and to find a series with leading term proportional to r^3 [131, 210]. Thus the field would be quadratic as desired. For the idealized case of cylindrical magnetic poles having surface magnetic potentials 0 and V that alternate among the six poles, the magnetic potential in the region between the poles is given by [211]

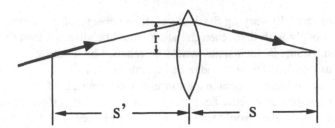

FIGURE 13.1. Schematic diagram of the magnetic lens described in the text. The path of a typical atom is indicated by the heavy arrows for entering and leaving the lens. Atoms whose paths cross the axis at s' are focused to retun to the axis at s.

$$\Phi(r, \theta) = \frac{4V}{\pi} \sum_{n=1}^{\infty} \left(\frac{r}{r_0}\right)^n \frac{\sin n\theta}{n} \sin\left(\frac{n\alpha}{2}\right) \sum_{\ell=1}^{N}(-1)^{\ell-1} \sin\left[\frac{n\pi}{2N}(2\ell - 1)\right],$$

(13.1)

where N is the number of pole pairs ($N = 3$ in this case), r_0 is the distance to the boundary (half of the lens aperture), and α is the axial angle subtended by each pole [211]. Thus the field for $N = 3$ and $\alpha = \pi/6$ is given by

$$B(r, \theta) = B_0 \left(\frac{r}{r_0}\right)^2 \left[1 - 2\left(\frac{r}{r_0}\right)^6 \cos 6\theta + \mathcal{O}\left(\frac{r}{r_0}\right)^{12}\right]^{1/2}.$$

(13.2)

The dominant term is the desired quadratic and the next term is smaller by about an order of magnitude for more than 3/5 of the total area of the aperture.

Atoms with magnetic moment μ and mass M oscillate in this harmonic potential with arbitrary A_0 according to $r = A_0 \cos \omega_{\text{lens}} t$, where ω_{lens} is given by

$$\omega_{\text{lens}}^2 = 2\mu B_0 / M r_0^2.$$

(13.3)

If the magnetic lens has a length L and atoms pass through it with longitudinal velocity v_0, then they undergo a phase change of the oscillatory motion of $\Delta\phi = \omega_{\text{lens}} L/v_0$.

Consider an atom that crossed the axis at a distance $s' \gg r_0$ from the lens with transverse velocity v_x and then enters the lens at a distance r from the axis (see Fig. 13.1). (The condition $s' \gg r_0$ corresponds to the small angle or paraxial domain for the lens.) Its initial phase in the oscillation is $v_x/A_0\omega_{\text{lens}}$, and its final phase is $v_x/A_0\omega_{\text{lens}} + \Delta\phi$. Thus it will be redirected to cross the axis again at a distance s from the lens that satisfies $(s + s')/ss' = 1/F$, where

$$F = \frac{v_0^2}{\omega_{\text{lens}}^2 L} = \frac{E_k}{E_{\text{mag}}} \frac{r_0^2}{L}.$$

(13.4)

Here E_k is the kinetic energy and $E_{\text{mag}} = \mu B_0$ is the magnetic energy at the boundary r_0. This is exactly the equation for a thin lens of focal length F, where s and s' are the image and object distances.

For ordinary thermal atoms it would require high fields and large L to get significant focusing, and even then, a practical lens would have a speed of only

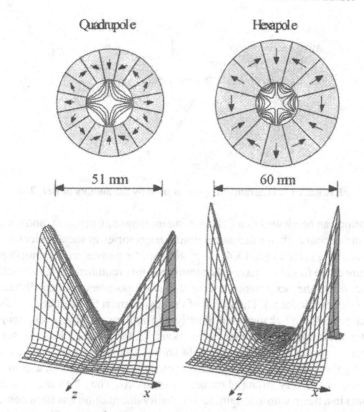

FIGURE 13.2. The design and construction of a magnetic quadrupole and hexapole lens made from permanent magnets. The lower part shows the potential seen by an atom whose magnetic moment is correctly oriented for focusing (figure from Ref. 213).

$\sim f/100$. In spite of this, such lenses were used routinely in early atomic beam experiments [11]. By contrast, laser-slowed atoms have E_k that is typically 10^{-7} times smaller, so a lens only a few cm long can have a speed of $\sim f/6$. The first use of a lens for slow atoms was reported in Ref. 212 using the lens described in Ref. 208.

Since then there have been much more sophisticated lenses built for slow atoms using permanent magnets made from modern ceramic rare-earth magnetic materials. The materials are cut to shape using electron erosion machining, and then magnetized just before assembly. An excellent description is in Ref. 213. Figure 13.2 shows both a quadrupole and hexapole lens made from permanent magnets.

The magnetic potential for atoms whose magnetic moment is correctly oriented for focusing is also shown in Fig. 13.2. Of course, both the magnetic potential and the focal length will be different for atoms with different orientations. If the fields are sufficiently strong, the nuclear magnetic moment will be decoupled from that of the electron, and only M_J will be important. For ground-state alkalis this requires fields around 0.1 T, and so atoms passing near the center of such a lens will behave differently from those near the edges where the field is stronger. This

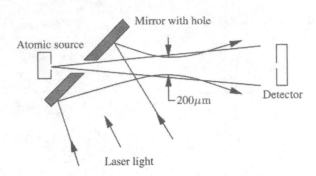

FIGURE 13.3. Diagram of apparatus used by the authors of Ref. 214.

decoupling can be viewed as a field-dependent magnetic moment, and needs to be taken into account when calculating atomic trajectories in such devices.

The formula given in Eq. 13.4 is very general. Its derivation does not depend on the nature of the force that acts on the atoms, but only requires that they pass through a region where they experience a force that is proportional to their distance from the axis (harmonic force). The length of the region must be short enough that their displacement doesn't change significantly during their passage, even though their transverse velocity does (thin lens). For example, the value of F for lenses whose focusing properties derive from the light shift is found by replacing the magnetic energy by the light shift. The dependence on v_0^2 is always present, and shows that such lenses have very strong chromatic aberrations. Thus they are suitable only for atoms in a beam whose longitudinal velocity distribution has been compressed considerably from thermal, for example, by laser slowing (see Chapter 6).

13.3.2   Optical Atom Lenses

Lenses that depend on the light shift have played an important role in a variety of atom optics applications. The first such experiment was performed in a beam of Na using a copropagating beam of light [214]. These authors placed a 45° mirror with a small hole in it in an atomic beam. The atoms passed through the hole, and a focused laser beam was reflected by the mirror from perpendicular to copropagating along the atomic beam (see Fig. 13.3). The Gaussian beam profile of the laser is approximately parabolic at its center, and at its focus the light shift is approximately harmonic, as required for a lens.

By changing the detuning of the laser frequency from above to below atomic resonance, the lens could be either converging or diverging. For red detuned light, the light shift of the ground state is negative, so atoms are attracted to strong field regions (see Sec. 11.2.1) and the lens focused the atomic beam into the movable detector. The data show strong focusing by their lens with speed of approximately $f/20$.

Another kind of atomic lens uses a two-dimensional MOT (see Sec. 11.4). Equation 11.5 therein shows that the total force on atoms in the combined magnetic and optical fields has both a damping and a harmonic component, but only the harmonic component is of interest here. It can dominate the damping force when the transverse component v_x of the velocity of atoms in a beam incident on such a lens satisfies $v_x \ll \mu' Az/\hbar k$, which constitutes the paraxial approximation (here A is the magnetic field gradient). Such a restriction is readily satisfied for easily achieved parameters for a lens used in the paraxial domain, as appropriate for the other lenses discussed in this section. The focal length can again be found from Eq. 13.4, using the relevant magneto-optic energy given by $m\omega_{lens}^2 A_0^2/2$, where the oscillation frequency ω_{lens} is found from Eqs. 11.5 and 11.6, and A_0 is the aperture radius.

13.4   Atomic Fountain

Not all atom optics devices have analogies to ordinary optics. In addition to dissipative elements, another obvious exception is those devices wherein the atomic trajectories are modified by gravity. One of the earliest suggestion to exploit the ballistic motion of atoms was the atomic fountain proposed by Zacharias. Authors often refer to the citation [215] but this is nothing more than the title only of a talk given by Zacharias at a conference. The only known published description of this experiment is in Ramsey's book [11], although there is further information in the proceedings of Zacharias' 61 birthday festschrift [216].

The idea was to make a fountain in which the slowest atoms emitted from a thermal source would rise only a small distance before gravity pulled them down. The objective was to lengthen the passage time of atoms through a microwave field that was driving the ground state hyperfine transition in order to reduce the associated transit time broadening to improve the precision of measurement. It was hoped that this would eventually lead to the establishment of an atomic time standard as envisioned by Rabi some 30 years earlier.

This experiment failed, because atomic collisions with fast atoms in the source aperture always speeded up the slowest atoms. The velocity distribution of atoms emerging from a typical thermal source does not resemble that calculated from kinetic theory (see Sec. 5.2) except for the case of extremely low densities.

However, laser cooling enables the production of samples with very slow atoms, and these can be vertically launched to make quite excellent fountains. Atoms are first loaded into a 3D optical molasses from a MOT (see Chapter 7 and Sec. 11.4) made with one pair of beams vertical. Then the frequency of one (or both) of the vertical beams is shifted so that the two beams have a frequency difference Δ. This results in an ordinary optical molasses in a frame of reference that is moving at velocity $\Delta/2k$, called the launch velocity.

If the sign of Δ is chosen so that this velocity is upwards, the atoms are quickly cooled to a low temperature in a frame moving upwards at the launch velocity. After a short time the molasses beams are shut off and the atoms are in free flight on trajectories that can take them to a height of $\Delta^2/8gk^2$ that can range from a few cm to a few m, depending on the height of the vacuum system. Of course, their residual horizontal velocities will cause the sample of atoms to spread out, and this, too, may limit the usable height.

Some interesting questions can arise in gravito-optics that aren't relevant in ordinary optics. When a sample of cold atoms is released from rest and is allowed to fall, atoms travel in parabolic trajectories. All possible parabolic trajectories are bounded by a parabolic caustic. Each point on a plane placed below the release point can be reached by two parabolic trajectories, one that starts in an upward direction and one that starts downward. Since these don't have the same path length, there can be atom interference fringes present on the plane [217]. By contrast, the light from a point source propagates radially and never folds back, so that the expanding spherical waves never show interference fringes.

13.5   Application to Atomic Beam Brightening

13.5.1   Introduction

In considering the utility of atomic beams for the purposes of lithography, collision studies, or a host of other applications, maximizing the beam intensity may not be the best option. Laser cooling can be used for increasing the phase space density, as described in Sec. 5.5, and this notion applies to both atomic traps and atomic beams. In the case of atomic beams, other quantities than phase space density have been defined as well, but these are not always consistently used. Many articles provide numbers to characterize their beams without specifying which of the defined quantities are being cited. Recently a summary of these beam properties has been presented in the context of phase space (see Fig. 13.4).

The geometrical solid angle occupied by atoms in a beam is $\Delta\Omega = (\Delta v_\perp/\bar{v})^2$, where $\bar{v} = \sqrt{(9\pi/8)}\ \tilde{v}$ is the average velocity of atoms in the beam (see Table 5.1) and $\Delta v_\perp$ is a measure of the width of the transverse velocity distribution of the atoms. The total current or flux of the beam is Φ, and the flux density or intensity is $\Phi/\pi(\Delta x)^2$ where Δx is a measure of the beam's radius. Then the beam brightness or radiance R is given by

$$R = \frac{\Phi}{\pi(\Delta x_\perp)^2\Delta\Omega}. \tag{13.5}$$

Optical beams are often characterized by their frequency spread, and, because of the deBroglie relation $\lambda = h/p$, the appropriate analogy for atomic beams is the longitudinal velocity spread. Thus the spectral brightness or brilliance B, is given by

$$B = R\frac{\bar{v}}{\Delta v_z}. \tag{13.6}$$

Note that both R and B have the same dimensions as flux density, and this is often a source of confusion. Finally, B is simply related to the 6D phase space density ρ.

One of the most important applications of these beams is for collision experiments. High-resolution studies of collisions between atoms in thermal beams were

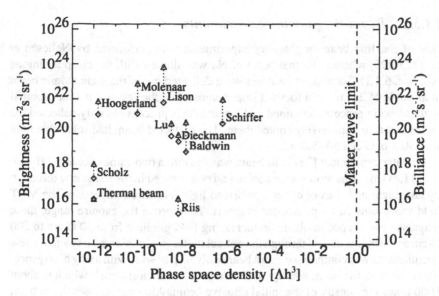

FIGURE 13.4. Plot of brightness (diamonds) and brilliance (triangles) *vs.* phase space density for various atomic beams cited in the literature. The lower-left point is for a normal thermal beam, and the progression toward the top and right has been steady since the advent of laser cooling. The experimental results are from Riis *et al.* [72], Scholz *et al.* [218], Hoogerland *et al.* [219], Lu *et al.* [220], Baldwin *et al.* [221], Molenaar *et al.* [65], Schiffer *et al.* [222], Lison *et al.* [223] and Dieckmann *et al.* [224]. The quantum boundary for Bose-Einstein condensation (see Chapter 17), where the phase space density is unity, is shown by the dashed line of the right (figure adapted from Ref. 223).

hampered in the past by the Maxwell-Boltzmann velocity distribution of effusive beams (see Sec. 5.2), so even in the simplest experiments the signals were always averaged over this distribution. Although some clever schemes have been devised to overcome this problem, they always suffer from loss of intensity. In addition to the longitudinal velocity compression discussed in Chapter 6, laser cooling can also provide intensity enhancement by transverse velocity compression. Most important, it enables collision experiments in the new regime of ultra-low temperatures (see Chapter 14).

Another application lies in the field of atom-surface scattering, where well-collimated atomic beams with large transverse deBroglie wavelengths can be used to study surface structures. Still another application is atomic nanofabrication discussed in Sec. 13.6. In this case, high brilliance beams are needed because of the strong velocity dependence of the focal length of atom lenses as discussed on pg. 184. Finally is the application to the area of precision measurements and atomic clocks (see Sec. 13.7). Many of the precision beam measurements of the 1950s and 60s were limited by the brightness of the atomic beam. These include several of fundamental importance, such as the electrical neutrality of matter and the search for dipole moments of elementary particles. This section focuses on techniques to obtain monochromatic, well-collimated, high brightness atomic beams.

13.5.2   Beam-Brightening Experiments

One of the first beam-brightening experiments was performed by Nellesen *et al.* [225, 226] where a thermal beam of Na was slowed with the chirp technique (see Sec. 6.2.1). Then the slow atoms were deflected out of the main atomic beam at an angle of 22° using a focused laser beam, while the fast atoms (not captured in the slowing process) remained undeflected. This process not only deflected the atoms, but also transversely cooled them. The deflected beam had selectable final velocities between 50–200 m/s.

In a later experiment [227] this beam was fed into a two-dimensional MOT (see Sec. 11.4) where the atoms were cooled and compressed in the transverse direction by an optical molasses of σ^+-σ^- polarized light. For this compression the MOT field was produced by permanent magnets. To improve the capture range, these magnets were shaped to obtain an increasing field gradient from 50 G/cm to 500 G/cm as the atoms moved through the optical molasses. In this way a beam of a few mm diameter was compressed into a beam only 43 μm wide with a tiny divergence. The density in the beam was approximately 10^9 cold atoms/cm^3, which is about 1000 times the density of the initial effusive beam. Although the density is high, the beam is still optically thin from the sides which makes it easy to manipulate it even further.

Another approach was used by Riis *et al.* who directed a slowed atomic beam into a hairpin-shaped coil that they called an "atomic funnel" [72]. The wires of this coil generated a two-dimensional quadrupole field that was used as a two-dimensional MOT as described before. Inside the trapping region, the beam of atoms is further slowed in the longitudinal direction by two counterpropagating laser beams of different frequencies, thereby forming a moving optical molasses, so that atoms moving at a certain selectable velocity experience zero force. In this way a monochromatic atomic beam with a velocity of 260 m/s, a diameter of 150 μm, and a flux of 10^9 atoms/s was produced, leading to a density of 2×10^6 atoms/cm^3. This is an increase of the density over chirped-cooled atomic beams by a factor of 40.

The authors of Ref. 223 have constructed a high-brilliance beam of Cs using a Zeeman slower for longitudinal velocity compression (see Sec. 6.2.2) and transverse collimation with optical molasses. They made two important improvements to the longitudinal phase space compression. First, they carefully tailored the field near the exit of the solenoid using extra magnets in a way explored in Ref. 65 but refined by them. Second, they improved the time-of-flight measurement system so that it could resolve only a few mm/s, nearly down to the recoil velocity for Cs. Thus Fig. 13.4 shows their beam with the highest brilliance to date, even though the phase space density is considerably lower than other beams. This beam is very well suited to nanofabrication.

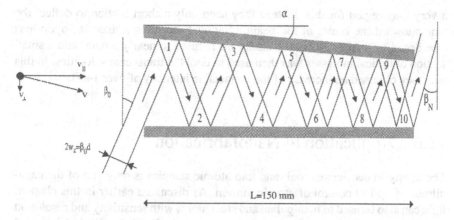

FIGURE 13.5. Schematic representation of a two-dimensional collimator. The incoming laser beam makes several bounces, and at each bounce the angle between the laser light and the normal to the atomic beam is reduced. In this way the light interacts with atoms coming from a much larger solid angle of the source. Furthermore, because of the recycling of the light, less laser power is required (figure from Ref. 219).

13.5.3   High-Brightness Metastable Beams

These approaches yield intense beams when the number of atoms in the uncooled beam is already high. However, if the density in the beam is initially low, for example in the case of metastable noble gases or radioactive isotopes, one has to capture more atoms from the source in order to obtain an intense beam. Aspect *et al.* [71] have used a quasi-standing wave of converging laser beams whose incidence angle varied from 87° to 90° to the atomic beam direction, so that a larger solid angle of the source could be captured. In this case they used a few mW of laser light over a distance of 75 mm.

One of the most sophisticated approaches to this problem has been developed for metastable Ne by Hoogerland *et al.* [219]. They used a three-stage process to provide a large solid angle capture range and produce a high brightness beam. The first stage of their beam brightener consists of two pairs of nearly parallel mirrors arranged so that multiply reflected beams of light cross the atomic beam at varying angles to provide a large capture range (see Fig. 13.5). The laser beams bounce between the mirrors 10 times, and the angle of the light with respect to the direction of the atomic beam increases by 0.5 mrad each time, so that at the end of the mirrors the light intersects at almost 90° with respect to the atomic beam. By recycling the laser light, the laser power consumption of this section is only 30 mW and the collimation of the beam could be extended over a large distance of 150 mm. The solid angle captured in the process is about 0.1 rad.

This region is followed by a magneto-optical lens to focus the atoms to a small space (see Sec. 13.3). This is required because the transverse collimation is necessarily accompanied by an increase in the diameter of the atomic beam to 20 mm in the last case. To overcome this problem, a two-dimensional MOT as discussed before could be used, but the high longitudinal speed of the atoms would require a very long region for this. Instead they used only a short section to deflect the atoms toward the center of the beam, effectively forming a magneto-optic lens (see Sec. 13.3). After 70 cm the atoms are focused to nearly a point and a small section of optical molasses was then used to cool the transverse velocities. In this way they obtained an increase of the metastable intensity of over 1400.

13.6   Application to Nanofabrication

The ability to decelerate, cool, and trap atomic samples is only one of the capabilities of optical control of atomic motion. As discussed earlier in this chapter, light can also be used to manipulate and steer atoms with sensitivity and resolution unimagined just a few years ago. Atomic beams can be focused, split, and delivered in complex patterns for the construction of objects on an incredibly small scale. Atomic beam nanofabrication has wavelength limits *much* smaller than similar optical processes because typical values of the deBroglie wavelength are 10^{-11} m in atomic beams. Furthermore, the neutrality of atoms removes the space charge limits and Coulomb repulsion effects associated with charged particle beams.

There are two fundamentally different methods for nanofabrication. The first consists of optically manipulating the atoms of interest from a beam or vapor directly into the desired pattern, typically on a substrate where the desired structures build up on the surface as the atoms strike and stick to it. Such direct fabrication is limited to those atoms having optical transitions that are convenient for laser manipulation and those substrates that are compatible with the atoms. The second method is atomic lithography, using atoms instead of light to expose the resist. There are many combinations of atoms and resists that have been shown to work, using either geometrical masks or optical fields to generate the distribution of atoms in the desired pattern to expose the resist. After exposure, the resist is "developed" by vapor etching, wet chemistry, or whatever is appropriate, just as in traditional lithography.

In the first experiments [228, 229] a pattern of lines of atoms was applied to a substrate. The Na atoms in a thermal beam traversed a standing wave of nearly resonant laser light, and the pattern of nodes and antinodes formed an array of microlenses resulting from the dipole force as discussed in Sec. 13.3.2. Even though the intensity distribution of such dipole force lenses is sinusoidal instead of the optimal parabolic, very good focusing can be achieved. In later experiments [230] a substrate placed in the "focal plane" of such an array of "cylindrical microlenses" was coated with an array of lines of Cr, as shown in Fig. 13.6. Although the Na pattern of the earlier experiments could not be exposed to air because metallic Na is unstable, Cr is quite durable and the pattern is very robust. There are now many laboratories seeking suitable combinations of atoms and substrates for such direct nanofabrication.

Atom lithography depends on having a thin layer of an appropriate resist for the atoms to react with. Perhaps the most successful types are self-assembled

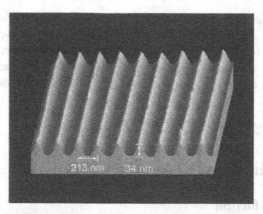

FIGURE 13.6. Chromium deposited directly onto a substrate of silica using a series of microlenses (figure from Ref. 230).

monolayers (SAM's) of polymers of various lengths. These long molecules have hydrophyllic radicals at one end and hydrophobic ones at the other end, so they tend to line up on a substrate in water like the vertical filaments of a plush carpet. Thus a chemically vulnerable layer of material, previously deposited on a suitable base, can be covered with a film of such a SAM. Then atoms arriving in a pattern formed by a mask or by optical steering or focusing can attack the polymer molecules somewhere along their length by penetrating a few atomic layers into the "carpet", compromising their ability to protect the underlying layer of material. One very common preparation is alkanethiolate polymers of various lengths between 5 and 10 units which form suitable SAMs on evaporated gold surfaces. After exposure, the broken alkanethiolate molecules undergo chemical removal and then the exposed gold is dissolved away. Then the remaining alkanethiolate molecules are removed, leaving the desired gold pattern.

Metastable rare gas atoms, carrying between 10 and 20 eV of energy, can effectively break these alkanethiolate chains with nearly 100% efficiency. The low energy (heavy) metastables are more suitable for shorter chains, whereas He* with its 20 eV energy can damage chains of length up to 12. Needless to say, the longer the chain of the SAM polymer, the better the undamaged molecules can protect the substrate and prevent unwanted etching. Some early experiments were done with He* atoms for this reason [231, 232], and now several laboratories are pursuing this technology.

Until 1997 all such atom lithography experiments were done by patterning the atomic beam with a physical mask. The authors of Ref. 233 used an optically focused Cs beam to attack the nine-element long chains of nonanethiole in a SAM on a thin layer of gold. The exposed sample was suitably etched and examined with both an optical and an atomic force microscope. Both images showed well-defined lines separated by 426 nm.

The potential applications of such atomic nanofabrication are manifold. Gratings made this way can serve as primary length standards in the μm range because the long-range order is not subject to systematic effects of the kind that arise with mechanical translations, and because the $\lambda/2$ separation between the grids is known to spectroscopic accuracy. By doing such nanofabrication in two dimensions, it may be possible to make arrays of microstructures with unprecedented ease and precision. These may be suitable for many diverse purposes such as diffractive optical devices or electronic chips such as computer memory or Josephson junction arrays.

13.7   Applications to Atomic Clocks

13.7.1   Introduction

Throughout history humans have tried to build devices for measuring time. There was a great increase in the development of clocks after the Europeans discovered the western hemisphere when good clocks became necessary for accurate navigation. Various governments offered large awards for the construction of clocks that could maintain accuracy through an ocean voyage. Many countries established Naval Observatories for this purpose, and several of these remain today as the principal national arbiters of time.

In the 16th century Galileo discovered the periodicity of the pendulum, and in the 17th century Huygens developed an escapement mechanism for both pendulum and spring-driven clocks that set the standard for 200 years. Time keeping experienced significant progress in the 19th century with the advent of the American railroads. The first clocks with accuracy much better than 1 sec/day were based on crystal quartz oscillators developed at the beginning of the 20th century.

Motivation for accurate time keeping comes from very many sources. For the purposes of scientific research, very accurate comparison of frequencies is necessary for testing basic theories, including relativity, QED, quantum mechanics, *etc.*. For the purposes of navigation, time keeping has been essential for hundreds of years. All commercial and military aircraft and spacecraft carry quartz or atomic clocks, and many even carry redundant systems. The new Global Positioning System (GPS), which has already begun to revolutionize travel, depends on atomic time, as do computer systems, radio and television broadcasting, telephone and communication systems, and a host of other contemporary technologies. As long as it is believed that all ^{133}Cs atoms are identical, then there is confidence that an atomic clock anywhere in the universe keeps the same time as the commercially available standards found in dozens of laboratories throughout the world.

The idea of atomic clocks grew out of the atomic beam research begun in the late 1930s. Rabi, Ramsey, Zacharias, and others promoted the idea after World War II, but it took 20 years more to become adopted. In 1967 the internationally accepted definition of the second changed from mechanical time pieces calibrated by the Earth's orbit to atomic time calibrated by the hyperfine structure splitting of the ground state of Cs. By definition, 1 second is exactly 9,162,631,770 cycles of the $(F, M_F) = (3, 0) \Leftrightarrow (4, 0)$ transition in ^{133}Cs, the natural stable isotope.

The limitation to both the accuracy and precision of atomic clocks is imposed by the thermal motion of the atoms (see Sec. 5.2). Both the nonzero speed and the thermal range of the speeds of different atoms from an atomic sample provide the ultimate limitation on high precision laboratory measurements and on clocks. One cause of this problem arises from the broadening of a spectral line caused by the small interaction time between the measuring equipment and rapidly moving atoms. At thermal velocities of typically 500 m/s, there are only a few ms to interact with a free atom in an apparatus of reasonable size (*i.e.*, a few meters).

The other source of this limit arises from a frequency shift caused by the relativistic time differences between reference frames in relative motion (sometimes called the second-order Doppler effect; the first-order Doppler effect is the familiar classical frequency shift between moving objects). If the velocity of the atoms with respect to the measuring apparatus were known, this effect could be calculated and accommodated as well. But the atoms have a velocity distribution, characterized by the temperature of their source. Although this too can be calculated as in Sec. 5.2, the details of the distribution at the low velocity end depend very sensitively on the details of the source, and sometimes cannot be adequately known (see Sec. 13.4). Thus a sample of laser-cooled atoms could provide a substantial improvement in atomic clocks and in spectroscopic resolution.

13.7.2   Atomic Fountain Clocks

The first attempts at providing slower atoms for better precision or clocks were by Zacharias in the 1950s, as discussed in Sec. 13.4. The advent of laser cooling changed this because the slow atoms far outnumber the faster ones. The first rf spectroscopy experiments in an atomic fountain using laser cooled atoms were reported in 1989 and 1991 [234, 235], and soon after that some other laboratories also reported successes.

Some of the early best results were reported by Gibble and Chu [236,237]. They used a MOT with laser beams 6 cm in diameter to capture Cs atoms from a vapor at room temperature. Their estimated capture velocity was 30 m/s, consistent with the estimates of Sec. 11.4.3. These atoms were launched upward at 2.5 m/s by varying the frequencies of the MOT lasers to form a moving optical molasses as described in Sec. 13.4, and subsequently cooled to below 3 μK. The atoms were optically pumped into one hfs sublevel, then passed through a 9.2 GHz microwave cavity on their way up and again later on their way down. The number of atoms that were driven to change their hfs state by the microwaves was measured *vs.* microwave frequency, and the signal showed the familiar Ramsey oscillatory field pattern. The width of the central feature was 1.4 Hz and the S/N was over 50 (see Fig. 13.7). Thus the ultimate precision was 1.5 mHz corresponding to $\delta v/v \cong 10^{-12}/\tau^{1/2}$ where τ is the number of seconds for averaging. Stability of the rf signals was maintained with a hydrogen maser.

The ultimate limitation to the accuracy of this experiment as an atomic clock was collisions between Cs atoms in the beam. Because of the extremely low relative velocities of the atoms, the cross sections are very large (see Sec. 14.3) and there

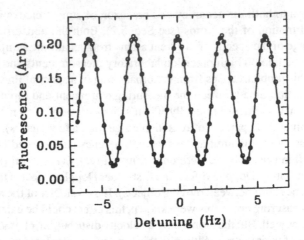

FIGURE 13.7. The central Ramsey fringes of the microwave clock transition $(F, M_F) = (3, 0) \Leftrightarrow (4, 0)$ in a 15 cm high Cs fountain. Each open circle data point represents approximately 1 s of data collection time (figure from Ref. 237).

is a measurable frequency shift [238]. By varying the density of Cs atoms in the fountain, the authors found frequency shifts of the order of a few mHz for atomic density of $10^9/cm^3$, depending on the magnetic sublevels connected by the microwaves. Extrapolation of the data to zero density provided a frequency determination of $\delta\nu/\nu \cong 4 \times 10^{-14}$. More recently the frequency shift has been used to determine a scattering length of $-400a_0$ [239] so that the expected frequency shift is 10^4 times larger than other limitations to the clock at an atomic density of $n=10^9/cm^3$. Thus the authors suggest possible improvements to atomic time keeping of a factor of 1000 in the near future. Even more promising are cold atom clocks in orbit (microgravity) where the interaction time can be very much longer than 1 s [240].

Another important approach to atomic clocks uses an optical transition frequency instead of a microwave frequency. A group at NIST is studying the 2 Hz wide transition in metastable Xe atoms driven by a two-photon transition between the $1s_5$ and $1s_3$ levels [241]. The energy difference corresponds to $\lambda = 1.1~\mu m$, but the angular momentum J of these levels differs by 2 so single-photon transitions are not allowed. Because the two-photon process at $\lambda = 2.19~\mu m$ is so weak, its natural width is very small, enabling very high spectral resolution. The atoms can be cooled and trapped on the allowed $1s_5 \Leftrightarrow 2p_8$ transition at $\lambda = 883$ nm whose natural width is ~ 5 MHz, an easily accessible wavelength for diode or Ti:Sapphire lasers.

13.8   Application to Ion Traps

The notion of laser cooling originates from attempts at improving the precision of measurements with trapped ions for high resolution spectrocopy and atomic clocks [37, 242]. The resolution limits are imposed by both residual first-order Doppler effects from "Doppler-free" spectroscopy, and from second-order Doppler effects (time dilation) that can not be completely characterized because of a lack of sufficently accurate information about the velocity distribution of ions in the traps. Although there was indeed early discussion about applications to neutral atoms as well [38], certainly the first experiments [242, 243] and the first quantitative description [39] were motivated by the ion trap groups. Several papers, but especially Ref. 39, laid down the fundamental ideas of resonant cooling that apply to both ions and neutral atoms.

These initial cooling ideas were related to Raman transitions (see Sec. 8.7.2) among the discrete bound states of trapped ions, as illustrated for optical lattices in Fig. 16.5. But the experiments described in Ref. 242 were in the domain where the discrete quantum states could not be resolved, and so the discussion here is more appropriate for this chapter than for Chapter 15 or 16.

In this first experiment [242], the authors used the currents induced in the trap electrodes as a measure of the motion of the ions, and assumed that the number of ions was fixed. They applied the cooling light from a frequency-doubled dye laser, and observed the time dependence of this induced voltage. They found that the temperatures were reduced to below their threshold of measurement of 40 K, although it's safe to assume that they had actually acheived a temperature below 100 mK.

It was not long before the experimenters could isolate a single ion [244] and laser cool it to extremely low energy. Such experiments were followed by dramatic technological progress in precision spectroscopy that is still an active subject in atomic clock research (see Sec. 13.7). Furthermore, the development of this new tool for one purpose was exploited for a plethora of fascinating experiments in the fundamentals of quantum mechanics. These include, but are not limited to the study of single quantum jumps, Wigner crystals, "shelved atoms", Schrödinger cat states, and laser cooling to the quantum limit of the zero-point motion in the trap. Although this book is about laser cooling of neutral atoms, its readers should pay careful attention to the extensive literature of beautiful experiments done with ion traps.

13.9   Application to Non-Linear Optics

One of the most widespread applications of the interaction between atoms and light is non-linear optics. Atomic absorption and scattering provide the appropriate interaction for multi-photon effects, Raman processes, and other related phenomena. In order to avoid interatomic effects in such studies, they are often done in a vapor where collisions are negligible. Since many cases require that the detuning of the light from atomic resonance be large enough to avoid resonant excitation, this detuning must generally exceed the Doppler width associated with the motion of the atoms in the atomic vapor. At ordinary temperatures, such Doppler widths

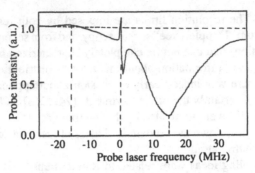

FIGURE 13.8. The absorption spectrum of a probe beam transmitted through a sample of Cs atoms in a MOT, where the probe frequency is measured relative to that of the MOT. The narrow dispersion-shaped feature at zero detuning corresponds to a stimulated Raman transition between ground state levels. Its 400 kHz width is less than 1/10 of the natural width, and is dominated by the Zeeman shifts from the inhomogeneous trap fields. Its spectroscopic linewidth limit is determined by the transit time broadening, the spread of the light shifts from the inhomogeneous optical field, and the residual population of the atomic excited state. The large, broad absorption feature centered at the atomic absorption frequency near 15 MHz, and the weak gain feature centered near -15 MHz, are not of interest here. (Figure adapted from Ref. 245.)

are a large fraction of a GHz, but laser cooling can provide atomic samples with much smaller values. With such samples, non-linear optics will enter a new domain where the Doppler widths are much smaller than the natural widths, and optical detunings can therefore be reduced to a few times the natural width. Thus one can expect enormous non-linear effects, allowing the exploration of previously known effects at much lower intensities, as well as the study of new effects that could not be observed at ordinary temperatures.

The earliest demonstration of non-linear optical spectroscopy in laser-cooled atomic vapors was in 1991 [245, 246]. Both groups studied the Raman transitions of atoms trapped in a MOT (see Sec. 11.4). The first experiments measured the transmission of an auxiliary probe laser beam through the atomic sample under conditions where it absorbed as much as 75% of the light. The experimenters observed three features in the spectrum, two broad and one narrow as shown in Fig. 13.8. The two broad features result from atomic fluorescence phenomena that are roughly independent of frequency, and hence are not affected by the Doppler shifts caused by atomic motion. But the narrow feature is a Raman transition involving light from each of the two laser beams, one from the probe and the other from the trap. Satisfying the Raman resonance condition for excitation by one beam and stimulated emission by the other demands that all atoms in the sample see both beams with the same frequency, and since they are not parallel, the atomic velocities must be small enough for the Doppler shifts to be negligible. This requires a sample of laser cooled atoms.

The signals arise from transitions between different magnetic sublevels of the ground state whose degeneracies have been lifted by the light shifts caused by the trap laser beams, and whose populations differ as a result of optical pumping, again by the trap laser beams. In later stages of the experiments, additional laser beams were added to test this hypothesis [245]. This test, along with careful modeling of all the experimental parameters, confirmed the scenario described here.

Other Raman transitions have provided further experiments in non-linear optics with cold, trapped atoms [247, 248]. In these experiments an optical cavity was placed around a sample of Cs atoms in a MOT, and tuned near the frequency of the gain peak near ω in Fig. 13.8. Since there is a population difference between the two ground states coupled by the Raman resonance, stimulated emission of radiation can occur, resulting in gain in the cavity. When the cavity is resonant with the emitted light, the process is enhanced and a laser beam emerges from the excited mode of the cavity. Again, careful tests verified this description of the origin of the strong gain reported [247, 248].

This is a most curious laser configuration, because the gain width of the active medium is very narrow. In the case of condensed matter lasers, the spectral range of the gain curve is typically many orders of magnitude larger than the width of the laser's optical cavity (the cavity width is approximately its free spectral range divided by its finesse, or several MHz in most cases). Even for gas lasers, the spectral width of the gain medium is the Doppler width, typically a large fraction of a GHz. In this unique gas phase laser, however, the Doppler width is well below the cavity width, and also below the atomic natural width. In fact, the spectral width of the gain medium is determined by the inhomogeneous laser and magnetic fields, as well as by the optical pumping rate, just as in the Raman spectroscopy experiment described above. Clearly there will be very many new and interesting studies on such non-linear systems.

Non-linear optical effects also play an important role in the experiments on atoms trapped in the periodic wells associated with the standing waves of optical molasses in one [162–164], two [249], or three dimensions [250,251]. That subject is discussed in Sec. 11.2.4 on microscopic optical traps and Chapter 16 on optical lattices. Here it is only mentioned for completeness, along with the suggestion that there will surely be non-linear experiments performed on such samples of atoms. In Ref. 250 there is a considerable discussion of Bragg reflection and four-wave mixing from atoms bound in a three-dimensional optical lattice. Further studies on this topic have been performed [252].

Still another multibeam effect on laser-cooled atoms is recoil-induced resonances. Here the resonance condition is calculated very precisely so that it includes both the energy *and* the momentum imparted to an atom that undergoes absorption followed by stimulated emission [253–256]. Since the recoil energy of the atom in the scattering process is included in the energy balance, the initial and final states are almost always non-degenerate, and a very sensitive dependence on atomic velocity enters the resonance condition. In many ways it is similar to the Compton effect, and may be appropriately described as stimulated optical Compton scattering. Thus the method lends itself to very high-resolution measurements of atomic velocities, well below the recoil velocity $\hbar k/M$. Since this corresponds to recoil energies in the kHz region, the limits of this spectroscopy will probably be dominated by interaction time and small field inhomogeneities.

One of the most interesting, and potentially useful, applications of non-linear optics is phase conjugate reflection. Two light beams prepare the atoms in a sample and a third incident beam is retroreflected, independent of its initial angle. The temporal phase of the beam is reversed, so that any aberrations or wavefront distortions may be removed. This has applications in processing images from satellites, airborne cameras, or other sources. The use of a laser-cooled sample for phase conjugate reflection will make enormous improvements in its sensitivity, and hence its utility. The first experiments of this kind have already demonstrated that it works [257], and further improvements are in progress.

14
Ultra-cold Collisions

14.1 Introduction

Laser-cooling techniques were developed in the early 1980s for a variety of reasons, such as high-resolution spectroscopy. During the development of the techniques to cool and trap atoms, it became apparent that collisions between cold atoms in optical traps was one of the limiting factors in the achievement of high density samples. Trap loss experiments revealed that the main loss mechanisms were caused by laser-induced collisions. Further cooling and compression could only be achieved by techniques not exploiting laser light, such as evaporative cooling in magnetic traps (see Chapter 12). Elastic collisions between atoms in the ground state are essential in that case for the rethermalization of the sample, whereas inelastic collisions lead to destruction of the sample. Knowledge about collision physics at these low energies is therefore essential for the development of high-density samples of atoms using either laser or evaporative cooling techniques.

At the end of the 1980s it became clear that laser-cooling techniques could also be used as a tool to study collision processes at low energies. Thermal collisions had been studied in laboratories since the beginning of the century, whereas high-energy collisions have been studied only since the development of accelerators in the 1930s. Laser cooling produces samples of atoms with temperatures below 1 mK and allows collision physicists to extend their energy range by more than six orders of magnitude. Moreover, the study of cold collisions in the presence of a light field became a fruitful subject from which high-resolution information on molecular structure could be obtained.

This chapter presents ultra-cold collisions from both points of view. Although excited-state collisions were studied in more detail before ground-state collisions, this discussion begins with ground-state collisions. It starts with potential scattering, which formed the heart of collision physics for many decades and which has been the subject of many books [258–260] and recent review articles [261–264] that will be cited for further details.

14.2   Potential Scattering

The interaction between two structureless particles is commonly described by the technique of potential scattering, where the interaction between the colliding partners is given by the interaction potential $V(R)$, where R is their separation. In the quantum mechanical description of the collision process, the incoming wavefunction of the relative motion of the two particles is expanded in partial waves, each having a well-defined angular momentum ℓ. In this so-called partial wave analysis, the scattering of each partial wave ℓ is calculated by solving the time-independent Schrödinger equation (SE) using the following Hamiltonian in spherical coordinates:

$$\mathcal{H} = -\frac{\hbar^2}{2\mu R^2}\frac{d}{dR}\left(R^2\frac{d}{dR}\right) + \frac{\hbar^2\ell(\ell+1)}{2\mu R^2} + V(R). \qquad (14.1)$$

Here $\mu = M_a M_b/(M_a + M_b)$ is the reduced mass of the interacting particles denoted by a and b. The second term denotes the centrifugal energy for a given partial wave ℓ.

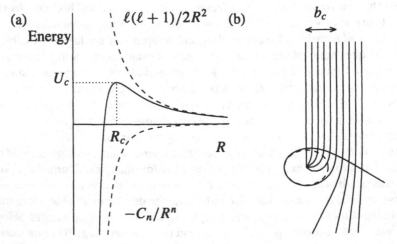

FIGURE 14.1. (a) Interaction potential of two colliding particles interacting by a C_n/R^n potential. (b) The trajectory of an atom in the potential of (a) for different impact parameters b. The critical impact parameter b_c for which the collision leads to a reaction in the inner region is indicated.

At small internuclear distances the potential $V(R)$ becomes appreciable compared with the total energy E, and the solutions of the SE are complicated functions. However, at large distance $V(R)$ becomes small and the solutions evolve toward simple oscillatory functions that are the solutions of the SE in the absence of the potential energy term $V(R)$. The only difference is that there is a phase shift δ_ℓ between the two solutions, and both the differential and total cross sections can be expressed in terms of these phase shifts [259]. The total cross section is

$$\sigma = \frac{4\pi}{k^2} \sum_{\ell=0}^{\infty} (2\ell + 1) \sin^2 \delta_\ell, \tag{14.2}$$

where the sum runs over all different partial waves ℓ and $k \equiv \sqrt{2\mu E}/\hbar$. The summation over ℓ can be truncated at a certain value of ℓ_{max}, where the centrifugal term in the Hamiltonian is so large compared to the total energy, that the incoming wave can no longer penetrate through the centrifugal barrier to small regions where the potential is appreciably different from zero. Then the phase shift δ_ℓ becomes small and therefore the contribution to the total cross section becomes vanishingly small. Although this partial wave analysis of the collision process is very powerful, it does not provide much insight into the reaction.

The connection between a quantum mechanical and semiclassical description of the collision process can easily be made by identifying the total orbital angular momentum $\hbar\sqrt{\ell(\ell+1)}$ as the magnitude of a classical angular momentum $\vec{L} = \vec{R} \times \vec{p}$. Since the magnitude of the angular momentum in a central potential $V(R)$ is conserved, its value is given by $L = \mu v_0 b$, where v_0 is the initial velocity and b is called the impact parameter (see Fig. 14.1b). In the semiclassical analysis the particle with reduced mass μ starts at infinity with a velocity v_0 and at each instant the classical equations of motion are solved using the interaction $V(R)$ and the centrifugal term. The quantum mechanical correspondence principle between quantum and classical mechanics requires that the semiclassical description becomes valid when the angular momentum in the process becomes large compared to $\hbar$, i.e., when many partial waves ℓ contribute to the cross section. At thermal collision energies a semiclassical description is normally sufficient.

The semiclassical description provides insight into the many special properties of ultra-cold collisions. The Langevin model that was first introduced in 1905 can be used to calculate mobility and diffusion coefficients [265]. In the simplest case, the interaction potential between two colliding particles can be approximated by the long-range interaction in combination with the centrifugal term:

$$U(R) = -\frac{C_n}{R^n} + \frac{\hbar^2 \ell(\ell+1)}{2\mu R^2}, \tag{14.3}$$

where n is the order of the interaction and C_n is the corresponding dispersion coefficient. Here the centrifugal term has been included in the definition of the interaction potential $U(R)$. Only the restricted case of collisions of identical atoms, either alkali-metal or metastable noble gas atoms is considered here. If the atoms collide in the S-state, the interaction is a van der Waals interaction and $n = 6$.

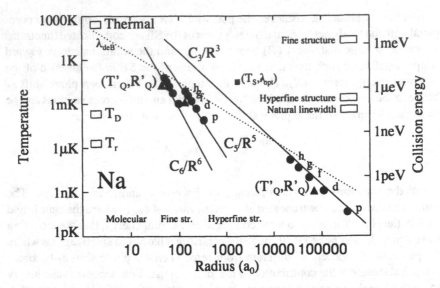

FIGURE 14.2. Potential energy of two Na atoms interacting with different interactions ($n=3$, 5, 6) indicated by solid straight lines. The ranges for which the molecular, fine structure and hyperfine structure interaction become important, are indicated in the bottom part of the graph. The dashed line indicates the deBroglie wavelength as a function of temperature. The symbols p, d, f, g, and h indicate the onset for scattering with the partial waves $\ell=1, 2, 3, 4$ and 5 for each of the three potentials. The temperature T'_Q indicates the temperature for which the WKB-approximation begins to fail. The temperatures T_D and T_r on the left side indicate the range of temperatures which can be reached respectively, by Doppler and sub-Doppler cooling (see Sec. 5.1). The energy ranges on the right indicate the typical atomic interaction energies. The temperature T_s is the temperature where spontaneous emission during the collision becomes important.

For two atoms colliding in the P-state, the interaction is a quadrupole-quadrupole interaction and $n = 5$. If one atom is in the S-state and the other atom is in a P-state, so the two states are coupled by an allowed dipole interaction, there is a dipole-dipole interaction and $n = 3$.

In the semiclassical model, the centrifugal barrier prevents reactions from taking place if the angular momentum is too large. For a low collision energy this cut-off becomes important even for small values of the angular momentum. However, in the quantum mechanical model there is no centrifugal barrier for $\ell = 0$ and the scattering of the lowest partial wave can always take place. So for sufficiently low energy, the scattering can be described in terms of only one partial wave and the regime is referred to as s-wave scattering. From the Langevin model, an estimate can be made for the energy where this regime becomes accessible.

The maximum of the interaction potential of Eq. 14.3 occurs at R_c, which can be found by setting $dU(R)/dR = 0$. Then (see Fig. 14.1a):

$$R_c = \left(\frac{\mu n C_n}{\hbar^2 \ell(\ell + 1)}\right)^{1/(n-2)} \tag{14.4a}$$

and

$$U(R_c) = \left(\frac{n-2}{2}\right) \left(\frac{\hbar^2 \ell(\ell+1)}{2\mu}\right)^{n/(n-2)} C_n^{2/(n-2)}. \tag{14.4b}$$

Reactive collisions can only take place if the collision energy is larger than $U(R_c)$ (see Fig. 14.1b). In order to estimate the cross section for the process, the maximum impact parameter b_c that contributes to the reaction needs to be found. Classically the first step consists of writing $L = \vec{R} \times \mu\vec{v} \cong \mu v b$. At threshold, taking the energy E equal to $U(R_c)$ and putting $L^2 = \hbar^2 \ell(\ell+1)$ yields

$$b_c = \frac{L}{\sqrt{2\mu E}} \propto \frac{E^{(n-2)/2n}}{E^{1/2}} = E^{-1/n}. \tag{14.5}$$

The cross section σ is then proportional to $b_c^2 \propto E^{-2/n}$.

Equation 14.4a also permits an estimate of the number of partial waves contributing to reactive collisions. For the highest partial wave ℓ_c contributing at a certain temperature T, the relation $U(R_c) = 3k_B T/2$ yields for ℓ_c:

$$\ell_c(\ell_c+1) = \left(\frac{n\mu C_n}{\hbar^2}\right) \left(\frac{3k_B T}{(n-2)C_n}\right)^{(n-2)/n}. \tag{14.6}$$

The onset of quantum behavior can be defined as the temperature T_Q where only s-waves can contribute, *i.e.*, the temperature for which $\ell_c = 1$ no longer contributes.

Figure 14.2 shows the R-dependence that this analysis predicts for the interaction energy of Na. Similar plots can be made for the other alkalis, but such plots are very similar. The onset for scattering with partial wave $\ell=1, 2, 3$, and 4 is indicated with the symbols p, d, f, and g for each of the three potentials indicated by the solid lines. The temperature T_Q is thus defined at the point where the p-wave starts to contribute. Notice that T_Q is of the order of the Doppler temperature (see Sec. 5.1) for an atomic interaction with an $n = 6$ potential, but that much lower temperatures are needed to observe the quantum threshold for dipole-dipole interactions ($n = 3$).

Julienne and Mies [266] suggested that the onset of quantum behavior can be found more rigorously by considering the temperature T'_Q, where the WKB-approximation fails. Defining the local deBroglie wave vector $k = 2\pi/\lambda_{dB} = \sqrt{2\mu[E - U(R)]}/\hbar$, the validity criterion for the use of WKB-methods becomes $d\lambda_{dB}/dR \ll 1$. Using the interaction potential of Eq. 14.3, they obtain the same scaling of the maximum angular momentum ℓ_c with temperature, but the overall onset of quantum behavior occurs at a somewhat higher temperature (see Fig. 14.2).

Both T_Q and T'_Q provide rigorous definitions of the transition point from semiclassical to quantum mechanical scattering. This transition point can also more loosely be defined as the point where the deBroglie wavelength becomes comparable to the size of the potential. Although this size is not strictly defined, for chemical reactions it is of the order of $10a_0$, where $a_0 = \hbar^2/me^2$ is the Bohr radius. However, as the deBroglie wavelength is lowered, the scattering probes different parts of the potential and as this chapter will show, cold collisions probe the long-range part of the potential as a result of the long deBroglie wavelength. Finally, in excited-state collisions the transition point can be defined as the temperature where the lifetime of the excited state becomes larger than the collision time. In that case the atoms can no longer be excited at long range and remain in the excited state during the collision. Since the lifetime of the first excited state in the alkalis and noble gases is of the order of tens of ns, this regime is characteristic of most optical traps.

Figure 14.2 summarizes this discussion in one plot for the different interactions important for cold collisions in optical traps. As the plot shows, the behavior for the different possible interactions is very different. For an $n = 6$ interaction, the regime of s-wave scattering appears at the Doppler temperature, so the scattering process for this potential has to be described quantum mechanically. However, for the $n = 3$ interaction many partial waves can still contribute at the Doppler temperature and the collision process can easily be described semiclassically. Therefore there is not *one* single transition point between quantum and semiclassical descriptions to be defined for a given system, but this point depends on the power n of the interaction. In this respect it is not practical to make a distinction between cold and ultra-cold collisions. Although this distinction is defined in the literature to be around 1 μK [264], the physics of the collision process below 1 μK is very different for S-S (n=6) or S-P (n=3) collisions.

14.3 Ground-state Collisions

Ground-state collisions play an important role in evaporative cooling (see Chapter 12). Elastic collisions are necessary to obtain a thermalization of the gas after the trap depth has been lowered, and a large elastic cross section is essential to obtain a rapid thermalization. These are therefore called "good" collisions in Chapter 12. Inelastic collisions, on the other hand, are called "bad" collisions, since the released energy accelerates the particles, which can then reach energies too high to remain trapped. A large good-to-bad collision ratio is essential for efficient evaporative cooling.

Ground-state collisions for evaporative cooling can be described by one parameter, the scattering length a. As Fig. 14.2 shows, ground-state collisions below or at the Doppler temperature are in the s-wave scattering regime and therefore only the phase shift δ_0 for $\ell = 0$ is important. Moreover, for sufficiently low energies, such collisions are governed by the Wigner threshold laws where the phase shift δ_0 is inversely proportional to the wavevector k of the particle motion. This can be understood as follows: The wavefunction in the inner range of the potential is no longer dependent on the energy of the collision, since in the inner range the potential energy is much larger than the collision energy. So the total accumulated phase in the inner region is independent of E. However, the phase of the unperturbed wavefunction is directly proportional to k, so the phase shift δ_0 is proportional to $1/k$. Taking the limit for low energy gives the proportionality constant, defined as

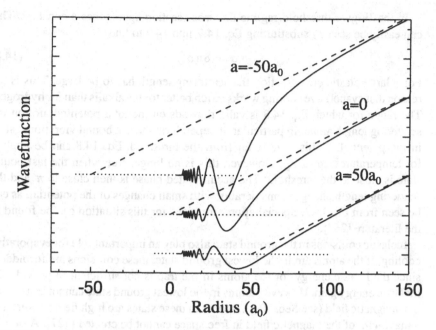

FIGURE 14.3. Wavefunctions of scattering states for Na for a temperature of 2 nK. The scattering length a is defined as the intersection of the unperturbed wavefunction, which is matched at long range to the "real" wavefunction. By making slight changes to the inner part of the potential, the scattering length can be changed over a wide range.

the scattering length a,

$$a = - \lim_{k \to 0} \frac{\delta_0}{k}. \qquad (14.7)$$

The scattering length not only plays an important role in ultra-cold collisions, but also in the formation of Bose-Einstein condensates (see Chapter 17).

The physical interpretation of the scattering length a can be inferred from Fig. 14.3 that shows a wavefunction plotted for three different values of a. In the inner region the wavefunctions are nearly identical in all three cases: the three cases only become different at long range. The scattering length is found by matching these wavefunctions at long range with the unperturbed wavefunctions, and this is indicated by the dashed lines. The intersection of this shifted wavefunction with the R-axis can now be identified as the scattering length a. For a negative scattering length, the shifted wavefunction has to be extended to negative R. Since at long range there is no difference between the original solution and the shifted, unperturbed wavefunction, it is clear that only the shift is important. Since the phase shift δ_0 for low energies is directly proportional to the scattering length a, the whole potential can be described by only one relevant parameter. Although cold collisions are very sensitive to the potential, the sensitivity is completely confined to changes in a. Details of the inner part of the potential cannot be obtained in low energy collisions.

In the Wigner threshold regime the cross section approaches a constant. This can easily be seen by substituting Eq. 14.7 into 14.2 to find

$$\sigma = 8\pi a^2. \tag{14.8}$$

For a large elastic cross section the scattering length has to be large. This is the reason that evaporative cooling works much better for the alkalis than for hydrogen. The range for which Eq. 14.8 is valid depends on the total potential, not on the scattering length alone. In particular, it depends on the last bound vibrational state in the potential. If this state is far from the threshold, Eq. 14.8 can be applied for temperatures up to T_Q. However, this is no longer true when the last bound state is close to the threshold. The accumulated phase is then close to π and the scattering length changes considerably with small changes of the potential, as can be seen from Eq. 14.7. Special approximations for this situation can be found in the literature [267].

Inelastic collisions in the ground state also play an important role for evaporative cooling. If the atoms are in the lowest ground state, these collisions are forbidden, since the kinetic energy for the atoms in the trap is not sufficient to provide the inelastic energy gain. However, atoms in the lowest ground state cannot be trapped in a magnetic field (see Sec. 10.1). Atoms in these states are high field seekers and a maximum of the magnetic field in free space cannot be created [127]. Atoms in a magnetic trap are therefore in higher magnetic sublevels of the ground state, and inelastic collisions to lower states are therefore possible.

The inelastic processes can be divided in two cases. In the first case, the total spin of the collision system is conserved and the process can only proceed if the spin of initial and final state are identical. Such exchange or relaxation processes are induced by the exchange potential between the two atoms, which is the difference potential between the singlet and triplet potential for the alkalis. Such process are generally strong and the rates are typically of the order of 10^{-11} cm^3/s. For a density of 10^{11} atoms/cm^3, which is rather low to reach quantum degeneracy, the sample will self destruct in the order of 1 s because of collisions. Exchange collisions should therefor be avoided. In the second case, the total spin is no longer conserved. Such transitions can only be induced by magnetic dipole-dipole interactions, which can be either electron-electron, electron-nuclear or nuclear-nuclear interactions. Since the magnetic dipole moment of the nucleus is 1000 times smaller than the magnetic dipole moment of the electron, the electron-electron interaction usually dominate. Rates for such processes are typically of the order of 10^{-15} cm^3/s for the alkalis.

In Fig. 4.2 the energies of the hyperfine ground states of Na in a magnetic field are shown. For atoms trapped in the high-field seeking state 7 exchange collisions lead to population of state 1. Such a reaction will not only release an amount of internal energy into kinetic energy, but also leads to trap loss, since state 1 is not trapped by the magnetic field. In order to avoid exchange collisions, atoms can be prepared in a doubly polarized state (state 8), which is generally used for BEC-experiments. Another possiblity is to trap in a sufficiently small magnetic field state 3, which is high-field seeking for a field up to 0.03 T. Although the trap cannot be very deep, the advantage is that the energy released in the reaction is small and therefore the inelastic rates are suppressed.

14.4   Excited-state Collisions

Although ground-state collisions are important for evaporative cooling and BEC, they do *not* provide a very versatile research field from a collision physics point of view. The situation is completely different for the excited-state collisions. For typical temperatures in optical traps, the velocity of the atoms is sufficiently low that atoms excited at long range by laser light decay before the collision takes place. Laser excitation for low-energy collisions has to take place during the collision. By tuning the laser frequency, the collision dynamics can be altered and information on the states formed in the molecular system can be obtained. This is the basis of the new technique of photo-associative spectroscopy, which for the first time has identified purely long-range states in diatomic molecules.

Excited-state collisions in optical traps play an important role. Since the laser cooling is done with nearly resonant light, a large fraction of the atoms are in the excited state. The cross section for excited-state collisions can be many orders of magnitude larger than for ground-state collisions, so inelastic excited-state collisions are the most dominant trap loss mechanism. In the rest of this section, information on excited-state collisions from trap losses is discussed first, then comes optical collisions where the collisions are induced by a probe laser, and finally the results obtained in photo-associative spectroscopy are presented.

14.4.1   Trap Loss Collisions

For atoms colliding in laser light closely tuned to the S-P transition, the potential is a C_3/R^3 dipole-dipole interaction when one of the atoms is excited. Since this potential has a much larger range than the C_6/R^6 ground-state potential, the ground-state potential can be considered flat at such ranges. Absorption takes place at the Condon point R_C given by

$$\hbar\delta = -\frac{C_3}{R_C^3} \quad \text{or} \quad R_C = \left(\frac{C_3}{\hbar|\delta|}\right)^{1/3}. \tag{14.9}$$

Note that the light has to be tuned below resonance, which is mostly the case for laser cooling. The Condon point for laser light detuned a few γ below resonance is typical 1000–2000 a_0.

Once the molecular complex becomes excited, it can evolve to smaller internuclear distances before emission takes place. Two particular cases are important for trap loss (see Fig. 14.5): (1) The emission of the molecular complex takes place at much smaller internuclear distance, and the energy gained between absorption and emission of the photon is converted into kinetic energy, or (2) the complex undergoes a transition to another state and the potential energy difference between the two states is converted into kinetic energy. In both cases the energy gain can

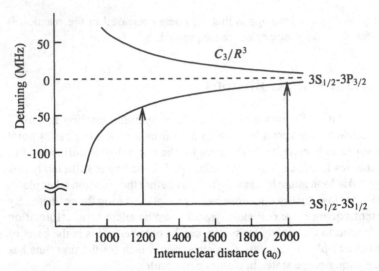

FIGURE 14.4. Molecular system of two Na atoms interacting in the ground and first excited state showing both attractive and repulsive potentials. Only excitation to attractive potentials leads to a close collision due to the small kinetic energy of ultra-cold atoms, whereas excitation to repulsive states leads to a breakup of the system. For light detuned more to the red, the excitation takes place at a smaller internuclear distance for which the excited complex has a larger probability to remain excited during the collision.

be sufficient to eject one or both atoms out of the trap. In the case of the alkalis, the second reaction can take place because of the different fine-structure states and the reaction is denoted as a fine-structure changing collision. The first reaction is referred to as radiative escape.

Trap loss collisions in MOT's have been studied to great extent, but results of these studies have to be considered with care. In most cases, trap loss is studied by changing either the frequency or the intensity of the trapping laser, which also changes the conditions of the trap. The collision rate is not only changed because of a change in the collision cross section, but also because of changes in both the density and temperature of the atoms in the trap. Since these parameters cannot be determined with high accuracy in a high-density trap, where effects like radiation trapping can play an important role, obtaining accurate results this way is very difficult.

The results of one trap loss study is shown in Fig. 14.6. The trap loss rate has a minimum at some modest value of light intensity, increases sharply at lower intensity, and increases more slowly at higher intensities. For low intensities, the trap depth becomes small so that even hyperfine-structure changing collisions in the ground state can lead to trap loss. This accounts for the sharp increase of the trap loss for small intensities. For high intensities, fine-structure changing collisions and radiative escape play a dominant role. Increasing the intensity increases the number of excited molecular complexes and thus the loss rate.

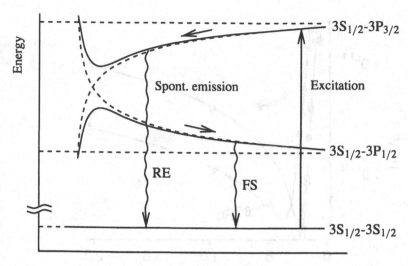

FIGURE 14.5. Schematic diagram of processes leading to trap loss in optical traps. In the radiative escape process (RE) the atoms gain kinetic energy from their mutual attraction and then a spontaneous photon is emitted that has less energy than the one that was initially absorbed. The energy difference appears as kinetic energy and can be enough to eject the atoms out of the trap. In a fine-structure changing collision (FS) the atoms gain kinetic energy from the transition, which is sufficient for ejection of one or both atoms out of the trap.

14.4.2 Optical Collisions

The previous section discusses experiments where trap loss is caused by collisions between atoms excited by the light of the trapping laser. More complete information on this type of collision can be obtained by using another laser to induce the losses. The benefits of a second, probe laser is that its intensity can be chosen to be so low that it will not perturb the trap. Furthermore, the detuning of the laser can be varied over a much larger range. In a more elaborate scheme, the trapping and probing can even be interchanged and the reaction products can be detected only in the probe phase. Such experiments are referred to as optical collisions, since the collisions take place in the presence of the optical field.

The first description of such processes was given by Gallagher and Pritchard [269]. In their semiclassical model (the GP-model) the first atom is located at the origin and the second atom is located at a certain distance R approaching the origin with a velocity v. The atoms are assumed to be distributed evenly over the reaction volume and the number of atoms between R and $R + dR$ is given by $n4\pi R^2 dR$, where n is the density of the atoms. The laser light is assumed to be weak enough that the excitation rate P_{exc} can be described by a quasi-static excitation probability (see Chapter 2):

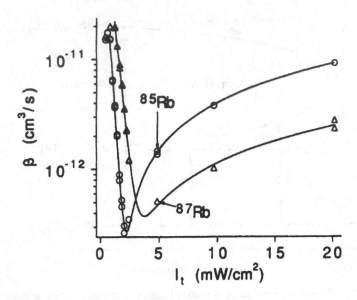

FIGURE 14.6. Trap loss as a function of the intensity of the trapping laser for two different isotopes of Rb [268].

$$P_{\text{exc}} = \frac{s_0 \gamma_m/2}{1 + (2\delta_m/\gamma_m)^2} \quad \text{with} \quad \delta_m = \omega_\ell - \omega_a + \frac{C_3}{R^3}. \tag{14.10}$$

Here δ_m is the detuning of the molecular system from resonance and γ_m the linewidth of the molecular transition, which is between 0 and 2γ depending on the molecular state [270]. Atoms in the excited state are accelerated toward the origin by the C_3/R^3 potential. In order to calculate the survival of the atoms in the excited state, the elapsed time t_{elap} between excitation and arrival at the origin is calculated. For low collision energies they find

$$t_{\text{elap}} \approx \left(\frac{\mu R^5}{2C_3}\right)^{1/2}. \tag{14.11}$$

The decay of the excited atomic state is assumed to be purely exponential and the survival rate becomes

$$P_{\text{surv}} = \exp(-\gamma_m t_{\text{elap}}). \tag{14.12}$$

The total number Q of collisions is then given by the number of atoms at a certain distance, the fraction of atoms in the excited state, and the survival rate, integrated over all distances:

$$Q = \frac{n^2 V}{2} \int_0^\infty dR \, 4\pi R^2 \, P_{\text{exc}} \, P_{\text{surv}}, \tag{14.13}$$

where V is the volume of the trap. To derive a rate coefficient, the authors define the collision rate coefficient k^* analogous to the rate for thermal, excited-state collisions, i.e.,

$$Q = k^* n^* n \, V, \tag{14.14}$$

where n^* is the density of excited-state atoms. Using this definition, the rate coefficient k^* becomes independent of the intensity of the laser light for low excitation rates. For small detunings, corresponding to large internuclear distances, the excitation rate is appreciable over a very large range of internuclear distances. However the excitation occurs at large internuclear distances so the survival rate of the excited atoms is small. For large detunings the excitation is located in a small region at small internuclear distances, so the total excitation rate is small, but the survival rate is large. As a result of this competition, the collision rate peaks at intermediate detunings. This is not true for the rate coefficient k^* as defined in Eq. 14.14. Since the atomic excitation rate becomes small for large detuning, the density of excited atoms n^* decreases for large detunings more strongly than the rate and thus the rate coefficient k^*, as defined in Eq. 14.14, increases for increasing detuning.

Another description of optical collisions is given by Julienne and Vigue [270]. Their description of optical collisions (JV-model) is quantum mechanical for the collision process, where they make a partial wave expansion of the incoming wavefunction. In order to determine the number of the incoming partial waves, they start with the statistical partition function and calculate the pair distribution function of two atoms with a relative collision energy E at an internuclear distance R. From this they obtain the flux F of atoms per unit of volume approaching each other with a given angular momentum ℓ and relative energy E:

$$dF = n^2 \frac{(2\ell + 1)}{h Q_{tr}} e^{-E/k_B T} dE, \tag{14.15}$$

with $Q_{tr} = (2\pi \mu k_B T / h^2)^{3/2}$ the translational partition function per unit volume. In their model dF describes the number of incoming collision pairs, which approach each other from infinity to small internuclear distance in the ground state.

In order to determine the number of optical collisions, the authors describe the excitation process in the same way as it was done in the GP-model. Thus the excitation is localized around the Condon point with a probability given by the quasi-static Lorentz formula of Eq. 14.10. The time between excitation and close collision is given by integration of the equation of motion given the interaction potential $U(R)$ with fixed angular momentum ℓ. The survival rate is given by Eq. 14.12. Note that the total number of partial waves that contribute to the collision cross section has to be considered carefully. The number of partial waves is in general much larger in the excited state than in the ground state, since the potential in the ground state has a much smaller range (see discussion in Sec. 14.2). However, for the collision to proceed in the excited state through a certain ℓ, the centrifugal barrier in the ground state at the Condon point has to be lower than the incoming kinetic energy, otherwise the atoms are repelled from each other before the Condon point is reached.

The total rate for excited-state collisions in the JV-model is given by

$$k^* = \frac{1}{N} \sum_{\beta=1}^{N} k_\beta I_\beta. \tag{14.16}$$

The sum runs over the total number N of channels. Here k_β is the rate of excited-state collisions in channel β, when one of the atoms is prepared in the excited state at infinite internuclear distance and no spontaneous decay takes place. This rate can be calculated with the method of potential scattering of Sec. 14.2. The factor I_β is the excitation and survival factor, which describes the combined effect of the modification of the collision because of the excitation and decay processes. This factor is a weighted average for each partial wave ℓ and energy E. Although the factor can be either larger or smaller than 1, it will be much smaller than 1 for most atoms as a result of the decay process.

The modification of the excitation process caused by the molecular interaction has been discussed in a large number of articles in the literature, which have been reviewed recently [264]. The problem in the case of optical collisions is related to the incompatible descriptions of the excitation and the collision process. For the excitation process a time-dependent description is used, for instance, solving OBE-equations (see Sec. 2.3). For the collision process the description is R-dependent. The relation between R and t is, of course, given by the velocity v, which depends on the potential $V(R)$. Since two states are coupled in the excitation process that have different potentials, this leads to an ambiguity regarding the choice of the potential. Although different choices have been considered, they do not lead to satisfactory results.

This problem can be resolved by going to a completely R-dependent description and treating the atom-laser interaction as a perturbation term in the Hamiltonian. The coupling between the two states can then be treated in the Landau-Zener model, where the excitation rate is given by

$$P_{LZ} = 1 - e^{-\pi\Lambda} \qquad \text{with} \qquad \Lambda = \frac{\hbar\Omega^2}{2\alpha v_C}. \qquad (14.17)$$

Here Ω is the Rabi-frequency, which is the coupling between the two states, α the gradient of the potential around the Condon point, and v_C the radial velocity of the system at the Condon point.

The validity of the Landau-Zener excitation rate for optical collisions is limited. First of all, the excitation has to be localized, which means that the detuning of the light from atomic resonance should not be too small. Furthermore, the effects of spontaneous emission are neglected in the Landau-Zener treatment, which means that it describes a single excitation. It therefore applies well for the low-intensity case. Excitation and spontaneous emission can then be treated separately and the survival rate in that case is again given by Eq. 14.12.

In still another approach, a completely semiclassical description of optical collisions has been given by Mastwijk et al. [271]. These authors start from the GP-model, but make several important modifications. First, the Lorentz formula is replaced by the Landau-Zener formula. Second, the authors consider the motion of the atoms in the collision plane. At the Condon point, where the excitation takes place, the trajectory of the atom in the excited state is calculated by integration of the equation of motion. Since the atom can cross the Condon point at different

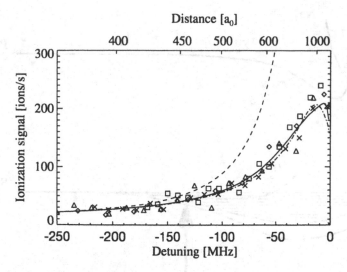

FIGURE 14.7. The frequency dependence for the associative ionization rate of cold He* collisions. The experimental results (symbols) is compared with the semiclassical model (solid line), JV-model (dashed line), and modified JV-model (dashed-dotted line). The axis on top of the plot shows the Condon point, where the excitation takes place.

angles in the collision plane, the trajectory is calculated for each angle and the outcome of the collision is determined. If this angle is large, the angular momentum L is too large, the atom is repelled by the rotational barrier in the excited state and no close collision occurs. If, on the other hand, the angle is small the excitation can lead to a collision. Finally, since the trajectory of the atoms is calculated, the elapsed time between excitation and close collision is known and Eq. 14.12 can be used to calculate the survival rate.

The results for their model are shown in Fig 14.7, and are compared with experiment and the JV-model. The agreement between the theory and experiment is rather good. For the JV-model two curves are shown. The first curve shows the situation for the original JV-model. The second curve shows the result of a modified JV-model, where the quasi-static excitation rate is replaced by the Landau-Zener formula. The large discrepancies between the results for these two models indicates that it is important to use the correct model for the excitation. The agreement between the modified JV-model and the semiclassical model is good, indicating that the dynamics of optical collisions can be described correctly quantum mechanically or semiclassically. Since the number of partial waves in the case of He* is in the order of 10, this is to be expected.

14.4.3 Photo-Associative Spectroscopy

The description of optical collisions in the previous section applies to the situation that the quasi-molecule can be excited for each frequency of the laser light. However, the quasi-molecule has well-defined vibrational and rotational states and the excitation frequency has to match the transition frequency between the ground and

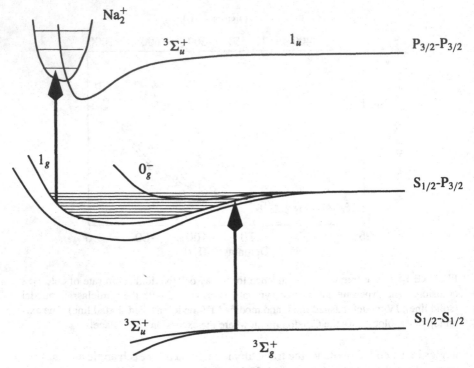

FIGURE 14.8. Photoassociation spectroscopy of Na. By tuning the laser below atomic resonance, molecular systems can be excited to the first excited state, in which they are bound. By absorption of a second photon the system can be ionized, providing a high detection efficiency.

excited rovibrational states. Close to the dissociation limit there are a large number rovibrational states with a small energy spacing, so the excitation can be treated as if there is a continuum. This is not true far from the dissociation limit, where well-resolved resonances are observed. This has been the basis of the method of photo-associative spectroscopy (PAS) for alkali-metal atoms, where detailed information on molecular states of alkali dimers have been obtained recently. Here photo-association refers to the process where a photon is absorbed to transfer the system from the ground to the excited state where the two atoms are bound by their mutual attraction.

The process of PAS is depicted graphically in Fig. 14.8. When two atoms collide in the ground state, they can be excited at a certain internuclear distance to the excited molecular state. If the excited state is attractive, the two atoms remain bound after the excitation and form a molecule. This so-called transient molecule lives as long as the systems remains excited, so after spontaneous emission of the molecule the atoms return to the ground state and in general dissociate again. The transition frequency is given by the difference between the total energy in the ground and excited state. The total energy in the ground state is well-determined, since the broadening from the kinetic energy is small because of the low temperature of

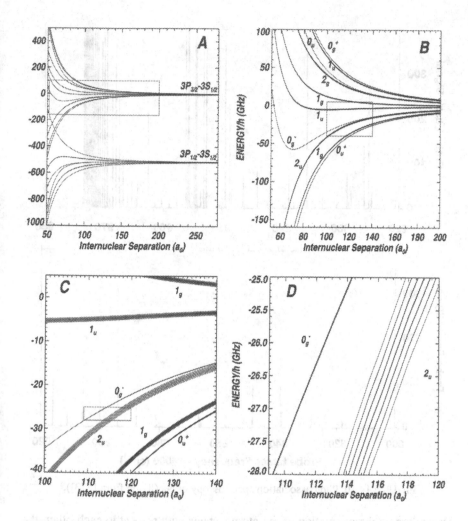

FIGURE 14.9. Molecular potentials connected to the 3S-3P asymptote of Na. The subplots b, c, d are sequential magnifications of the indicated regions. The curves are labeled with Hund's case (c) notation (see Ref. 272).

the atoms. The total energy of the bound excited state is given by the energy of the rovibrational state, and thus PAS yields precise information on the position of these states. The number of rotational states that can contribute to the spectrum is small for low temperature. The resolution is only limited by the linewidth of the transition, which is comparable to the natural linewidth of the atomic transition. With PAS, molecular states can be detected with a resolution of ≈ 10 MHz, which is many orders of magnitude better than traditional molecular spectroscopy.

The molecular potentials for alkali-metal dimers are very complex. In Fig. 14.9 the potentials connected to the 3S-3P asymptote of Na are shown. Note that the fine structure interaction results in two asymptotes, either 3S-3P$_{1/2}$ or 3S-3P$_{3/2}$.

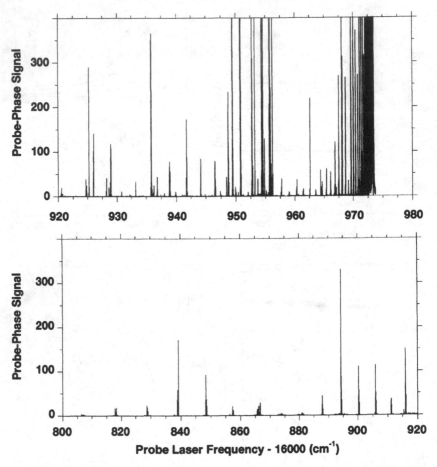

FIGURE 14.10. Photoassociation spectroscopy of Na (figure from [273]).

Depending on the orientation of the atomic states with respect to each other, the molecular states split up in energy when the two atoms come closer together. The region around $100a_0$ in Fig. 14.9a shows 10 states that asymptotically connect to the 3S-3P$_{3/2}$ state. These states are identified by their molecular labels (for details on the labeling of molecular states, see Ref. 272). Further magnification of the scale shows a broadening of each molecular state caused by the hyperfine interaction. This hfs interaction differs strongly for different molecular states, as can be seen in Fig. 14.9d. The state labeled as 2_u shows a splitting between the different hyperfine state of ≈ 100 MHz, whereas the hyperfine splitting of the 0_g^- can not be observed on this scale.

In Fig. 14.10 the excitation spectrum for Na$_2$ transient molecules is shown. The formation of the molecules is probed by absorption of a second photon of the same color, which can ionize the molecule. This way ions can be detected, which provides a high detection efficiency of the process. In the spectrum a typical vibrational structure can be observed, where the spacing between the states becomes smaller close to resonance. Although it is to be expected that all five attractive states connected to the $3S-3P_{3/2}$ asymptote would contribute to the spectrum, only the 1_g state is clearly observable because of the small excitation or ionization probability of the other states.

A more flexible way to study PAS is the use of a second color to ionize the transient molecules. The frequency and intensity of this second ionizing laser can be controlled independently, which has a number of benefits. Since the ionization process has a smaller cross section than the association process, the intensity of the association laser can be lowered compared to the one-color case without decreasing the signal strength. Furthermore, since the frequency of the ionizing laser can be tuned independently, the ionization process can proceed via different channels. In the case of one color, the total energy of the two photons will always be smaller than twice the energy of the resonance frequency, since the first step requires a frequency below the resonance frequency. This is no longer the case for two-color spectroscopy, and in this way doubly excited states can be reached, which have a much larger probability for ionization compared to direct photo-ionization.

In this way new molecular states have been observed that are now referred to as long-range molecular states. These states have been predicted in 1977 by Movre and Pichler [274] and are shown in Fig. 14.9 labeled as 0_g^- and 1_u. In Na the 0_g^- state has been detected in two-color spectroscopy [275,276]. The state connects to the $3S-3P_{3/2}$ asymptote, is attractive at very large internuclear distance, but already becomes repulsive at long range because of an avoided crossing with a similar state connected to the $3S-3P_{1/2}$ asymptote. The state is bound by only 55 GHz and has an inner turning point of $55a_0$, which is very large compared to "normal" molecular states. Therefore such states are referred to as long-range molecular states.

Since the total potential is determined by long-range interaction, it can be calculated with high accuracy. For the first excited state the interaction at long-range is a dipole-dipole interaction and only depends on the one parameter C_3. This dispersion coefficient in turn depends only on the dipole moment of the atom, which is related to the natural lifetime τ of the atomic state involved. Accurate determination of the C_3 parameter thus leads to an accurate determination of τ. Jones et al. [277] have measured 10 vibrational states of the 0_g^- state of Na_2 with high accuracy and compared their results with accurate calculation of these states. Although the calculation relies on a careful analysis of several effects, such as higher-order dispersion, non-adiabatic effects and retardation, the only fit parameter in the analysis is the C_3 coefficient. The lifetime derived from their analysis agrees with the most recent theoretical values.

Finally, PAS has been discussed in the literature as a technique to produce cold molecules. The methods discussed employ a double resonance technique, where the first color is used to create a well-defined rovibrational state of the molecule and a second color causes stimulated emission of the system to a well-defined vibrational level in the ground state. Although such a technique has not yet been shown to work experimentally, cold molecules have been produced in PAS recently using a simpler method [278]. The 0_g^- state in Cs_2 has a double-well structure, where the top of the barrier is accidentally close to the asymptotic limit. Thus atoms created in the outer well by PAS can tunnel through the barrier to the inner well, where there is a large overlap of the wavefunction with the vibrational levels in the ground state. These molecules are then stabilized against spontaneous decay and can be observed. The temperature of the cold molecules has been detected and is close to the temperature of the atoms. This technique and similar techniques will be very important for the production and study of cold molecules.

14.5   Collisions Involving Rydberg States

Most studies dealing with cold collision with excited states only consider collisions with one atom in the first excited state. Although such studies have already produced a wealth of new information on molecular states, atoms have many more than one excited state. The situation becomes extremely interesting when using atoms in Rydberg states [279]. Such studies are currently under way and the first preliminary results have recently been discussed in the literature [280, 281].

Rydberg atoms are atoms in highly excited states close to the dissociation level. The radius of Rydberg atoms is proportional to n^2 with n the principal quantum number of the state. For very high n the radius of the atom can become in the order of a few μm. The cross section for a collision between two Rydberg atoms becomes very large and thus for already a moderately high density of the cold atomic sample, such collisions dominate its evolution.

The situation is rather complicated. Since the radius of the Rydberg atoms is so large and the velocity of the atoms is rather small, the atoms react with a large number of other atoms without changing their position appreciably. This situation is referred to as a "frozen Rydberg gas", much like the situation of interactions between atoms in a solid. Collisions are then no longer dominated by the relative velocity between two atoms, but by the diffusion of the atoms through the gas. This leads to novel phenomena that will surely be studied in the near future.

15
deBroglie Wave Optics

15.1 Introduction

One of the major developments of laser cooling of neutral atoms in the 1990s has been the evolution toward quantization of the atomic center-of-mass motion (external coordinates). For Newtonian atom optics, as defined in Chapter 13, the motion of atoms is described in a perfectly classical way, assuming they have arbitrary position and momentum, and that both of these quantities can be known simultaneously. This classical picture of atoms moving as particles without regard to their overall wavelike character has been of great use, but some recent experiments have come into the range where the center-of-mass motion of atoms must be viewed quantum mechanically. The first discussion of such ideas has already appeared in the calculations of quantum states in a magnetic trap in Sec. 10.4 and in optical traps in Sec. 11.2.4. A collection of articles on related topics is to be found in Ref. 282.

Thus it becomes necessary to consider atomic position and momentum as quantum mechanical variables, replete with wavepacket spreading and non-commuting operators. A deBroglie wave field occupies allowed states of a region of space that may have a spatially varying potential, which defines modes of the field. For example, these may be eigenstates in the optical potentials created by the laser fields. By contrast, if the potential is uniform, it can then be set to zero and a classical description of the motion may be used.

In analogy with optics, the occupation of particular modes of this field can result in spatial interference, and the entire field of atom interferometry emerges as a subset of this way of thinking. However, there are several important differences between this and the optical case. First, atoms have internal structure that can be manipulated by the same fields that determine the light shifts they experience as they move in the light fields. The deBroglie wave view of this phenomenon leads to the equivalent of multiple refractive indices in the same region of space. Second, atoms have mass that can be affected by the earth's gravitational field and that also leads to a different dispersion relation for deBroglie waves than for light. Third, unlike photons, atoms are not all bosons. With fermionic atoms there will be cases where only a single atom can occupy a particular mode of the deBroglie wave field.

Thus there is a new view of optical control of atomic motion in terms of its quantum mechanical behavior. Of course, atoms at ordinary velocities are distributed over thousands of quantum states, so laser cooling is intimately involved in these studies. But the subject is evolving toward a quantum description of optical control of atomic motion, and may also involve rf or microwave transitions to prepare the desired internal states. After all, it is the internal states of atoms that determine the magnitude and nature of the electromagnetic forces on them.

Among the most obvious manifestations of the wave nature of atoms are interferometry, Bose-Einstein condensation (BEC), and the band structure that results from the movement of atoms in periodic potentials. The first of these is discussed in some detail in Sec. 15.7. As for the second topic, when atoms are sufficiently slow and dense that their deBroglie wavelengths are large enough to overlap one another, BEC occurs and an entirely new domain of phenomena appears. The whole of Chapter 17 is devoted to this special topic of deBroglie wave optics. The third topic, band structure, becomes important when sufficiently cold atoms move in a periodic potential. Then their kinetic energy levels exhibit band structure in accordance with the Kronig-Penney model to be found in many quantum mechanics texts. Reference 161 presents a full quantum treatment of laser-cooled atoms moving in a periodic potential, and the topic is discussed further in Chapter 16.

15.2   Gratings

Perhaps the simplest atom optical device is the mask or slit that passes only those atoms in a beam that are incident on its openings, and blocks the others. Although its role in Newtonian atom optics is clear and simple, for example, collimating atomic beams as described in Chapter 13, its role in deBroglie wave optics needs further discussion. An array of parallel slits constitutes a grating, and atoms in a sufficiently collimated beam that pass through such a grating can undergo interference that is constructive in the favored directions, given by the grating equation

$$n\lambda_{dB} = d \sin \theta. \qquad (15.1)$$

Here n is an integer that gives the order of the diffraction, $\lambda_{dB} = h/p$ is the deBroglie wavelength of the atoms, and d is the grating spacing.

In this case "sufficiently collimated" means that the atoms must sample more than one of the grating's slits, and this statement only has meaning in the context

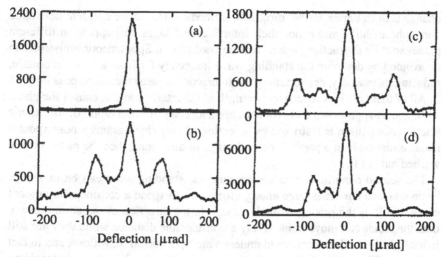

FIGURE 15.1. The measured profile of the atomic beam after diffraction from the material grating described in Ref. 283.

of deBroglie wave atom optics: localizable particles cannot possibly pass through more than one slit. Sampling more than one of the grating's slits requires that the atoms' transverse deBroglie wavelength is larger than d, which means that the angular spread of the beam incident on the grating must be less than θ. This corresponds to the obvious geometrical criterion that diffraction fringes at angles smaller than those of the incident beam spread can not be observed.

The diffraction of waves from such an amplitude modulator grating is an old problem whose solutions for optical beams are Bessel functions [31] that appear in many textbooks. The first detailed study of atomic beams diffracted by a material grating was reported in 1988 [283]. The authors used a thermal beam of Na that was carefully collimated by two 10 μm slits placed ~1 m apart. Their early results are shown in Fig. 15.1. The manufacturing, mounting, and handling of such gratings obviously requires great skill and care. Sec. 13.3 describes the use of atom optics for lithographic fabrication of such nanostructures, but here is an example of the complementary process. A device prepared by electron beam lithography is used for atom optics.

Another kind of periodic structure that can function as an atom grating is a phase rather than amplitude modulator. This means that *all* the atoms pass through, but the phase of the deBroglie wavefront is altered from place to place. Atoms in a beam traversing a perpendicular optical standing wave experience position-dependent light shifts. If the atomic beam is sufficiently collimated, its transverse deBroglie wavelength is larger than the $d = \lambda/2$ periodicity of the standing wave, and so various parts of the transmitted deBroglie wavefront lead or lag adjacent regions. This is manifest as a diffraction pattern that also obeys Eq. 15.1.

Such optical gratings provide two complications that are absent in material gratings. The first one arises because the atoms can be excited by the light and then undergo spontaneous emission to the ground state. Since the recoil momentum change is in an unpredictable direction, different atoms suffer different deBroglie wave phase shifts, and hence their interference fringes will appear in different places so their diffraction pattern will be washed out. Spontaneous emission can be avoided by detuning the standing wave frequency far from atomic resonance, reducing its intensity, or shortening the interaction time as the atoms pass through it. All of these choices reduce the strength of the interaction that causes the phase shift, and so appropriate compromises are necessary. If observation of the atomic interference pattern is restricted by selection of only those atoms whose spontaneous emission is in a particular small range of directions, then the pattern is not washed out [284].

The second complication arises because the standing wave can be in a laser beam whose diameter is large enough for atoms to spend a considerable amount of time in it. If the grating is thick enough in the direction of the atomic beam, then the atoms can move transversely a considerable distance while they are still in the light field. The atoms could undergo multiple optical transitions, and in fact are Bragg diffracted by the "lattice" comprised of the periodic potential associated with the standing wave. This long interaction time domain is called the "Bragg regime", and is discussed further in Sec. 15.3 and in Chapters 16 and 18. It is precisely this kind of interaction that is most effective for the atom lithography described in Sec. 13.3. The converse of this thick grating Bragg regime is called the Raman-Nath regime, and corresponds to a thin grating similar to the very thin structures that constitute the material gratings discussed above.

Such gratings necessarily change the transverse velocity of atoms that are diffracted, so the atomic momentum is changed. Momentum conservation is easily described for optical gratings because atoms can absorb a photon from one beam, and then undergo stimulated emission to provide an additional photon in the other beam. Thus their transverse momentum would be changed by $\pm 2\hbar k$, and their angle of travel would be changed by $\pm 2\hbar k/p = \pm 2\lambda_{dB}/\lambda$, exactly the condition given by Eq. 15.1 for $d = \lambda/2$.

It is important to realize, however, that the sign of the transverse momentum change does not vary from one atom to another. Instead, *all* atoms are driven into superposition states having momentum components of *both* signs, and it is only after a measurement is made that one component of the superposition is determined. This view helps to provide understanding of momentum conservation for the material grating case where there is no photon exchange picture. As with the light grating, atoms enter a superposition of opposite transverse momentum states. In some sense they are thus "split into two parts" having opposite transverse momenta, and so the net momentum transfer is zero.

Energy conservation is not so easily handled. For material gratings, there is no problem with either energy or momentum conservation because there is always a large mass present whose recoil can provide for both. But for optical gratings in the Raman-Nath regime, it is necessary to consider that a light beam confined to a small region of space, such as the focused standing waves under consideration here, are not plane waves. Thus they are superpositions of many spatial modes with the same frequencies but various k-vectors. Energy conservation then requires that longitudinal velocity is diminished by the absorption-stimulated emission sequence because the photon momenta that are exchanged are neither exactly opposite to one another nor exactly perpendicular to the incident atomic beam. Instead, the nearly π angle between them must slow the atoms by just the right amount to conserve energy. This can be pictured most easily in terms of the spherical Huygen's wavelets emerging from each slit in a water wave (ripple tank) grating.

For optical gratings in the Bragg regime, energy and momentum must be strictly conserved. Thus an incident atomic beam can be reflected only if both the incident *and* the emerging deBroglie wavefront satisfy the Bragg condition. This places severe limits on the range of angles and velocities that can be used. Nevertheless, several important atom optics experiments are described in Secs. 15.3 and 15.7 that have been done using such Bragg diffraction.

15.3 Beam Splitters

One of the most elementary devices for deBroglie wave optics is a coherent beam splitter. It must divide an atomic sample or beam into two or more spatially or temporally separated parts in such a way that each atom undergoes interactions exactly the same as all the others. This precludes such inhomogeneous events as spontaneous emission or collisions, because the wavefunction of the external state of motion of such atoms that suffer such interactions will necessarily undergo uncontrolled and therefore different phase shifts. On the other hand, if atoms undergo absorption followed by stimulated emission in a transverse standing wave geometry, the wavepacket of the atoms splits up coherently and such a beam splitter might serve in an atomic interferometer.

To minimize the effects of spontaneous emission, one has either to reduce the number of spontaneous emissions by detuning far from resonance ($\delta\tau \gg 1$) or reduce the interaction time τ_{int} to much shorter then the spontaneous lifetime τ. In an early experiment, Moskowitz et al. [285] showed that a beam of sodium atoms crossing a standing wave is deflected into two symmetric peaks, where the scattered atoms acquire momentum in multiples of $2\hbar k$ from the combined absorption and stimulated emission of photon momenta $\hbar k$. Later Gould et al. [286, 287] showed that the rms momentum gained by the atoms in the case of large detuning is proportional to $\Omega\tau_{int}/\delta$, where Ω is the Rabi frequency.

Another demonstration of an atomic beam splitter was shown by Sleator et al. [288]. They created a large period standing wave by reflecting a laser beam at a small angle from a mirror. The interference field in front of the mirror forms a standing wave with a spatial period much larger than λ, in their case 15 μm. As discussed in Sec. 9.3, the proper description for a two-level atom in a strong near-resonant radiation field is in terms of dressed states. Since the dressed states have opposite light shifts, they feel opposite forces as atoms traverse the light field, and thus an incoming deBroglie wavepacket splits in two. They achieved a splitting of $4\hbar k$ in their experimental setup. In a similar arrangement they observed the focusing of an atomic beam, where they used the approximately harmonic force of the light shift near the antinode of this large period standing wave [289]. Since the thickness of such a lens is given by the beam waist of the laser beam in the direction of the atomic beam (in their case only 40 μm), one can produce thin atomic lenses this way (see Sec. 13.3).

A significant increase of the transverse momentum can be obtained by using a novel magneto-optical force, as was first described by Pfau et al. [290]. In this scheme the atomic beam is crossed transversely by two counterpropagating laser beams, where both laser beams are linearly polarized and the polarizations make an angle ϕ with respect to each other. To obtain this new magneto-optical force, the authors discuss a V-type level scheme with one ground state g coupled to two excited states e_+ and e_-, which are split by a magnetic field along the laser beam direction. By matching the Larmor precession frequency caused by the magnetic field with the Rabi oscillation frequency, it is possible to achieve efficient absorption of light from one beam followed by stimulated emission into the other beam. In this way a large number of photon momenta $\hbar k$ can be transferred from the light field to the atom. A dressed atom picture of this magneto-optical force has been given [291]. In such an experimental arrangement, Pfau et al. showed that an atomic beam can be split in two, where the splitting between the two peaks was as large as 42 $\hbar k$ [292].

Still another scheme for an atomic beam splitter, that used the bichromatic force of Ref. 118 was proposed in 1994. In the configuration as described in Sec. 9.4.3, which was optimized for beam slowing and collimation, the relative phase of the standing waves of the two frequencies was chosen to be $\pi/2$ in order for spontaneous emission to provide a bias for the force direction. However, if the phase is chosen instead to be π, then there is no preferred direction and thus atoms may experience a force in either direction [293]. The dressed atom states of Sec. 9.4.3 become coherent superpositions of motion in both directions. Because of the large size and velocity range of the bichromatic force, atoms entering such a laser field are coherently split into two beams with very large momentum differences. It is only necessary to choose the interaction time τ_{int} to be small enough so that $\gamma \tau_{int} \ll 1$, thereby minimizing the chance for spontaneous emission to destroy the atomic coherence.

15.4   Sources

Any spatially extended source of deBroglie waves can be characterized by its spatial modes, and a finite size atomic sample having a finite temperature occupies many such modes. In analogy to light sources, laser-cooled atomic samples can be made very monochromatic just as highly filtered classical light sources. But single mode light can only come from a process of stimulated emission into one single mode of an optical cavity, and correspondingly, single mode deBroglie wave sources are necessarily Bose-Einstein condensates (BEC's) as described in

Chapter 17. Eventually, atom optics experiments will all start with BEC's as their source.

Nevertheless, there is still a large class of deBroglie wave optics experiments that can be done with laser-cooled atomic samples. Some of these are described in Sec. 15.7 on atom interferometry, or in other places scattered throughout this book. These are analogous to optics experiments performed with white light or multimode lasers. For example, most HeNe lasers operate in several modes, and yet interferometry, metrology, spectroscopy, holography, and a host of other experiments are readily done with these lasers. This is because each mode interferes only with itself, and as long as there are not too many independent interference patterns overlapping one another so that the fringes wash out, the finite coherence length of HeNe lasers poses only limitations but not prohibitions. Thus laser-cooled atoms that are not in BEC's can be used for lithography, interferometry, inertial measurements such as atom gyroscopes, and many other experiments.

Another form of experiments in atom optics that can be done with such "classical" atom sources are in reduced dimensions (see Sec. 13.5.1). Atoms in a beam can be collimated by slits sufficiently well that they occupy just a single transverse spatial mode, and one-dimensional experiments can be performed. Furthermore, certain kinds of subrecoil laser cooling such as the velocity selective coherent population trapping described in Secs. 18.2 and 18.3 as well as the Raman cooling of Sec. 8.7.2 can also provide appropriate samples for experiments in deBroglie wave optics. In addition, both VSCPT and Raman cooling can be done in 3D [294].

15.5 Mirrors

In the domain of deBroglie wave optics, atomic mirrors must be designed to preserve atomic coherence. This means that atoms must not suffer inhomogeneous events such as collisions, spontaneous emissions, or any other interactions that will alter either their internal or external states in a way that is different from one atom to the next. In view of this, perhaps the ideal atomic mirror is produced using Bragg reflection as discussed in Sec. 15.3 and in Chapters 16 and 18. The atoms experience a plane optical standing wave whose only function is to reflect the incident deBroglie waves using momentum exchanges from absorption followed by stimulated emission. All the photons absorbed by the atoms are returned to the standing wave laser field with the same frequency and polarization because there is only stimulated emission, and thus neither the internal energy nor the internal states of the atoms can be changed. Therefore the external energy of the atoms must also be unchanged, and so their final momenta must be either identical to or opposite to their initial momenta. Such a mirror may also be viewed as reflection at the edge of a Brillouin zone formed by the optical lattice of the standing wave light field (see Chapter 16).

Other atomic mirrors have been discussed in Sec. 13.2. None of these is assuredly coherent, but some could be made so by careful choices of their parameters. For example, reflection by the evanescent waves near the surface of a dielectric can be made coherent if the light is detuned sufficiently far from resonance. Similarly, the Rydberg atom mirror could be made coherent if the atoms experienced π pulses for both excitation and de-excitation from the Rydberg state, and these were spaced much closer together than the natural lifetime of the chosen Rydberg state to avoid spontaneous decay. Needless to say, if the pulse timing were determined by passage of the atoms through the laser beams, the mirror would be coherent only for atoms within a small velocity range.

By contrast, the reflection of atoms from the inhomogeneous magnetic fields above the surface of magnetic recording media may be incoherent. This is because various atoms experience the fields from different microscopic domains, and so their Zeeman sublevels are precessed differently. Their external motion suffers no inhomogeneous effects caused by spontaneous emission, but their internal states do (see Sec. 13.2).

15.6   Atom Polarizers

Polarized light is usually characterized by its transmission through various polarizers at various angles. Such polarization is attributed to the state of the light field or the spin orientation of the photons, and a similar property exists for neutral atoms. The ground and excited states of two-level atoms cannot be considered as appropriate polarization components, even though the wavefunction is often written as a spinor, simply because the excited state eventually decays to the ground state. Therefore the appropriate coordinates are restricted to ground states, and one suitable choice is the orientation M_F of the total angular momentum F (J for $I = 0$). The principle difference is that optical polarization has only two states (helicity ± 1 or two coordinates on the Poincaré sphere), but atomic polarization can have as many as $2F + 1$ coordinates. For $F = 1/2$ the systems are formally identical.

The simplest example of an atomic polarizing component is an inhomogeneous magnetic field in the spirit of the Stern-Gerlach experiment. Atoms with $F = 1/2$ are deflected either up or down depending on their M_F value. If atoms that were deflected in a particular direction subsequently pass through a second magnet, their transmission depends on the relative field orientation as $\cos^2 \theta$ just as does light traveling through successive polarizers. Similar descriptions apply for atoms with $F > 1/2$.

The orientation of linearly polarized light may be rotated by half wave plates or optically active materials, and analogously the magnetization of atoms may be altered by Raman transitions among the ground states. For $F = 1/2$ a π-pulse on a Raman transition completely inverts the polarization, just as a half wave plate whose axes are at $\pm \pi/4$ to a linearly polarized light beam. More complicated descriptions apply to other Raman transitions or more complicated atoms. Superpositions of multiple M_F states can be described as a pure M_F state in a different coordinate system if they are in phase, corresponding to linearly polarized light at an angle to the chosen coordinate axes. Superpositions with different phases correspond to circularly or elliptically polarized light.

15.7   Application to Atom Interferometry

Atom interferometry is an important part of deBroglie wave optics. Just as Dirac said about photons, atoms can only interfere with themselves. In order to see the fringes, at least two components of the same internal and external atomic states must evolve differently and then be recombined. These shared wavefunction components do not need to involve 100% of the total atomic wavefunction, but the contrast of the fringes may be reduced proportionately. What is required is that there be no way to distinguish which of the two (or more) paths was followed by some parts of the recombined wavefunction. In this sense, Ramsey oscillations, spin and photon echoes, and quantum beats are simply interference in the time domain rather than in space.

Some interference experiments have been done in the spatial domain by deflecting the coherently split atomic beams away from one another and then recombining them (transverse), while others have been done in the time domain by delaying or phase shifting one of the two states that form an atomic superposition (longitudinal). The transverse experiments are much more similar to the familiar optical interferometers, and the most commonly used configuration resembles a Mach-Zehnder type. The longitudinal experiments are much more similar to the familiar Ramsey separated oscillatory fields method, and these experiments may be regarded simply as a reinterpretation of the Ramsey oscillations. Longitudinal interference experiments have been used in atomic fountains to study possible atomic clocks [236,295], as discussed in Sec. 13.7. Transverse experiments have been used to measure the gravitational acceleration [296] or photon recoil [297].

The first atom interferometers were reported in 1991. There were two of the transverse or spatial kind, and two of the longitudinal or temporal kind. The transverse ones [298, 299] were both made with material atom optical elements for beam splitting. In Ref. 298 the authors used a single slit followed by a double slit so it was analogous to a Young's double slit experiment. The authors of Ref. 299 used multislit gratings to split and recombine the beams so the interferometer was analogous to a Mach-Zehnder. Within a few months, the authors of Ref. 298 also described the focusing of atoms with a material fabricated Fresnel zone plate [300]. This is an example of an atomic lens in the deBroglie wave optics regime.

The first demonstration of spatial atom interferometry with an all optical beam splitter was reported in 1995 [301,302]. In the case of Ref. 301, the atomic beam was collimated by narrow slits spaced far apart so that the transverse velocity distribution of the atoms was sufficiently narrow. Atoms were Bragg diffracted by optical standing waves tuned several hundred linewidths from resonance to avoid spontaneous emission. The angle between the incident atomic beam and the standing wave field was carefully adjusted to satisfy the Bragg condition, and the laser parameters were chosen so that the beam was divided in half. The atomic beam was thus split into two coherent components separated by 58 μrad, traveled freely for 31 cm, passed a second standing wave field for a second Bragg reflection, and then the two components crossed 31 cm further downstream. At that crossing point they passed a third standing wave Bragg region and were recombined. The interference fringes were observed by scanning a detector downstream from the third Bragg region.

The efficiency of Bragg diffraction depends on the laser and atomic parameters, and can be chosen to be 50% for beam splitters. However, atoms passing through such an optical field can be Bragg reflected with nearly 100% efficiency after sufficiently long time. Of course, the newly redirected beam can be Bragg reflected again back to the original direction if it is still in the optical field. This oscillation of the direction of Bragg reflection is called "pendulösung", and was first studied in neutron diffraction. Pendulösung has also been observed in the Bragg diffraction of neutral atoms in 1996 [303].

The temporal interferometers went beyond the demonstration of fringes, and were actually used for inertial measurements. In Ref. 295 the entire apparatus was mounted on a rotatable table, and the authors were able to measure the shift of the fringes arising from rotation. This fringe shift is called the Sagnac effect [304], and for any interferometer in a non-inertial reference frame (*e.g.*, rotating), there is a phase shift of the fringes $\Delta \Phi = 4\pi \vec{\Omega} \cdot \vec{A}/\lambda v$, where $|\vec{\Omega}|$ is the rotation frequency, λ is the wavelength of the interfering entity, and v is its velocity of propagation. Also, $\vec{A}$ is a vector whose magnitude is the area of the plane enclosed by the two paths through the interferometer and whose direction is normal to this plane.

For an optical interferometer $v = c$, and for an atom interferometer, $\lambda = h/p = h/Mv$. Thus the Sagnac phase shift is larger for atoms by a factor $Mc^2/\hbar\omega$ which is typically a few times 10^{10}. This huge factor is somewhat compromised because optical interferometers can easily have $|\vec{A}|$ hundreds of times larger than atomic ones, and the S/N is surely better, too. Nevertheless, the possibility of atomic inertial sensors, navigational gyroscopes, and other devices is very much a possibility. This first demonstration of the Sagnac effect for atoms is a precursor of possible atomic gyroscopes for navigational purposes, especially in space where the atomic trajectories would not be influenced by gravity.

More recently, another group has used a similar apparatus with several improvements that achieved a factor of 100 better sensitivity [305]. They used Raman transitions between the two ground hfs states of Cs to double the momentum transfer, and an atomic beam apparatus nearly 2 m long to increase $|\vec{A}|$. The ultimate resolution was a rotation of $|\vec{\Omega}| = 2 \times 10^{-8}$ rad/s, corresponding to 2×10^{-4} of the Earth's rotation rate.

The opposite point of view was the objective of the other temporal interference experiments. In Refs. 296, 306 the authors exploited the extreme sensitivity to inertial frames to measure the acceleration of gravity to the extraordinary precision of one part in 3×10^8. Such exquisite sensitivity has been further exploited to measure the recoil frequency $\omega_r \equiv \hbar k^2/M$ from the atomic motion, and with the help of other fundamental constants and well-measured quantities, provide a new measurement of the ratio h/m_e where m_e is the electron mass. Further refinements of the technique, using two atomic samples separated vertically by 1 m, allowed measurement of the gradient of the gravitational acceleration resulting from the inverse square law [307]. The the measured $3.3 \times 10^{-6}/s^{-2}$ is within 10% of the expected value.

There have been several other experiments that demonstrated the superb capabilities of atom interferometry for various measurements. For example, the MIT group has measured both the real and imaginary parts of the "index of refraction" of various gases for deBroglie waves of Na [308]. These were found to differ from unity by only ~ 1 part in 10^8, demonstrating again the high sensitivity of atom interferometry. They also applied an electric field to Na vapor in one arm of their interferometer and measured the polarizability of the atomic ground state to better than $\pm 0.25\%$ [309].

Atom interferometry has also been used to study such fundamental subjects as spatial topological effects. There have been measurements of both Berry's phase for atoms [310] and the Aharonov-Casher effect [311]. Thus atom interferometry is fulfilling its promise of providing physicists with new tools to make measurements that could not otherwise be possible.

16
Optical Lattices

16.1   Introduction

In 1968 V.S. Letokhov [312] suggested that it is possible to confine atoms in the wavelength size regions of a standing wave by means of the dipole force that arises from the light shift, as discussed in Chapters 1, 9, and 11. This was first accomplished in 1987 in one dimension with an atomic beam traversing an intense standing wave [313]. Since then, the study of atoms confined in wavelength-size potential wells has become an important topic in optical control of atomic motion because it opens up configurations previously accessible only in condensed matter physics using crystals.

The basic ideas of the quantum mechanical motion of particles in a periodic potential were laid out in the 1930s with the Kronig-Penney model and Bloch's theorem, and optical lattices offer important opportunities for their study. For example, these lattices can be made essentially free of defects with only moderate care in spatially filtering the laser beams to assure a single transverse mode structure. Furthermore, the shape of the potential is exactly known, and doesn't depend on the effect of the crystal field or the ionic energy level scheme. Finally, the laser parameters can be varied to modify the depth of the potential wells without changing the lattice vectors, and the lattice vectors can be changed independently by redirecting the laser beams.

Because of the transverse nature of light, any mixture of beams with different $\vec{k}$-vectors necessarily produces a spatially periodic, inhomogeneous light field. The importance of the "egg-crate" array of potential wells arises because the associated atomic light shifts can easily be comparable to the very low average atomic

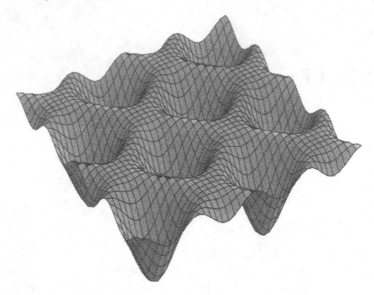

FIGURE 16.1. The "egg-crate" potential of an optical lattice shown in two dimensions. The potential wells are separated by $\lambda/2$.

kinetic energy of laser-cooled atoms. Thus the picture of an atomic vapor of laser cooled atoms moving in an optical molasses with no spatial order in a homogeneous region eventually fails, and the atomic motion is instead strongly influenced by the unavoidable periodic potential. A typical example projected against two dimensions is shown in Fig. 16.1.

The name "optical lattice" is used rather than optical crystal because the filling fraction of the lattice sites is typically only a few percent (as of 1998). The limit arises because the loading of atoms into the lattice is typically done from a sample of trapped and cooled atoms, such as a MOT for atom collection, followed by an optical molasses for laser cooling. The atomic density in such experiments is limited to a few times $10^{11}/cm^3$ by collisions and multiple light scattering as discussed on p. 27. Since the density of lattice sites of size $\lambda/2$ is a few times $10^{13}/cm^3$, the filling fraction is necessarily small. Experiments are underway in several laboratories to confine atoms with far-off-resonant lattices using long wavelength lasers such as the $\lambda \approx 10 \ \mu m$ CO_2 laser so that the density of lattice sites is comparable to readily achieved atomic densities of a few times $10^9/cm^3$.

16.2   Laser Arrangements for Optical Lattices

The simplest optical lattice to consider is a 1D pair of counterpropagating beams of the same polarization as was used in the first experiment [313]. A variation of this uses beams of different polarization, where the conservative part of the force arising from the polarization gradient provides the periodic potential for the lattice

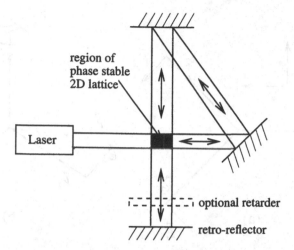

FIGURE 16.2. A single mode of a standing wave cavity is folded by mirrors to make an optical field consisting of two perpendicular phase-stable standing waves. Their polarizations can be different by placement of a retarder at the indicated position [316].

(see Chapter 8). Such 1D lattices have produced much new knowledge about atoms moving in periodic potentials as discussed below.

The "crystallography" of optical lattices in 2D and 3D is a large subject, well beyond the scope of this book. Nevertheless, some of the most important features are summarized here. The reader is directed to an excellent article on the subject [314], as well as a more general review article [315].

At first thought it would seem that a rectangular 2D or 3D optical lattice could be readily constructed from two or three mutually perpendicular standing waves. However, a sub-wavelength movement of a mirror caused by a small vibration could change the relative phase of the standing waves. This could make dramatic changes in the local polarization, for example, converting linearly polarized light to circular, and hence to the depth and nature of the wells, just as in optical molasses. In 2D this problem can be partially avoided by choosing all the beams to have the same linear polarization, perpendicular to the plane containing the beams' $\vec{k}$-vectors, but this can't be done in 3D because no such plane exists.

One way to avoid problems arising from such phase fluctuations is to exert careful control of the relative phases of the standing waves. This has been done to interferometric precision [249], but presents a technical challenge because many of the mirrors and other optical components are necessarily mounted far from the solid surface of an optical table.

Clearly the phase problem can also be controlled by making the optical field from the single mode of a multiply folded cavity, as shown for example in Fig. 16.2 for 2D [316]. The light field is a standing wave formed by reflection from the mirror, and therefore has a fixed temporal phase everywhere. The scheme of Fig. 16.2 suffers if the mirrors are outside the vacuum system because the returning horizontal beam will have passed through six more windows and undergone three more reflections than the incident horizontal beam, and so balancing the intensities can be quite

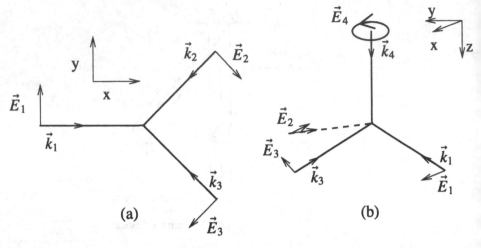

FIGURE 16.3. Arrangement for a stable optical lattice. In (a) the three beams all have the same polarization plane and propagate at 120° to one another. In (b) the four beams still share the same polarization plane and the vertically traveling one is circularly polarized. The other three no longer have coplanar wavevectors (figure adapted from Ref. 250).

a task. Therefore the authors of Ref. 316 devised a slightly different geometry that produces the same fixed-phase standing wave but eliminates the intensity balancing problem. They also described other similar configurations in both 2D and 3D [316]. In such cases, vibrations or other phase changes would displace the optical wells in space, but would not make major changes in the character of the optical field [249].

In 1993 a very clever scheme was described [250]. It was realized that an n-dimensional lattice could be created by only $n + 1$ traveling waves rather than $2n$. Instead of producing optical wells in 2D with four beams (two standing waves), these authors used only three. The $\vec{k}$-vectors of the co-planar beams were separated by $2\pi/3$, and they were all linearly polarized in their common plane (not parallel to one another) as shown in Fig. 16.3a. As in the folded cavity scheme discussed above, vibrations or other phase changes would only displace the optical wells in space [249].

The same immunity to vibrations was established for a 3D optical lattice by using only four beams arranged in a quasi-tetrahedral configuration. The three linearly polarized beams of the 2D arrangement described above were directed out of the plane toward a common vertex, and a fourth circularly polarized beam was added (Fig. 16.3b). All four beams were polarized in the same plane [250]. The authors showed that such a configuration produced the desired potential wells in 3D.

Other four-beam configurations for 3D optical lattices were also devised and studied. For example, the Paris group realized that a more symmetric laser beam configuration could be made by altering the standard 1D lin ⊥ lin configuration to make a 3D lattice [317]. They replaced each of the beams by two beams of the same linear polarization traveling at an angle with respect to one another.

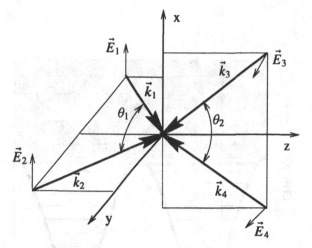

FIGURE 16.4. The four beams of the Paris lattice can be thought of as originating from a 1D lin ⊥ lin configuration by replacing each of the beams by two of the same polarization traveling in a plane perpendicular to the polarization (figure adapted from Ref. 314).

If the original lin ⊥ lin configuration had horizontal $\vec{k}$-vectors with horizontal and vertical polarization, then the lattice was formed by replacing the originally horizontally polarized beam by two other horizontally polarized beams, traveling at an angle θ_2 with respect to one another in one vertical plane, and by replacing the originally vertically polarized beam by two vertically polarized beams at an (possibly different) angle θ_1 with respect to one another traveling in one horizontal plane (see Fig. 16.4).

The NIST group studied atoms loaded into this type of lattice using Bragg diffraction of laser light from the spatially ordered array [318]. They cut off the laser beams that formed the lattice, and before the atoms had time to move away from their positions, they pulsed on a probe laser beam at the Bragg angle appropriate for one of the sets of lattice planes. The Bragg diffraction not only enhanced the reflection of the probe beam by a factor of 10^5, but by varying the time between the shut-off of the lattice and turn-on of the probe, they could measure the "temperature" of the atoms in the lattice. The reduction of the amplitude of the Bragg scattered beam with time provided some measure of the diffusion of the atoms away from the lattice sites, much like the Debye-Waller factor in X-ray diffraction.

16.3 Quantum States of Motion

Laser cooling has brought the study of the motion of atoms into an entirely new domain where the quantum mechanical nature of their center-of-mass motion must be considered. Such exotic behavior for the motion of whole atoms, as opposed to electrons in the atoms, has not been considered before the advent of laser cooling simply because it is too far out of the range of ordinary experiments. A series of

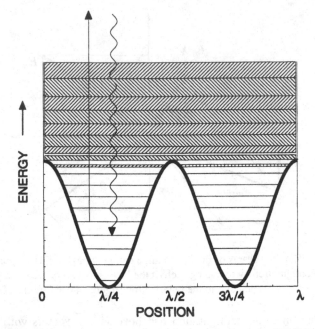

FIGURE 16.5. Energy levels of atoms moving in the periodic potential of the light shift in a standing wave. There are discrete bound states deep in the wells that broaden at higher energy, and become bands separated by forbidden energies above the tops of the wells. Under conditions appropriate to laser cooling, optical pumping among these states favors populating the lowest ones as indicated schematically by the arrows (see Sec. 16.5).

experiments in the early 1990s provided dramatic evidence for these new quantum states of motion of neutral atoms, and led to the debut of deBroglie wave atom optics (see Chapter 15).

The limits of laser cooling discussed in Sec. 8.7 suggest that atomic momenta can be reduced to a "few" times $\hbar k$. This means that their deBroglie wavelengths are equal to the optical wavelengths divided by a "few". If the depth of the optical potential wells is high enough to contain such very slow atoms, then their motion in potential wells of size $\lambda/2$ must be described quantum mechanically, since they are confined to a space of size comparable to their deBroglie wavelengths. Thus they do not oscillate in the sinusoidal wells as classical localizable particles, but instead occupy discrete, quantum mechanical bound states as shown in the lower part of Fig. 16.5.

Optical lattices can be used for the study of many effects normally associated with solid state physics. One example is the Bragg diffraction discussed in the previous section. Another example arises because atoms bound in states near the tops of the potential wells can readily tunnel to adjacent wells. The result is that bound states near the tops of the wells are broadened into bands, and of course, there are also forbidden energy bands even above the tops of the wells. Figure 16.5 shows this energy level structure for the simplest 1D case. The generalization to more complicated 1D cases, as well as higher dimensions, is obvious. In addition

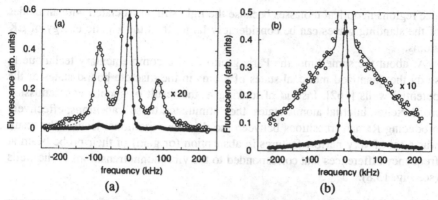

FIGURE 16.6. (a) Fluorescence spectrum in a 1D lin $\perp$ lin optical molasses. Atoms are first captured and cooled in an MOT, then the MOT light beams are switched off leaving a pair of lin $\perp$ lin beams. Then the measurements are made with $\delta = -4\gamma$ at low intensity. (b) Same as (a) except the 1D molasses is σ^+-σ^- which has no spatially dependent light shift and hence no vibrational motion (figure from Ref. 163).

to the presence of the discrete bound states shown by the sidebands in Fig. 16.6 (see discussion below), the existence of such energy bands is also a direct manifestation of the quantum mechanical nature of the atomic center-of-mass motion in the lattice, since tunneling is a purely quantum phenomenon.

The very low temperatures of optical molasses described in Chapter 8 produced severe limits on the ballistic technique described in Sec. 7.4 used for measuring the atomic velocity distribution. The group at NIST therefore developed a new method that superposed a weak probe beam of light directly from the laser upon some of the fluorescent light from the atoms in a 3D optical molasses, and directed the light from these combined sources onto on a fast photodetector [319]. The resulting beat signal carried information about the Doppler shifts of the atoms in the optical molasses [163]. These Doppler shifts were expected to be in the sub-MHz range for atoms with the previously measured 50 μK temperatures. The observed features confirmed the previous measurements done by atomic ballistics [82].

The results of such experiments, shown in Fig. 16.6, display a central peak whose width corresponds to the average velocity along the direction of the weak probe beam, and two sidebands that correspond to transitions that raise or lower the energy of atoms in the wavelength size traps by one vibrational quantum (see Fig. 16.5). Their unequal strength reflects the unequal populations of the vibrational levels, and allows extraction of the "temperature" from the Boltzmann factor. The dependence of this measured temperature on the laser trap parameters is consistent with the theories of laser cooling previously described in Chapter 8 above.

The experiments also showed a much narrower peak atop the sub-MHz central peak whose width corresponded to velocities much less than a single atomic recoil. Since there was no evidence to support the existence of such a narrow velocity distribution, the sharp peak was attributed to Dicke narrowing [320], a suppression of Doppler shifts when radiators or scatterers are confined to a space smaller than a wavelength. It is not surprising to expect atoms to be confined to sub-wavelength size regions in optical molasses because the light shifts associated with antinodes of the standing waves can be considerably larger than the kinetic energy of μK atoms.

At about the same time, the Paris group used a complementary technique to study the quantized motional states of atoms in the discrete bound states of the potential wells [162]. Instead of looking at the result of spontaneous emission from excited internal atomic states, they stimulated these transitions, effectively producing Raman transitions between the discrete vibrational levels. The result was a set of very narrow features in absorption (or gain) of their probe beam at frequency differences that corresponded to the vibrational transitions in the wells (see Fig. 13.8).

16.4   Band Structure in Optical Lattices

The band structure associated with the motion of atoms in standing waves was first studied theoretically in 1991 [161, 321], and experimentally in 1994 [322]. The experimenters used the $2^3S \rightarrow 2^3P$ transition in He* at $\lambda = 1.083~\mu$m in various 1D polarization configurations. The ratio of its high recoil energy ($\approx$ 42 kHz) to its small natural width ($\approx$ 1.6 MHz) combine to produce a relatively large value of $\varepsilon \approx 0.025$ as defined in Eq. 5.5, and thereby maximize the observable effects. It is easy to see why a large value of ε maximizes such quantum effects by considering the ratio r of the energy level spacing to the optical excitation rate γ_p given by Eq. 2.26. Classically this ratio gives approximately the Q of the oscillations, since the lifetime of the ground states is given by γ_p. Using the harmonic approximation to the shape of the bottom of the sinusoidal wells given by $U = U_0(1+\cos 2kx)/2$, the ratio r becomes

$$r = \frac{1}{\gamma_p}\sqrt{\frac{1}{M}\frac{d^2U}{dx^2}} = \sqrt{\varepsilon\frac{4U_0\gamma}{\hbar\gamma_p^2}} \propto \sqrt{\varepsilon\left(\frac{|\delta|}{\gamma}\right)^3} \qquad (16.1)$$

for $|\delta| \gg \gamma$. Of course, $|\delta|$ can never be made too large in any experiment because the laser cooling rate will be considerably decreased, so a large value of ε is indeed important.

The results of the measurements showed features that depended on the existence of bands with finite energy widths, rather than simply the presence of discrete states in the wells. For example, the Bloch states within a band have a continuum of energy levels whose populations may vary as a result of optical pumping. This variation can influence the velocity distribution of the entire population of the band in a calculable way, and this is precisely what these authors observed [322].

In the 1930s Bloch realized that applying a uniform force to a particle in a periodic potential would not accelerate it beyond a certain speed, but instead would result in Bragg reflection when its deBroglie wavelength became equal to the lattice period. Thus an electric field applied to a conductor could not accelerate electrons to a speed faster than that corresponding to the edge of a Brillouin zone, and that

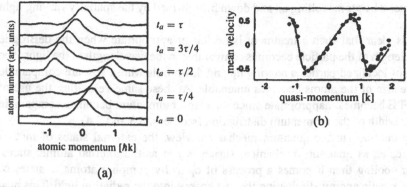

(a)

(b)

FIGURE 16.7. Plot of the measured velocity distribution *vs.* time in the accelerated 1D lattice. The atoms accelerate only to the edge of the Brillouin zone where the velocity is $+v_r$, and then the velocity distribution appears at $-v_r$ (figure from Ref. 323).

at longer times the particles would execute oscillatory motion. Ever since then, experimentalists have tried to observe these Bloch oscillations in increasingly pure and/or defect-free crystals.

Atoms moving in optical lattices are ideally suited for such an experiment, as was beautifully demonstrated in 1996 [323]. The authors loaded a 1D lattice with atoms from a 3D molasses, further narrowed the velocity distribution, and then instead of applying a constant force, simply changed the frequency of one of the beams of the 1D lattice with respect to the other in a controlled way, thereby creating an accelerating lattice. Seen from the atomic reference frame, this was the equivalent of a constant force trying to accelerate them. After a variable time t_a the 1D lattice beams were shut off and the measured atomic velocity distribution showed beautiful Bloch oscillations as a function of t_a. The centroid of the very narrow velocity distribution was seen to shift in velocity space at a constant rate until it reached $v_r = \hbar k / M$, and then it vanished and reappeared at $-v_r$ as shown in Fig. 16.7. The shape of the "dispersion curve" allowed measurement of the "effective mass" of the atoms bound in the lattice.

16.5 Quantum View of Laser Cooling

In the semiclassical picture of laser cooling used up to now, the motion of atoms is treated as if they were point particles whose positions and velocities can be known simultaneously. The optical damping force on them is calculated from the force operator $\nabla \mathcal{H}$, and the average over the mixture of states created by the light field is done using $F = -\text{Tr}(\rho \nabla \mathcal{H})$ (see Eq. 3.4). Here ρ is the atomic density matrix found from its equation of motion, Eq. 2.21, called the optical Bloch equations. The velocity distribution evolves in classical phase space according to the Fokker-Planck Equation (FPE), given in Eq. 5.20. Solution of these equations results in a formal quantitative description of the Sisyphus cooling that has been modelled in

Chapter 8 by atoms rolling up and down hills formed by the spatially varying light shifts.

It is clear that such a treatment is no longer appropriate when the deBroglie wavelength of the particle becomes comparable to the wavelength of the light. The view of localized particles moving up and down hills whose size are comparable to the size of the atoms becomes untenable. At these same velocities, the use of the FPE becomes inappropriate since the recoil momentum becomes comparable to the width of the momentum distribution (see Secs. 5.3 and 5.4).

By contrast, in the quantum mechanical view, the external states of motion are treated as quantum mechanical variables just as the internal atomic states. Laser cooling then becomes a process of optically pumping atoms to states of lower kinetic energy, dissipating the lost energy into the radiation field. This new theoretical approach requires a different explanation of laser cooling than that of a damping force competing with momentum diffusion (see Sec. 8.7), because stationary quantum states rather than classical trajectories are involved. Optical excitation and spontaneous emission deplete the more energetic quantum states faster than the low energy states because the transition rates are asymmetric in total energy. Thus laser cooling becomes an optical pumping process among external states of motion as well as among internal atomic states [324].

Figure 16.5 shows a model of how this works [325]. Atoms occupy discrete levels in the wells formed by the interference of the laser beams that do the cooling. They absorb light and undergo spontaneous emission from the excited states as indicated by the wavy lines. When the laser parameters correspond to those appropriate for cooling, the spatial overlap of the wavefunction of the excited and ground states favors excitation to those particular vibrational levels of the excited state whose decay paths preferentially lead to lower ground state vibrational energies.

This is an especially interesting description because it depends on the light-shift potential caused by stimulated emission to make the wells, and also on the optical pumping caused by the spontaneous emission to enable the cooling. Both of these quite different processes are caused by the same light field.

17
Bose-Einstein Condensation

17.1 Introduction

The basis underlying the great advance of the Planck distribution law for black body radiation in 1901 was a mystery in the era before the development of quantum mechanics in the late 1920s. Early attempts to calculate this spectrum using classical statistical mechanics had failed dismally, resulting in the catastrophe of the Rayleigh-Jeans law. In 1924 S. Bose found the correct way to evaluate the distribution of identical entities such as Planck's radiation quanta that allowed him to calculate the Planck spectrum using the methods of statistical mechanics. Within a year Einstein had seized upon this idea, and generalized it to identical particles with discrete energies. The result was the Bose-Einstein (BE) statistical mechanics of identical particles, even before the idea of wavefunctions had appeared. The Fermi-Dirac statistics, and their contrast with BE statistics, came after the advent of quantum mechanics and the Pauli exclusion principle (antisymmetrization postulate). The BE distribution is

$$N(E) = \frac{1}{e^{\beta(E-\mu)} - 1}, \qquad (17.1)$$

where $\beta \equiv 1/k_B T$ and μ is the chemical potential that vanishes for photons: Eq. 17.1 with $\mu = 0$ is exactly the Planck distribution.

Einstein observed that this distribution has the peculiar property that for sufficiently low average energy (*i.e.*, low temperature), the total energy could be minimized by having a discontinuity in the distribution for the population of the lowest allowed state. That is, at sufficiently low temperatures, the total energy of a sample of atoms would be minimized if there were a significant fraction of its atoms in the ground state, and infinitesimal fractions of the atoms in each of the discrete excited states.

The condition for this Bose-Einstein condensation (BEC) in a gas can be expressed in terms of the deBroglie wavelength λ_{dB} associated with the thermal motion of the atoms as

$$n\lambda_{dB}^3 \geq 2.612\ldots, \tag{17.2}$$

where n is the spatial density of the atoms. In essence, this means that the atomic wavefunctions must overlap one another.

The most familiar elementary textbook description of BEC focuses on non-interacting particles. The photons of the Planck distribution is the most widely studied example, but BEC for massive, non-interacting particles is also discussed. However, particles *do* interact and the lowest order approximation that is widely used to account for the interaction takes the form of a mean-field repulsive force. It is inserted into the Hamiltonian for the motion of each atom in the trap (*n.b.*, not for the internal structure of the atom) as a term V_{int} proportional to the local density of atoms. Since this local density is itself $|\Psi|^2$, it makes the Schrödinger equation for the atomic motion non-linear, and the result bears the name "Gross-Pitaevski equation". For N atoms in the condensate it is written

$$\left[-\frac{\hbar^2}{2M}\nabla_{\vec{R}}^2 + V_{trap}(\vec{R}) + N V_{int}|\Psi(\vec{R})|^2\right]\Psi(\vec{R}) = E_N\Psi(\vec{R}), \tag{17.3}$$

where $\vec{R}$ is the coordinate of the atom in the trap, $V_{trap}(\vec{R})$ is the potential associated with the trap that confines the atoms in the BEC, and $V_{int} \equiv 4\pi\hbar^2 a/M$ is the coefficient associated with strength of the mean field interaction between the atoms. Here a is the scattering length (see Chapter 14), and M is the atomic mass.

For $a > 0$ the interaction is repulsive so that a BEC would tend to disperse. This is manifest for a BEC confined in a harmonic trap by having its wavefunction somewhat more spread out and flatter than a Gaussian. By contrast, for $a < 0$ the interaction is attractive and the BEC eventually collapses. However, it has been shown that there is metastability for a sufficiently small number of particles with $a < 0$ in a harmonic trap, and that a BEC can be observed in vapors of atoms with such negative scattering length as ^{7}Li [326–328]. This was initially somewhat controversial.

Solutions to this highly non-linear equation 17.3, and the ramifications of those solutions, form a major part of the theoretical research into BEC. Note that the condensate atoms all have exactly the same wavefunction, which means that adding atoms to the condensate does not increase its volume, just like the increase of atoms to the liquid phase of a liquid-gas mixture makes only an infinitesimal volume increase of the sample.

The consequences of this predicted condensation are indeed profound. If atoms in a container, or bound in a trap, satisfy the condition of Eq. 17.2, then a significant fraction of them will be in the lowest bound state, whose wavefunction spans a large fraction of the accessible volume. For example, in a harmonic trap, the lowest state's wavefunction is a Gaussian. With so many atoms having *exactly* the same wavefunction they form a new state of matter, unlike anything in the familiar experience.

17.2   The Pathway to BEC

The earliest interpretation of an experiment to observe BEC is usually associated with studies of the Λ transition in liquid ^{4}He in the late 1930s. At the temperature 2.2 K the conditions of Eq. 17.2 are satisfied, there is clearly a phase transition, and the properties of the fluid change dramatically. However, because of the very strong interaction between the atoms, only a small fraction ($\sim 9\%$) of the atoms are in the condensate. In fact, the behavior of the system is sufficiently complicated that neither the simple Bose distribution given by the Planck spectrum nor the Gross-Pitaevski equation is adequate.

In the late 1970s there were several attempts to achieve BEC in spin-polarized atomic hydrogen gas. This gas does not liquefy even at $T = 0$ K, and the BEC could be made at densities sufficiently low that the interactions associated with collisions could be neglected. The major problem was recombination of two H atoms into a molecule of H_2, and this was thought to be impossible since the required three-body collisions would be quite rare. The process was doomed to failure by the existence of long-range two-body interactions that were predicted soon after the experiments began. These caused the atomic spins to invert as a result of the dipole magnetic field associated with their spin magnetic moments, and this greatly enhances recombination into H_2 molecules. After considerable development of forced evaporative cooling techniques (see Sec. 12.5), BEC in H was finally reported in 1998 without using laser cooling [329].

A claim to BEC in condensed matter was made based on the motion of excitons. These sparse electron excitations in certain materials have some of the properties of atoms, and are sufficiently delocalized that their wavefunctions can overlap. But the system is rather restricted, and has not attracted as much attention as the BEC in laser-cooled alkali vapors.

Achieving the conditions required for BEC, namely satisfying Eq. 17.2 in a low-density atomic vapor, requires a long and difficult series of cooling steps. First, note that an atomic sample cooled to the recoil limit would need to have a density of a few times 10^{13} atoms/cm^3 in order to satisfy Eq. 17.2. However, atoms can not be optically cooled at this density because the resulting vapor would have an absorption length for on-resonance radiation approximately equal to the optical wavelength. Thus atoms would form an opaque vapor that would not permit the cooling light in or the fluorescent light to escape it. Furthermore collisions between ground and excited state atoms that can heat them by energy transfer among fine or hyperfine states have such a large cross section, that at this density the optical cooling would be extremely ineffective.

In fact, the practical upper limit to the atomic density for laser cooling in a 3D optical molasses (see Sec. 8.2) or MOT (see Sec. 11.4) corresponds to $n \sim 10^{10}$ atoms/cm^3. In order to satisfy Eq. 17.2 at this density, the temperature would have to be a few nK, clearly out of the range of any of the usual optical processes that have been described up to now (Sec. 8.7). (Note that the Raman cooling described in Sec. 8.7.2, and the VSCPT process described in Sec. 18.3 are optical processes that can achieve such sub-recoil temperatures, but these are inefficient so that cooling enough atoms to make a BEC with these techniques is quite a formidable task.) Thus it is clear that the final stage of cooling toward a BEC must be done either in the dark or in a far-off-resonance trap (Sec. 11.2.3).

The process typically begins with a MOT for efficient capture of atoms from a slowed beam or from the low-velocity tail of a Maxwell-Boltzmann distribution of atoms at room temperature. The MOT fields are then shut off in a very short time (few ms), a process that requires very great care in the electronics that power the MOT coils. Then a polarization gradient optical molasses stage is initiated (see Sec. 8.2) that cools the atomic sample from the mK temperatures of the MOT to a few times T_r. For the final cooling stage, the optical molasses beams are quickly switched off and the magnet coils are turned back on again so that the cold atoms are confined in the dark in a purely magnetic trap. Then a forced evaporative cooling process, described in some detail in Chapter 12, is used to cool the atoms.

The observation of BEC in trapped alkali atoms in 1995 has been the largest impetus to research in this exciting field. In each reported case the atoms were trapped, laser-cooled, and then evaporatively cooled. At first there were only three reported successful experiments, and then there was a hiatus for over a year. Slowly other labs developed the technology and reports of successful BEC's began to appear more rapidly. As of this writing (1999), the only atoms that have been condensed are Rb [138], Na [330], Li [331], and most recently, H [329]. The case of Cs is special because, although BEC is certainly possible, the presence of a near-zero energy resonance severely hampers its evaporative cooling rate. The case of H is also special because it was done using standard cryogenic methods followed by evaporative cooling, without laser cooling.

17.3   Experiments

17.3.1   Observation of BEC

The apparatus for producing the conditions necessary for BEC has been described in several parts of this book, and is summarized in the previous section. By contrast, the apparatus for observing BEC is quite different from the usual setups for the study of laser cooling. Nevertheless, the basic principles apply, namely to devise a scheme to measure the spatial distribution of atoms some delay time after the atoms have been released from whatever trap holds them. Simple classical ballistic calculations then are used to infer the velocity distribution of atoms in the original sample from the measured spatial distribution.

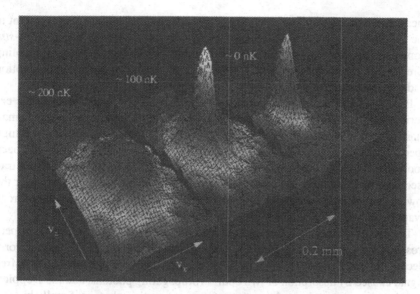

FIGURE 17.1. Three panels showing the spatial distribution of atoms after release from the magnetostatic trap following various degrees of evaporative cooling. In the first one, the atoms were cooled to just before the condition of Eq. 17.2 was met, in the second one, to just after this condition, and in the third one to the lowest accessible temperature consistent with leaving some atoms still in the trap (figure taken from the JILA web page).

The first observations of BEC were in Rb [138], Li [331], and Na [330], and the observation was done with such destructive ballistic techniques. The results from one of the first experiments are shown in Fig. 17.1. The three panels show the spatial distribution of atoms some time after release from the trap. From the ballistic parameters, the size of the BEC sample, as well as its shape and the velocity distribution of its atoms could be inferred. For temperatures too high for BEC, the velocity distribution is Gaussian but asymmetrical. This is because the magnetic trap is necessarily anisotropic, as outlined in Sec. 10.2, so the Gaussian wavefunction of the lowest state is narrower in one direction than in the other. Of course, the momentum distribution is wider in the direction where the spatial distribution is narrower. For temperatures below the transition to BEC, the distribution is also not symmetrical, but now shows the distinct peak of a disproportionate number of very slow atoms corresponding to the ground state of the trap from which they were released. The effect is most dramatic, and was described as not the usual kind of "clue hidden in the data", but more like "Venus rising from the sea, fully formed" [332]. As the temperature is lowered further, the number of atoms in the narrow feature increases very rapidly, a sure signature that this is truly a BEC and not just very efficient cooling.

Such destructive ballistic measurement methods have been complemented by direct optical imaging in many cooling and trapping experiments. However, the physical size of a typical BEC is only $\sim 10~\mu m$, so direct imaging strains the capability of optical design for viewing a BEC that is several cm from the nearest window of a vacuum system. Nevertheless, both absorptive and dispersive measurements, with limited resolution, have become more and more important in the experiments. The destructive methods have recently been replaced by *in-situ* non-destructive dispersive measurements that have enormous advantages, enabling direct observation of the growth of the BEC, as well as of its collective motion under the influence of various kinds of perturbations.

BEC in H was first reported in 1998 [329] and is quite different from the experiments in the alkalis. The temperature is as high as 50 μK because the small mass of H results in a sufficiently large deBroglie wavelength, even at such high speed. It is produced by evaporative cooling of a sample of atomic H that has been cooled in a conventional cryostat and then magnetically confined in a Ioffe trap (see Sec. 10.2). At first, the sample is allowed to evaporate spontaneously over the low end of the magnetic trap, cooling the gas to about 120 μK, and then forced evaporation using rf is applied (see Sec. 12.5).

The detection scheme is also quite different. The authors use two-photon spectroscopy to excite the metastable 2S state of H at $\lambda = 243$ nm [333]. With retroreflected beams of light, the spectrum shows both the very narrow, Doppler-free signal caused by absorption of two counterpropagating photons, and the Doppler-broadened spectrum from absorption of co-propagating photons. Small shifts or broadening of the narrow feature are easily observable, and the signature of the BEC is a strong shift arising from the strong interatomic interactions associated with the condensate. The magnitude of the shift is found from the non-linear term of the Gross-Pitaevski equation, Eq. 17.3. BEC in H enjoys the special advantage that the atomic structure is so well known that calculation of V_{int} in Eq. 17.3 can be done much more accurately.

17.3.2   First-Order Coherence Experiments in BEC

One of the earliest experiments performed with a BEC was to observe deBroglie wave interference fringes [334]. A sample of Na atoms was collected and trapped in a MOT, cooled in molasses, and confined in an elongated magnetic trap related to the Ioffe trap (see Sec. 10.2). Then an intense beam from an Ar ion laser was focused into the center of the trap, and even though the resulting light shift was small because of the huge detuning, it was large enough to form a barrier that could split the very cold trapped atoms into two separated samples. Then these were evaporatively cooled to form independent BEC's, and released from the trap to fall under gravity. It took ~ 40 ms for the BEC's to expand sufficiently to overlap one another, and during this time they fell ~ 8 mm under gravity.

The atomic cloud was imaged dispersively so as not to disturb it, and it displayed bright interference fringes with a spatial period appropriate to the experimental geometry (see Fig. 17.2). The phase of the fringes varied unpredictably from trial to trial as might be expected from independent coherent wave sources. This can be understood by considering the pulse-to-pulse variation of the phase of interference fringes of the light from two independent pulsed lasers. Furthermore, if the barrier between the BEC's was made small enough to permit tunneling between them,

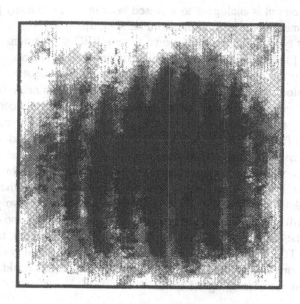

FIGURE 17.2. DeBroglie wave interference fringes produced by allowing two independent BEC's to run into one another (figure from Ref. 335). The atoms in the BEC's are dropped from the trap, and as they fall they expand and overlap, producing the fringes.

then a relatively stable phase feature was observed (but not measured because of limitations imposed by vibration).

In summary, the atoms in a BEC have a spatial coherence that derives from their momentum coherence, and this can be observed by interfering two independent BEC's. This experiment was the first step toward coherent atom optics using a BEC as a source.

The next step in the study of atoms in a BEC was the demonstration of an output coupler for them. This was done by the same group [336], using the same kinds of rf transitions that produced the evaporative cooling (see Chapter 12). Instead of completely inverting the atomic spins with a π-pulse of rf radiation, the experimenters applied a pulse that gave a much smaller inversion. This means that the atoms were placed in a superposition consisting mostly of the trapping state, but with a small component of a non-trapping state. Of course, this small fraction of the atoms was removed from the BEC, but most of the atoms remained trapped in the condensate.

These pulses of atoms could be repeated many times, slowly depleting the population of the BEC. The trapping state was the $M_F = -1$ sublevel of the $F = 1$ hfs ground state of Na, and the rf transitions populated both the $M_F = 0$ and $+1$ sublevels. The first of these simply fell away and was easily photographed, but the second is a strong field seeker and was therefore repelled quickly from the trapping region. But each separate pulse of atoms was internally coherent and this coherence could enable interference so that a collision of two pulses should show fringes. This output is analogous to a pulsed laser output, and many people refer to this experiment as the birth of the atom laser [336].

A different type of extraction of atoms from a BEC was demonstrated by the NIST group [337]. They employed the recoil-induced resonances of the type described near the end of Sec. 13.9, but using counterpropagating laser beams [253–256]. Atoms undergo a single cycle of absorption from one beam followed by stimulated emission from the other, and thus have a precisely controlled momentum change. By choosing the strength of the interaction, the authors could select a small fraction of the atoms in the condensate to be ejected, and then repeat the process periodically. Thus atoms were kicked by $\pm 2\hbar k$, and were ejected in opposite directions from the condensate. Furthermore, repetition of the laser pulse could split each of the ejected packets of atoms into two others, one with momentum $\pm 4\hbar k$ and one with 0. A third pulse would allow two packets that had followed different sequences of momentum kicks to overlap, and evidence for interference between them was observed. In a related experiment, the magnetic field of their TOP trap (see Sec. 10.4.3) caused these packets of atoms to follow curved trajectories, and stroboscopic-like traces of their orbits could be followed with the technique of imaging the atoms.

17.3.3   Higher-Order Coherence Effects in BEC

There have been a number of tests that dramatically confirm the bosonic character of the particles in the observed BEC's, and show excellent corroboration of the theoretical pictures. One of these derives from the collisional losses that occur in the BEC by the formation of alkali metal dimers that are no longer trapped. This can only happen in three-body collisions, but such collisions are strongly suppressed for atoms in a BEC because all the atoms have the same wavefunctions and thus are not "moving" relative to one another to have collisions.

The correlations of the positions of atoms in a BEC is expressed in terms of the same correlation functions that are used to describe the intensity correlations of light, for example, in the famous Hanbury-Brown and Twiss experiment. The second order correlation coefficient $g^{(2)}$ is a measure of such intensity correlations for atoms, but dimerization cannot occur in two-body collision whose rate would be proportional to $g^{(2)}$. Thus the observation of atomic density correlations by loss from a BEC would have to depend on the third order coefficient, $g^{(3)}$, corresponding to three-body collisions [338].

The density-dependent loss of atoms from a BEC is relatively easy to sort out from the density-independent part by measurement of the decay rate of the sample. The measurements of Ref. 338 showed that the density-dependent part indeed derived almost solely from three-body collisions. The presence of atomic density correlations in a BEC was then extracted from the data by comparing the loss rates for a sample of atoms both in and not in a condensate, but having the same density. This was accomplished by cooling Rb atoms to a temperature just above the condition of Eq. 17.2, measuring the density-dependent loss rate, and then comparing this with a sample of atoms that had been cooled well into the BEC regime. The ratio of the two measured values of the decay rates was found to be 7.4 ± 2.6, vastly different from unity, and this is in agreement with various theoretical models that give $g_{BEC}^{(3)}/g_{normal}^{(3)} = 3! = 6$ [338].

Another confirmation of the theory for BEC comes from a measurement of the dynamics of the formation of the condensate from a vapor of super-cooled atoms [339]. In this experiment, a vapor of Na was brought to nearly the threshold of the condition of Eq. 17.2, and then suddenly cooled to well below this. The atoms began forming a condensate, and the rate of populating it was not limited to the cooling rate as in previous experiments. Instead it depended only on the dynamics of BEC formation.

When bosons interact with one another, there is a tendency for them to undergo transitions that result in growth of the population of the states that already have higher populations. It is said that bosons "like to clump together" in the same state, and stimulated emission of light in a laser is one often cited example of this. In a similar way, it might be expected that cold atoms that can occupy only discrete states of a trap would collide in such a way that they would be driven into the most populated trap state. Thus if a small BEC had formed by some nucleation process, then the rate of atoms going into the BEC ought to be enhanced.

In the experiments of Ref. 339, the population of the condensate was found to grow rapidly and then level off, but the approach to steady state was definitely *not* exponential. Instead it followed part of an "$\tan^{-1}$"-shaped curve, starting out slowly, growing rapidly, and then leveling off. This suggests that population of the BEC is not a relaxation recovery from the transient of the sudden cooling, but in fact arises from bosonic stimulation.

17.3.4 Other Experiments

One of the more interesting questions pertaining to BEC is how such a sample of material moves under the influence of a force. From the first experiments, it's clear that a BEC falls ballistically under gravity. It's also expected that it would oscillate harmonically as a rigid body if perturbed in a harmonic trap, and this too has been tested [140, 340]. In general, the nature of higher order modes of motion, such as breathing or deformation, can be found from solutions of the highly non-linear equation Eq. 17.3, which are generally not simple. For example the frequency of the quadrupole deformation oscillation (*e.g.*, from vertical to horizontal ellipsoid) depends on the number of atoms in the BEC because the potential term in Eq. 17.3 is multiplied by N.

In a series of experiments [140, 340–343] several authors in two groups have made extensive studies of the collective excitations of BEC's. The excitations were produced by shaking or squeezing the trap in a phase sensitive way, at or near the frequencies corresponding to the various modes of oscillation. It was found that the oscillation frequencies of various modes can indeed be predicted by Eq. 17.3, and these do not follow any simple, general rules. The measurements have compared the motion of BEC's, vapors of trapped atoms that were too hot to condense, and mixtures. The damping of the oscillations of BEC's was found to be much less than that of ordinary vapors, as expected.

There have also been experiments that test the propagation of sound waves in a BEC, and in particular the observation of the purely quantum mechanical phenomenon called second sound. For normal sound the superfluid and normal components of the density fluctuations in the condenstate are in the same direction. For second sound, these components are in opposite directions and cancel. There are no density fluctuations, but there is still transport of energy from the second sound through the sample. Several observations were found to be inconsistent with direct applications of the theory, and the measurements currently present a challenge to the calculations of the detailed behavior of BEC's.

In a totally different series of experiments, unrelated to the excitations discussed above, BEC's of atoms in two different magnetic substates have been formed in the same trap. Each of the two ground state hfs levels of Rb has a magnetic sublevel with maximum trapping coefficient, and both of these can be trapped simultaneously. In a recent experiment with such a double population in a trap, only one of these was evaporatively cooled and it condensed into a BEC [344]. Because of the different g-factors (see Eq. 4.4), these rf transitions can be selective for one or the other state. Measurements on the other population showed that it, too, had condensed because it had been cooled by interaction with the first population. Such sympathetic cooling had previously been demonstrated with trapped ions, but not with neutral atoms. The experimenters could observe the two BEC's simultaneously in the trap because they were separated in space by the imbalance between the same gravitation force but different magnetic trapping forces on both of them. Such experiments pave the way for making BEC with atoms that can be trapped but not cooled easily, as well as for molecules.

18
Dark States

18.1 Introduction

As the techniques of laser cooling advanced from a laboratory curiosity to a tool for new problems, the emphasis shifted from attaining the lowest possible steady-state temperatures to the study of elementary processes, especially the quantum mechanical description of the atomic motion. In the completely classical description of laser cooling, atoms were assumed to have arbitrary position and momentum that could be known simultaneously. However, when atoms are moving sufficiently slowly that their deBroglie wavelength precludes their localization to less than $\lambda/2\pi$, these descriptions fail and a quantum mechanical description is required, as discussed in Chapter 15.

One special area of interest is the study of dark states, atomic states that can not be excited by the light field. Some atomic states are trivially dark, that is, they can't be excited because the light has the wrong frequency or polarization. The more interesting cases are superposition states created by coherent optical Raman coupling. A very special case are those superpositions whose excitable component vanishes exactly when their external (deBroglie wave) states are characterized by a particular momentum. Such velocity selective coherent population trapping (VSCPT) has been a subject of considerable interest since its first demonstration in 1988 [345]. VSCPT enables arbitrarily narrow momentum distributions and hence arbitrarily large delocalization for atoms in the dark states [346].

The quantum description of atomic motion requires that the energy of such motion be included in the Hamiltonian. The total Hamiltonian for atoms moving in a light field would then be given by

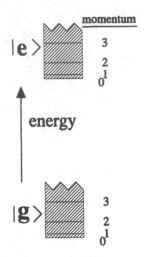

FIGURE 18.1. The kinetic energies accessible to atoms moving freely (no spatially varying light shift). The horizontal lines mark the momentum values that correspond to integer values of the recoil momentum $\hbar k$.

$$\mathcal{H} = \mathcal{H}_{\text{atom}} + \mathcal{H}_{\text{rad}} + \mathcal{H}_{\text{int}} + \mathcal{H}_{\text{kin}}, \tag{18.1}$$

where $\mathcal{H}_{\text{atom}}$ describes the motion of the atomic electrons and gives the internal atomic energy levels, $\mathcal{H}_{\text{rad}}$ is the energy of the radiation field and is of no concern here because the field is not quantized, $\mathcal{H}_{\text{int}}$ describes the excitation of atoms by the light field and the concomitant light shifts, and $\mathcal{H}_{\text{kin}}$ is the kinetic energy E_k of the motion of the atoms' center of mass. This Hamiltonian has eigenstates of not only the internal energy levels and the atom-laser interaction that connects them, but also of the kinetic energy operator $\mathcal{H}_{\text{kin}} \equiv \mathcal{P}^2/2M$. These eigenstates will therefore be labeled by quantum numbers of the atomic states as well as the center of mass momentum p. For example, an atom in the ground state, $|g;\ p\rangle$, has energy $E_g + p^2/2M$ which can take on a range of values. Figure 18.1 shows the continuum of kinetic energy values for both ground and excited states, with the integer momentum values marked as lines. Because $p^2 = (-p)^2$, all these kinetic energy states are doubly degenerate in 1D except for $p = 0$. (In this chapter, momentum is measured in units of $\hbar k$.)

18.2   VSCPT in Two-Level Atoms

To see how the quantization of the motion of a two-level atom in a monochromatic field allows the existence of a velocity selective dark state, consider the states of a two-level atom with single internal ground and excited levels, $|g;\ p\rangle$ and $|e;\ p'\rangle$. Two ground eigenstates $|g;\ p\rangle$ and $|g;\ p''\rangle$ are generally not coupled to one another by an optical field except in certain cases. For example, in oppositely propagating light beams (1D) there can be absorption-stimulated emission cycles

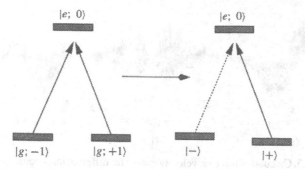

FIGURE 18.2. Schematic diagram of the transformation of the eigenfunctions from the internal atomic states $|g; \ p\rangle$ to the eigenstates $|\pm\rangle$. The coupling between the two states $|g; \ p\rangle$ and $|g; \ p''\rangle$ by Raman transitions mixes them, and since they are degenerate, the eigenstates of $\mathcal{H}$ are the non-degenerate states $|\pm\rangle$.

that connect $|g; \ p\rangle$ to itself or to $|g; \ p \pm 2\rangle$, depending on whether the stimulated emission is induced by the beam that excited the atom or by the other one.

In the first case, the states of the atom and field are left unchanged, but the interaction shifts the internal atomic energy levels thereby producing the light shift (see Sec. 1.2.1). In the second case, the initial and final E_k of the atom differ by $\pm 2(p \pm 1)/M$ so energy conservation requires $p = \mp 1$ (the energy of the light field is unchanged by the interaction since all the photons in the field have energy $\hbar\omega_\ell$). Thus energy conservation corresponds to Raman resonance between the distinct states $|g; \ -1\rangle$ and $|g; \ +1\rangle$, and is therefore velocity selective. The coupling of these two degenerate states by the light field produces off-diagonal matrix elements of the total Hamiltonian $\mathcal{H}$ of Eq. 18.1, and subsequent diagonalization of it results in the new ground eigenstates of $\mathcal{H}$ given by (see Fig. 18.2).

$$|\pm\rangle \equiv (|g; \ -1\rangle \pm |g; \ +1\rangle)/\sqrt{2}. \qquad (18.2)$$

The excitation rate of the eigenstates $|\pm\rangle$ given in Eq. 18.2 to $|e; \ 0\rangle$ is proportional to the square of the electric dipole matrix element $\vec{\mu}$ given by

$$|\langle e; \ 0|\vec{\mu}|\pm\rangle|^2 = |\langle e; \ 0|\vec{\mu}|g; \ -1\rangle \pm \langle e; \ 0|\vec{\mu}|g; \ +1\rangle|^2/2. \qquad (18.3)$$

This vanishes for $|-\rangle$ because the two terms on the right-hand side of Eq. 18.3 are equal since $\vec{\mu}$ does not operate on the external momentum of the atom (dotted line of Fig. 18.2). Excitation of $|\pm\rangle$ to $|e; \ \pm 2\rangle$ is much weaker since it's off resonance because its energy is higher by $4\hbar\omega_r = 2\hbar^2 k^2/M$, so that the required frequency is higher than to $|e; \ 0\rangle$. The resultant detuning is $4\omega_r = 8\epsilon(\gamma/2)$, and for $\epsilon \sim 0.5$, this is large enough so that the excitation rate is small, making $|-\rangle$ quite dark. Excitation to any state other than $|e; \ \pm 2\rangle$ or $|e; \ 0\rangle$ is forbidden by momentum conservation. Atoms are therefore optically pumped into the dark state $|-\rangle$ where they stay trapped, and since their momentum components are fixed, the result is VSCPT.

A useful view of this dark state can be obtained by considering that its components $|g; \ \pm 1\rangle$ have well defined momenta, and are therefore completely delocalized. Thus they can be viewed as waves traveling in opposite directions but having

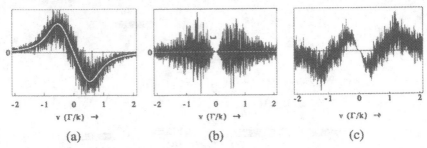

(a)   (b)   (c)

FIGURE 18.3. Calculated force *vs.* velocity curves for different laser configurations showing both the average force and a typical set of simulated fluctuations. Part (a) shows the usual Doppler cooling scheme that produces an atomic sample in steady state whose energy width is $\hbar\gamma/2$. Part (b) shows VSCPT as originally studied in Ref. 345 with no damping force. Note that the fluctuations vanish for $\wp = 0$ because the atoms are in the dark state. Part (c) shows the presence of both a damping force and VSCPT. The fluctuations vanish for $\wp = 0$, and both damping *and* fluctuations are present at $\wp \neq 0$.

the same frequency, and therefore they form a standing deBroglie wave. The fixed spatial phase of this standing wave relative to the optical standing wave formed by the counterpropagating light beams results in the vanishing of the spatial integral of the dipole transition matrix element so that the state cannot be excited. This view can also help to explain the consequences of p not exactly equal ± 1, where the deBroglie wave would be slowly drifting in space. It is common to label the average of the momenta of the coupled states as the family momentum, $\wp$, and to say that these states form a *closed family*, having family momentum $\wp = 0$ [345,347].

In the usual case of laser cooling, atoms are subject to both a damping force *and* to random impulses arising from the discrete photon momenta $\hbar k$ of the absorbed and emitted light. These can be combined to make a force *vs.* velocity curve as shown in Fig. 18.3a. Even in the present case, atoms with $\wp \neq 0$ are always subject to the light field that optically pumps them into the dark state and thus produces random impulses as shown in Fig. 18.3b. There is no damping force in the most commonly studied case of a real atom, the $J = 1 \rightarrow 1$ transition in He*, because the Doppler and polarization gradient cooling cancel one another as a result of a numerical "accident" for this particular $J = 1 \rightarrow 1$ case.

Figures 18.3a and b should be compared to show the velocity dependence of the sum of the damping and random forces for the two cases of ordinary laser cooling and VSCPT. Note that for VSCPT the momentum diffusion vanishes when the atoms are in the dark state at $\wp = 0$, so they can collect there. In the best of both worlds, a damping force would be combined with VSCPT as shown in Fig. 18.3c. Such a force was predicted Ref. 348 and was first observed in 1996 [349].

18.3   VSCPT in Real Atoms

Real atoms have multiple internal levels that include the effects of the magnetic, hyperfine, and other sublevels. The strength of their optical interactions depends

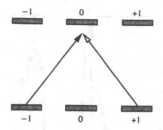

FIGURE 18.4. The magnetic sublevels of the $J = 1 \rightarrow 1$ transition can be coupled by circularly polarized light in four different ways. However, the $M_J = 0 \rightarrow 0$ transition is forbidden, and therefore optical pumping quickly empties the $J = 0$ sublevel. Thus the only remaining transitions are those two indicated by the arrows, and they are the ones involved in VSCPT.

on the light polarization and on these other quantum numbers (see Sec. 4.5). For the present purpose, only the multiplicity of the magnetic sublevels associated with $J \neq 0$ will be important. A particularly beautiful example of dark states appears in the $J = 1 \rightarrow 1$ transition, where the optical selection rules associated with ΔM_J produce an analog of the hypothetical state $|-\rangle$ that is perfectly dark.

18.3.1 Circularly Polarized Light

The most well-studied example occurs in this $J = 1 \rightarrow 1$ transition with counterpropagating beams of opposite circular polarization. This is designated the σ^+-σ^- configuration because the light induces $\Delta M_J = \pm 1$ transitions when the quantization axis is chosen parallel to $\vec{k}$ of one of the light beams so that there is no z-component of the optical electric field. Decay from the excited $M_J = 0$ state to the ground $M_J = 0$ state is forbidden by the selection rules, so the ground $M_J = 0$ state is emptied by optical pumping, and the only populated ground states are $M_J = \pm 1$. Then the $\Delta M_J = \pm 1$ transitions can populate only the excited $M_J = 0$ state, thus forming a "Λ" system of levels as shown in Fig. 18.4. Although this optical arrangement is similar to that used in the magneto-optical trap (MOT) discussed in Sec. 11.4, there is a *very* important difference: the transition scheme for a MOT generally uses a $J \rightarrow J + 1$ angular momentum scheme, carefully chosen to exploit the Zeeman effect, and this $J \rightarrow J + 1$ scheme precludes dark states [350].

The two ground states having $M_J = \pm 1$ can be coupled by a Raman transition requiring the participation of *both* light beams, and thus their momenta must be different by ± 2 in units of $\hbar k$. Moreover, the resulting superposition states, designated $|\pm\rangle$ as above, constitute *entangled states*, a case of very special importance to be discussed later in this chapter. For the case where the σ^+ (σ^-) beam propagates in the positive (negative) z-direction, the superposition states are given by

$$|\pm\rangle = \left[|g_{-1}; -1\rangle \pm |g_{+1}; +1\rangle \right]/\sqrt{2}, \qquad (18.4)$$

where the subscripted quantum number denotes M_J and the other one denotes the atomic momentum. As for the two-level atom case discussed above, one of

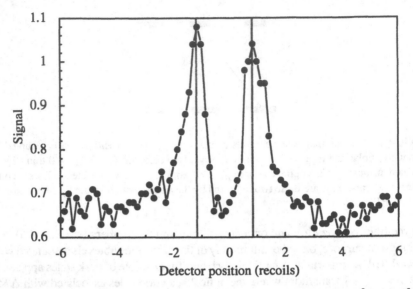

FIGURE 18.5. Measured velocity distribution of He* atoms driven on the $2^3S_1 \to 2^3P_1$ transition at $\lambda = 1.083$ μm by counterpropagating beams, linearly polarized at 90° to one another, for a 20 μs interaction time. The intensity was 120 μW/cm^2 (each beam) and the laser was on resonance.

the states given in Eq. 18.4 is dark, and in this case it is $|+\rangle$ because the Clebsch-Gordan coefficients for the two transitions that couple the ground states of different M_J have opposite signs.

The state $|g_{+1};\ +1\rangle$ cannot be excited by the σ^+ light because there is no excited state with $M_J = +2$. Thus excitation must be by the σ^- beam that can only change its momentum to zero, and correspondingly for the state $|g_{-1};\ -1\rangle$. Therefore these two ground states can be coupled only through the excited state $|e_0;\ 0\rangle$ because the positive-going σ^+ light increases both the linear momentum *and* M_J by one unit, and vice versa.

Clearly the two states that are mixed together to form $|\pm\rangle$ in this case have the same total energy because they are both degenerate ground states, and their equal but opposite momenta result in the same E_k. Thus the mixture is also a stationary state of the total Hamiltonian $\mathcal{H}$ of Eq. 18.1. If the family momentum $\wp$ is not zero, then the two states that are mixed do not have the same E_k, and so neither of the states $|\pm\rangle$ are stationary states of $\mathcal{H}$. An atom originally in the dark state $|+\rangle$ would then evolve into $|-\rangle$ from which it can be excited. The subsequent spontaneous emission would change its momentum in an unpredictable way, but possibly toward a value of $\wp$ closer to zero. As a result, atoms with $\wp \neq 0$ continue to interact with the light field until they are optically pumped by random walk in momentum space into the dark eigenstate $|+\rangle$, and then the optical excitation ceases. Measuring the velocity distribution of an ensemble of atoms in $|+\rangle$ produces two distinct peaks at $p = \pm 1$ as shown in Fig. 18.5, and this is the usual VSCPT as it was first observed [345].

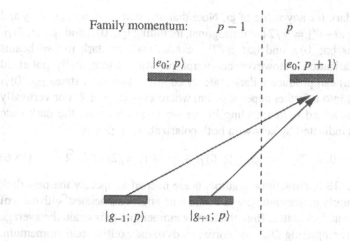

Family momentum: $p - 1$ p

$|e_0; p\rangle$ $|e_0; p + 1\rangle$

$|g_{-1}; p\rangle$ $|g_{+1}; p\rangle$

FIGURE 18.6. Construction of the dark state for linearly polarized light from the states $M_J = \pm 1$ and $p = \pm 1$.

18.3.2   Linearly Polarized Light

Dark states in linearly polarized light are more complicated because each beam can drive both $\Delta M_J = \pm 1$ transitions. Of course, one could always simplify the description by choosing a quantization axis parallel to one of the optical electric fields, but this results in an additional complication. It arises because for a single linearly polarized light beam driving a $J = 1 \rightarrow 1$ transition, there is always an uninteresting, velocity-independent dark state corresponding to the forbidden $M_J = 0 \rightarrow 0$ transition. To avoid confusion with this dark state, the z-axis is chosen parallel to $\vec{k}$ as before. The selection rules now dictate $\Delta M_J = \pm 1$ because $\hat{z} \perp \vec{E}$.

Then the velocity-independent dark state of the beam traveling in the $+z$ direction can be visualized by considering that such light mixes the states $|g_{-1}; p\rangle$ and $|g_{+1}; p\rangle$ only via a single excited state $|e_0; p + 1\rangle$ as shown in Fig. 18.6 (the excited state $|e_0; p - 1\rangle$ is not coupled by a beam traveling in the $+z$ direction). The mixing forms two new superposition states, and one of them is dark just as above. Needless to say, the darkness property of this state is velocity independent, and corresponds to the forbidden $M_J = 0 \rightarrow 0$ transition.

Such a superposition state that is dark to linearly polarized light at some particular angle θ needs to be indicated by a different ket that must also include the family momentum $\wp$ in the specification of the state. Thus the superposition dark state is no longer simply $|+\rangle$ (or $|-\rangle$ for a two level atom), but instead the state that is dark to vertically polarized light is denoted by $|\wp; \theta = 0\rangle_D$, where the subscript "D" denotes a dark state, and for a horizontally polarized beam, the corresponding dark state is $|\wp; \pi/2\rangle_D$. More generally, the state that is dark to a single beam of light linearly polarized at an angle θ from the vertical is

$$|\wp; \theta\rangle_D = \left(e^{-i\theta}|g_{+1}; \wp\rangle + e^{i\theta}|g_{-1}; \wp\rangle\right)/\sqrt{2}. \qquad (18.5)$$

Such a state is dark for any value of $\wp$. Note that any pair of states $|\wp; \theta\rangle_D$ and $|\wp; \theta'\rangle_D$ having $|\theta - \theta'| = \pi/2$ are orthogonal, including $|\wp; 0\rangle_D$ and $|\wp; \pi/2\rangle_D$.

States such as $|\wp; 0\rangle_D$ and $|\wp; \pi/2\rangle_D$ cannot each be dark to two beams of different polarizations. However, counterpropagating orthogonally polarized beams (lin $\perp$ lin) can produce a dark state by coupling two such states $|\wp; 0\rangle_D$ and $|\wp'; \pi/2\rangle_D$ into a further superposition, where $\wp - \wp' = \pm 2$. For vertically (horizontally) polarized light traveling in the $+z$ ($-z$) direction, the dark state superposition is indicated by kets with both polarization angles as

$$D_{\text{lpl}} \equiv |\wp; 0, \pi/2\rangle_D \equiv (|\wp + 1; 0\rangle_D + |\wp - 1; \pi/2\rangle_D)/\sqrt{2} \qquad (18.6)$$

as shown in Fig. 18.6. Now three arguments are needed to specify the new dark state D_{lpl}: the family momentum $\wp$ and both of the angles associated with each of the two component dark states. Here the family momentum $\wp$ is again the average of the two states comprising D_{lpl}, and corresponds to the excited state momentum. The state D_{lpl} can not be excited, independent of the value of $\wp$. Its first component $|\wp + 1; 0\rangle_D$ can be excited only by the horizontally polarized beam (chosen to be traveling in the $-z$ direction) so absorption causes the momentum to be decreased to $\wp$, and correspondingly for the second component, $|\wp' - 1; \pi/2\rangle_D$.

The velocity dependence arises because the superposition D_{lpl} can be a stationary state of the Hamiltonian $\mathcal{H}$ of Eq. 18.1 only if $(\wp + 1)^2 = (\wp - 1)^2$ and thus $\wp = 0$. For this case there is again a closed family of states because atoms cannot be transferred out of the three orthogonal states, $|\wp + 1; 0\rangle_D$, $|\wp - 1; \pi/2\rangle_D$, and the unmixed excited state having $p = 0$ and $M_J = 0$ except by spontaneous emission. Thus the stationary dark state D_{lpl} represents a velocity selective trapped state, and atoms collect in it [347].

18.4   VSCPT at Momenta Higher Than $\pm\hbar k$

A most interesting effect occurs with two linearly polarized counterpropagating beams with their electric fields at an angle θ to one another (one beam is still vertically polarized). The state $|\wp; 0\rangle_D$ is still dark to the vertically polarized beam, but it is not orthogonal to the state that is dark to the other beam, $|\wp; \theta\rangle_D$ (orthogonality requires $|\theta - \theta'| = \pi/2$ as discussed above). Nevertheless, there is a dark state superposition

$$|\wp; 0, \theta\rangle_D = \frac{|\wp + 1; 0\rangle_D + |\wp - 1; \theta\rangle_D}{\sqrt{2 - \cos\theta}}, \qquad (18.7)$$

where the normalization constant is not $\sqrt{2}$ because the states are not orthogonal. The components of this superposition are not part of a closed family because $|\wp; 0, \theta\rangle_D$ is a stationary state of the Hamiltonian only for $\wp = 0$. Such superposition states offer new insights into VSCPT and related phenomena.

The lin-angle-lin optical field considered here can also couple the two independent dark (but not stationary) states that have $\wp = \pm 1$

$$|+1;\ 0,\theta\rangle_D = \frac{|2;\ 0\rangle_D + |0;\ \theta\rangle_D}{\sqrt{2-\cos\theta}} \qquad (18.8a)$$

and

$$|-1;\ 0,\theta\rangle_D = \frac{|0;\ 0\rangle_D + |-2;\ \theta\rangle_D}{\sqrt{2-\cos\theta}} \qquad (18.8b)$$

that have the same average energy $\wp^2/2M$. The superposition of these two dark states is also dark, and is given by

$$|0^{(2)};\ 0,\theta\rangle_D \equiv \frac{|+1;\ 0,\theta\rangle_D - |-1;\ 0,\theta\rangle_D}{\sqrt{2}}. \qquad (18.9)$$

The momentum distribution of this state $|0^{(2)};\ 0,\theta\rangle_D$ consists of peaks at $p = 0$ and ± 2, a total of three peaks. The relative phase of the superposition in Eq. 18.9 is chosen to be π (negative sign) because this allows near cancellation of the middle peak at $p = 0$, leaving just the two side peaks at $p = \pm 2$ [351]. Even though the two states $|\pm 1;\ 0,\theta\rangle_D$ may each be readily pumped to the excited state through their mixing that arises because they're not stationary, $|0^{(2)};\ 0,\theta\rangle_D$ has a far lower mixing rate because it is nearly an eigenstate having $E_k = 4\hbar\omega_r$. This is because its largest components have $p = \pm 2$ and therefore the same energy. Exact cancellation of the $p = 0$ component is only possible for $\theta = 0$, but that case has only velocity-independent dark states.

The two momentum states having $p = \pm 2$ comprising the superposition in Eq. 18.9 cannot be coupled by a Raman transition involving a single excitation-stimulated emission cycle because their momenta differ by $\pm 4\hbar k$. Instead, a four-photon Raman transition is required to conserve momentum, corresponding to a higher-order process in VSCPT. This is a rare example of using higher-order non-linear optical effects to produce dark states and laser cooling.

Population accumulates in the state given by Eq. 18.9, producing peaks in the momentum distribution at ± 2. Also in this lin-angle-lin laser field, the state $|\wp;\ 0,\theta\rangle_D$ given in Eq. 18.7 is perfectly dark for $\wp = 0$, and population also accumulates in it producing the well-studied peaks at $p = \pm 1$ [348]. Both of these long-lived states are populated by a random walk in momentum space, and each of them can be readily observed in an experiment with appropriate interaction time. Thus there would be four very narrow (FWHM $\leq \hbar k$) peaks expected in the measured momentum distribution. Such a distribution has been observed in metastable He driven on the $2^3 S_1 \rightarrow 2^3 P_1$ transition using light of wavelength $\lambda = 1.083\ \mu$m as shown in Fig. 18.7 [351].

18.5   VSCPT and Bragg Reflection

A completely new view of VSCPT has emerged from more careful consideration of the motion of such dark state atoms in the spatially periodic field of oppositely propagating light beams [352]. As Fig. 18.8 shows, dark state atoms traveling with longitudinal momentum p_ℓ make an angle ϕ with the optical wavefronts, and their deBroglie wavelength is

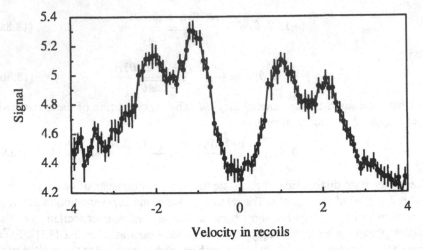

FIGURE 18.7. Measured He* velocity distributions after 20 μs interaction time for $I = 1.6$ mW/cm^2 and zero detuning with counter-propagating beams linearly polarized at 30°. The four-peaked structure is quite evident, and each peak has width less than $\hbar k$.

$$\lambda_{dB} = \frac{h}{\sqrt{p_\ell^2 + 1}}. \tag{18.10}$$

Since $\sin \phi = 1/\sqrt{p_\ell^2 + 1}$, it is clear that

$$\lambda_{dB} = 2d \sin \phi, \tag{18.11}$$

where $d \equiv \lambda/2$ is the spatial periodicity of the light field, and momentum is in units of $\hbar k$. Equation 18.11 is exactly the equation for Bragg reflection, but its *interpretation in this context is indeed most astounding* [352]. Here the deBroglie "matter" wave is Bragg reflected by the spatially periodic optical field: matter and field have been interchanged from the usual case of Bragg reflection of an electromagnetic field by crystalline planes of atoms!

The usual case of X-ray Bragg reflection can be viewed as arising from multi-center scattering of radiation by atoms at each lattice site in a crystal. It follows that propagation of the reflected wave can occur only in the preferred direction defined by Eq. 18.11. Such waves are the only ones not diffusively scattered by atoms in the lattice. The equivalent view of atoms in dark states is simply that the deBroglie wave fields propagate without scattering (*i.e.*, no spontaneous emission) in the light field only when the atoms are indeed in dark states.

Finally, note that this Bragg reflection description of dark states is enhanced by the high velocity states described in Sec. 18.4. Equation 18.11 describes the lowest order Bragg reflection, but for a state such as $|0^{(2)}; 0, \theta\rangle_D$, it is only necessary to put a factor of 2 on the left-hand side. The notion is readily generalizable to any order n of Bragg reflection. Consider atoms entering a light field with any integer value of transverse momentum p_n (in units of $\hbar k$). The dark state associated with this motion can not undergo spontaneous emission, so the only interactions with

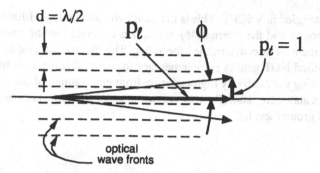

FIGURE 18.8. The relationship between dark states and Bragg reflection is shown by considering the trajectories of atoms with transverse momenta $\pm\hbar k$ as shown. The total atomic momentum is the vector sum of this transverse momentum with the longitudinal momentum p_ℓ and the corresponding deBroglie wavelength relation is just Eq. 18.11 (see Ref. 352).

the light field are stimulated emissions. This leaves the frequency, and hence the energy of the light field unchanged, so that conservation of energy requires that p_n^2 be unchanged. Therefore the trajectories of atoms are either unchanged or reflected. If the longitudinal momentum is zero, the reflection corresponds exactly to the edge of a Brillouin zone. This type of atomic mirror is discussed in Sec. 15.5.

18.6   Entangled States

One of the most interesting aspects of dark state physics arises from the entanglement of motional and internal states. This leads to the opportunity for fundamental studies of many topics whose basis is at the heart of quantum mechanics, such as the well-known Einstein-Podolsky-Rosen paradox, Schrödinger's cat, quantum communications, and quantum computing. The key feature of entangled states is embodied in the form of Eq. 18.4. Here $|\pm\rangle$ is written as a sum of products, and it can be shown that is not possible to find a basis where this state can be described as an outer product. Moreover, a generalization of Eq. 18.4 in which the two components of $|\pm\rangle$ are combined with arbitrary relative amplitudes and a phase factor, $e^{i\alpha}$, can never be measured exactly because it requires two independent parameters to specify the state (e.g., longitude and latitude on the Bloch sphere), and a single measurement perturbs the system so that a second one is unreliable.

Clearly dark state entanglements with multilevel neutral atoms offer several advantages over related optical experiments. First, atoms arrive as discrete objects, unlike optical fields with the notorious difficulties of producing Fock states. Perhaps more important, the number of Hilbert spaces that are available, as well as their dimensionality, can each be larger than two.

As an example, note that the primary element in quantum computing is the quantum controlled NOT gate because it can be combined with rotations to enable any computational operation. Such a gate can be realized directly with the states that are entangled in VSCPT. This is because two independent Hilbert spaces, the external motion and the internal M_J levels, are entangled in the state. Therefore a measurement of one determines the other. This neutral atomic beam version of a controlled NOT gate is complementary to one realizable with trapped ions, while retaining the relatively high isolation from environmental decoherence (the momentum states are naturally very robust, and the internal states are composed entirely of ground levels).

Part IV

Appendices

Appendix A
Notation and Definitions

TABLE A.1: Notation and Definitions

Parameter	Definition	Description	Section	Page
A	dB/dz	Magnetic field gradient	10.2	139
$A_{l'm'lm}$	$\langle l'm'|Y_{1q}|lm\rangle$	Angular part, dipole moment	4.5.1	51
α		Identification of state, apart from angular momentum	4.4	48
β		Damping coefficient	3.3.1	35
$D(E)$		Density of states	12.3	168
δ	$\omega_l - \omega_a$	Laser detuning	1.2	5
E_0		Electric field amplitude	1.2	5
$\vec{\mathcal{E}}(r,t)$		Electric field operator	1.2	5
$\hat{\varepsilon}$		Polarization of the light field	1.2	5

continued on next page

continued from previous page

Parameter	Definition	Description	Section	Page
ε	ω_r/γ	Ratio recoil frequency to natural width	5.1	59
F	$J + I$	Total angular momentum, atom	4.1	40
F		Force on two-level atom	3.1	30
$\mathcal{F}$	$-d\mathcal{H}/dr$	Force operator	3.1	30
$\mathcal{F}$	$\vec{\nabla}(\vec{\mu} \cdot \vec{E})$	Force operator	8.5	107
γ	$1/\tau$	Spontaneous decay rate, linewidth	2.16	22
γ'	$\gamma\sqrt{1+s_0}$	Linewidth, power-broadened	2.4	26
γ_p	$\gamma\rho_{ee}$	Scattering rate	2.4	25
γ	$\log(T'/T)/\log\nu$	Decrease in temperature	12.3	168
Γ	β/M	Damping rate	3.3.1	35
$\mathcal{H}'_{jk}$	$\langle\phi_j\vert\mathcal{H}'\vert\phi_k\rangle$	Coupling matrix element between states j and k	1.1	4
I		Nuclear spin	4.1	40
I_s	$\pi hc/3\lambda^3\tau$	Saturation intensity	2.4	25
j	$\ell + s$	Total angular momentum, electron	4.1	40
J	$L + S$	Total angular momentum, electrons	4.1	40
L		Orbital angular momentum, electrons	4.1	40
ℓ		Orbital angular momentum, electron	4.1	40
ℓ		Partial wave	14.2	200
λ	$2\pi c/\omega_l$	Optical wavelength	1.2	5

continued on next page

continued from previous page

Parameter	Definition	Description	Section	Page
$f(\vec{v})$		Maxwell-Boltzmann distribution	5.2	61
μ'	$(g_e M_e - g_g M_g)\mu_B$	Effective magnetic moment	6.2.2	77
μ_{eg}	$e\langle e|r|g\rangle$	Dipole moment	4.5.1	50
n		Principal quantum number, electron	4.1	40
n	N/V	Atom density	12.1	165
ν	N'/N	Fraction of atoms remaining in trap	12.3	168
η		Design parameter, Zeeman slower	6.2.2	77
η	$U_{\text{trap}}/k_B T$	Ratio trap depth to kinetic energy	12.3	168
ω_a	ω_{eg}	Atomic resonance frequency	1.2	5
ω_{jk}	$\omega_j - \omega_k$	Transition frequency	1.1	4
ω_l		Laser frequency	1.2	5
ω_D	$-\vec{k}\cdot\vec{v}$	Doppler shift	1.2.4	13
ω_r	$\hbar k^2/2M$	Recoil frequency	5.1	59
Ω	$-eE_0\langle e|r|g\rangle/\hbar$	Rabi frequency	1.2	5
Ω'	$\sqrt{\Omega^2 + \delta^2}$	Generalized Rabi frequency	1.2	6
$\Phi(r,\theta)$		Magnetic potential	13.3.1	181
R_c	$(C_3/\hbar|\delta|)^{1/3}$	Condon point	14.4.1	207
ρ	$|\Psi\rangle\langle\Psi|$	Density operator	2.1	17
ρ_{ij}	$\langle\phi_i|\rho|\phi_j\rangle$	Density matrix element	2.1	18
$\rho(\vec{r},\vec{p},t)$		Phase space density	5.5	68

continued on next page

continued from previous page

Parameter	Definition	Description	Section	Page
ρ	$n\lambda_{dB}^3$	Phase space density	12.1	165
$\vec{r}$		Electron coordinate	1.1	3
$\mathcal{R}$	$\vec{\mu}_{eg} \cdot \vec{E}_+/\hbar$	Rabi operator	8.5	107
$\mathcal{R}_{l'm'lm}$	$\langle n'l' \| r \| nl \rangle$	Radial part, dipole moment	4.5.1	51
s		Spin angular momentum, electron	4.1	40
S		Spin angular momentum, electrons	4.1	40
s	$s_0/(1 + (2\delta/\gamma)^2)$	Saturation parameter, off-resonance	2.4	25
s_0	$2|\Omega|^2/\gamma^2$	Saturation parameter, on-resonance	2.4	25
τ	$1/\gamma$	Lifetime of excited state	2.2	22
T	$M\langle v^2 \rangle/k_B$	Temperature	5.1	58
T_D	$\hbar\gamma/2k_B$	Doppler temperature	5.1	58
T_e	$\pi M k_{\text{dip}}^2/16\sigma^2 k_B$	Evaporative cooling limit	12.4.3	174
T_r	$\hbar^2 k^2/M k_B$	Recoil temperature	5.1	59
$\bar{v}$	$\sqrt{3k_B T/M}$	Thermal velocity	5.2	63
v_c	γ/k	Capture velocity	5.1	58
v_D	$\sqrt{\hbar\gamma/2M}$	Doppler velocity	5.1	58
v_r	$\hbar k/M$	Recoil velocity	5.1	59
v_{vsr}	$\pm\omega_Z/2k$	Velocity for VSR-resonances	8.9	118
ξ	$1/s_1 + 1/s_2 + 1/s_3$	Scale trapping potential	12.3	167

Appendix B
Review Articles and Books on Laser Cooling

Special Issues of Regularly Published Journals

Special Issue	Ref.
W.D. Phillips, Ed. *Laser Cooled and Trapped Atoms* — Probably the first special issue, filled with original ideas	353
P. Meystre and S. Stenholm, Eds., *The Mechanical Effects of Light* — One of the earliest major special issues	40
S. Chu and C. Wieman, Eds., *Laser Cooling and Trapping of Atoms* — An early special issue of major importance	41
H.C.W. Beijerinck and B. J. Verhaar, Eds., *Dynamics of Inelastic Collisions of Electronically Excited Atoms* — Proceedings of a Dutch conference	354
J. Mlynek, V. Balykin, and P. Meystre, Eds., *Optics and Interferometry with Atoms* — A special issue on Atom Optics	282
E. Arimondo and H-A. Bachor, Eds., *Special Issue on Atom Optics* — A collection of articles spanning a broad range of topics	355

Books or Large Conference Proceedings

Book	Ref.
V. Minogin and V. Letokhov, *Laser Light Pressure on Atoms* — A wide-ranging text that treats many high-intensity phenomena	356
A. Kazantzev, G. Surdutovich, and V. Yakolev, *Mechanical Action of Light on Atoms* — Thorough, formal treatment of force calculations and other topics	357
L. Moi, S. Gozzini, C. Gabanini, E. Arimondo, and F. Strumia, Eds., *Light Induced Kinetic Effects on Atoms, Ions, and Molecules* — Conference proceedings, commonly called LIKE	358
A. Arimondo, W. Phillips, and F. Strumia, Eds., *Proceedings of the Fermi School CXVII* — Proceedings of a major summer school with very many articles	359
V.I. Balykin and V.S. Letokhov, *Atom Optics with Laser Light* — Introduction to collimation, focussing, channeling and reflection by laser light	360
A. Aspect, W. Barletta, and R. Bonifacio, Eds., *Proceedings of the Fermi School CXXXI* — Proceedings of a major summer school with very many articles	361
P. Berman, Ed., *Atom Interferometry* — Several long excellent articles on atom optics	362

General Review Articles

Article	Ref.
C. Cohen-Tannoudji, *Laser Cooling and Trapping of Neutral Atoms — Theory* — A short, elegant theoretical summary of selected topics	363
H. Metcalf and P. van der Straten, *Cooling and Trapping of Neutral Atoms* — This article was the predecessor of this book	364
C.S. Adams, M. Sigel, and J. Mlynek, *Atom Optics* — Theoretical discussion of several topics in deBroglie wave optics	365
C.S. Adams and E. Riis *Laser Cooling and Trapping of Neutral Atoms* — A wonderful article with crystal-clear descriptions and hundreds of references	366

Specialized Review Articles

Article	Ref.
G. Nienhuis, *Impressed by Light: Mechanical Action of Radiation on Atomic Motion* — Very early formal treatment of optical forces, including light-induced drift	367
V. Balykin and V. Letokhov, *Laser Optics of Neutral Atomic-Beams* — A compact early description of several phenomena and ideas	205
P.S. Julienne, A.M. Smith, and K. Burnett, *Theory of Collisions Between Laser Cooled Atoms* — Cold collisions	262
T. Walker and P. Feng, *Measurements of Collisions between Laser-Cooled Atoms* — Cold collisions, emphasis on trap loss collisions	261
J. Weiner, *Advances in Ultracold Collisions: Experimentation and Theory* — Cold collisions, emphasis on excited-state collisions	263
A. Aspect, *Manipulation of Neutral Atoms — Experiments* — Short summary of some experimental results	368
J. Thomas and L. Wang, *Precision Position Measurement of Moving Atoms* — Atomic position measurement, but important for optical forces	369
H. Wallis, *Quantum-Theory of Atomic Motion in Laser-Light* — Thorough formal theoretical treatment of several important topics	370
J.P. Dowling and J. Gea-Banacloche, *Evanescent Light-Wave Atom Mirrors, Resonators Wave-Guides, and Traps* — Detailed description of evanescent wave effects	105
W. Ketterle and N.J. van Druten, *Evaporative Cooling of Trapped Atoms* — Evaporative cooling	199
P.S. Jessen and I.H. Deutsch, *Optical Lattices* — Optical Lattices	315

There are also many review articles to be found in the proceedings of major serial conferences. The most notable of these are "International Conference on Atomic Physics, ICAP" whose proceedings are named **Atomic Physics n**, and the "Laser Spectroscopy Conference" whose proceedings are named **Laser Spectroscopy n**.

Appendix C
Characteristic Data

In principle laser cooling can be used for any atom. In practice, it is necessary that the atom to be cooled has a transition for which enough laser power can be generated. Furthermore, one needs to scatter many thousands of photons on this transition, so spontaneous emission out of the excited state should be entirely to the ground or metastable state. Until now laser cooling has been used primarily for the following atoms:

metastable noble gas atoms	^{4}He*	^{20}Ne*	^{40}Ar*	^{84}Kr*	^{132}Xe*
alkali-metal atoms	^{7}Li	^{23}Na	^{39}K	^{85}Rb	^{133}Cs
alkaline-earth atoms		^{24}Mg	^{40}Ca	^{88}Sr	^{138}Ba

Characteristic values for these elements together with H and Cr are given in the following tables.

Atom	transition	I	λ (nm)	$\hbar\omega_a$ (eV)	τ (ns)	$\gamma/2\pi$ (MHz)
^{1}H	$1^2S_{1/2} - 2^2P_{3/2}$	$1/2$	121.57	10.199	1.60	99.58
^{4}He*	$2^3S_1 - 2^3P_2$	0	1083.33	1.144	98.04	1.62
^{4}He*	$2^3S_1 - 3^3P_2$	0	388.98	3.187	106.83	1.49
^{7}Li	$2^2S_{1/2} - 2^2P_{3/2}$	$3/2$	670.96	1.848	26.87	5.92
^{8}Be	$2^1S_0 - 2^1P_1$	0	234.93	5.277	1.46	108.88
^{20}Ne*	$3^3P_2 - 3^3D_3$	0	640.40	1.936	18.79	8.47
^{23}Na	$3^2S_{1/2} - 3^2P_{3/2}$	$3/2$	589.16	2.104	15.90	10.01
^{24}Mg	$3^1S_0 - 3^1P_1$	0	285.30	4.346	1.97	80.95
^{40}Ar*	$4^3P_2 - 4^3D_3$	0	811.75	1.527	27.09	5.87
^{39}K	$4^2S_{1/2} - 4^2P_{3/2}$	$3/2$	766.70	1.617	26.13	6.09
^{40}Ca	$4^1S_0 - 4^1P_1$	0	422.79	2.933	4.60	34.63
^{52}Cr	$a^7S_3 - z^7P_4$	0	425.55	2.913	31.77	5.01
^{84}Kr*	$5^3P_2 - 5^3D_3$	0	811.51	1.528	28.63	5.56
^{85}Rb	$5^2S_{1/2} - 5^2P_{3/2}$	$5/2$	780.24	1.589	26.63	5.98
^{88}Sr	$5^1S_0 - 5^1P_1$	0	460.86	2.690	4.98	31.99
^{132}Xe*	$6^3P_2 - 6^3D_3$	0	882.18	1.405	33.03	4.82
^{133}Cs	$6^2S_{1/2} - 6^2P_{3/2}$	$7/2$	852.35	1.455	30.70	5.18
^{138}Ba	$6^1S_0 - 6^1P_1$	0	553.70	2.239	8.68	18.33

TABLE C.1. Spectroscopic data for optical transitions used for laser cooling. Given are the atomic mass M, the wavelength λ, the transition energy $\hbar\omega_a$, the lifetime of the upper state τ, and the linewidth γ. In the case of metastable helium, values for two transitions are given.

Atom	σ_{ge} (10^{-15} m^2)	I_s (mW/cm^2)	a_{max} (10^6 m/s^2)	$\omega_r/2\pi$ (kHz)	ε (10^{-3})
^{1}H	7.1	7244.	1019.	13391.	134.5
^{4}He*	560.4	0.17	0.469	42.46	26.2
^{4}He*	72.2	3.31	1.199	329.35	221.0
^{7}Li	215.0	2.56	1.577	63.15	10.7
^{9}Be	26.4	1097.	64.444	400.98	3.68
^{20}Ne*	195.8	4.22	0.829	24.33	2.87
^{23}Na	165.7	6.40	0.926	24.99	2.50
^{24}Mg	38.9	455.	14.824	102.17	1.26
^{40}Ar*	314.6	1.44	0.227	7.57	1.29
^{39}K	280.7	1.77	0.256	8.71	1.43
^{40}Ca	85.3	59.9	2.569	27.92	0.81
^{52}Cr	86.5	8.50	0.284	21.20	4.23
^{84}Kr*	314.4	1.36	0.102	3.61	0.65
^{85}Rb	290.7	1.64	0.113	3.86	0.65
^{88}Sr	101.4	42.72	0.990	10.68	0.33
^{132}Xe*	371.6	0.92	0.052	1.94	0.40
^{133}Cs	346.9	1.09	0.057	2.07	0.40
^{138}Ba	146.4	14.12	0.301	4.72	0.26

TABLE C.2. Characteristic values for the excitation of different elements with laser light. Given are the cross section for absorption $\sigma_{ge} = 3\lambda^2/2\pi$ (see Eq. 2.28b), the saturation intensity $I_s = \pi hc/3\lambda^3\tau$ (see Eq. 2.24c), the maximum acceleration a_{max}, the recoil frequency ω_r and the ratio $\varepsilon = \omega_r/\gamma$. The values for the cross section and the saturation intensity apply for the strongest transition between magnetic sublevels. In most cases the maximum obtainable acceleration is of the order of 10^5–10^6 m/s^2.

Atom	Capture limit		Doppler limit		Recoil limit	
	v_c	T_c	v_D	T_D	v_r	T_r
	(m/s)	(mK)	(cm/s)	(μK)	(cm/s)	(μK)
^{1}H	12.11	17.77	443.	2389.	325.	1285.
^{4}He*	1.76	1.49	28.44	38.95	9.200	4.075
^{4}He*	0.58	0.16	27.25	35.75	25.6	31.61
^{7}Li	3.97	13.33	41.03	142.11	8.474	6.061
^{9}Be	25.58	709.4	155.23	2612.	18.8	38.48
^{20}Ne*	5.43	70.80	29.07	203.29	3.116	2.335
^{23}Na	5.90	96.18	29.47	240.18	2.945	2.399
^{24}Mg	23.09	1539.	82.04	1942.	5.830	9.80
^{40}Ar*	4.77	109.33	17.12	140.96	1.230	0.727
^{39}K	4.67	102.23	17.66	146.16	1.335	0.836
^{40}Ca	14.64	1031.	41.57	831.	2.361	2.680
^{52}Cr	2.13	28.41	13.87	120.23	1.805	2.035
^{84}Kr*	4.51	205.47	11.50	133.40	0.586	0.346
^{85}Rb	4.66	222.12	11.85	143.41	0.602	0.370
^{88}Sr	14.74	2299.	26.94	768.	0.985	1.025
^{132}Xe*	4.25	286.83	8.54	115.64	0.343	0.186
^{133}Cs	4.42	312.14	8.82	124.39	0.352	0.198
^{138}Ba	10.15	1710.	16.28	439.96	0.522	0.453

TABLE C.3. Limiting values for the velocity and temperature for laser cooling of different elements. Values for the velocity v and temperature T are given for the capture, Doppler and recoil limit.

Element	Abundance	I	F_g ($J_g=1/2$)	A (MHz)	ΔE_{fs} (GHz)	F_e ($J_e=1/2$)	A (MHz)	F_e ($J_e=3/2$)	A (MHz)	B (MHz)
^{1}H	99.985	1/2	0,1	1420.405	10.968	0,1	59.18	1,2	23.67	—
^{6}Li	7.5	1	1/2,3/2	152.137		1/2,3/2	17.375	1/2,3/2,5/2	−1.155	−0.10
^{7}Li	92.5	3/2	1,2	401.752	10.091	1,2	45.914	0,1,2,3	−3.055	−0.221
^{23}Na	100	3/2	1,2	885.813	515.53	1,2	94.3	0,1,2,3	18.69	2.90
^{39}K	93.26	3/2	1,2	230.859	1730.4	1,2	28.85	0,1,2,3	6.06	2.83
^{40}K	0.0117	4	7/2,9/2	−285.731		7/2,9/2	—	5/2,7/2,9/2,11/2	−7.59	−3.5
^{41}K	6.73	3/2	1,2	127.007		1,2	—	0,1,2,3	3.40	3.34
^{85}Rb	72.17	5/2	2,3	1011.910	7123.0	2,3	120.72	1,2,3,4	25.009	25.88
^{87}Rb	27.83	3/2	1,2	3417.341		1,2	406.2	0,1,2,3	84.845	12.52
^{133}Cs	100	7/2	3,4	2298.157	16611.8	3,4	291.90	2,3,4,5	50.34	−0.38

TABLE C.4. Fine- and hyperfine structure constants for the various alkali-metal atoms. The parameters A and B can be used in Eqs. 4.2 and 4.3 to calculate the shift and the splitting from the hyperfine interaction. The values for A and B are from Ref. 28.

Element	Metastable level	Exc. energy (eV)	IP (eV)	Lifetime (s)
He	2^3S_1	19.82	24.580	7900
Ne	$3s[\frac{3}{2}]_2(^3P_2)$	16.62	21.559	20
Ar	$4s[\frac{3}{2}]_2(^3P_2)$	11.55	15.755	60
Kr	$5s[\frac{3}{2}]_2(^3P_2)$	9.915	13.996	85
Xe	$6s[\frac{3}{2}]_2(^3P_2)$	8.315	12.127	150

TABLE C.5. Constants for the various metastable noble gas atoms.

Appendix D
Transition Strengths

The following diagrams show the transition strength for alkali-metal atoms for optical transitions from the ground state $n^2 S_{1/2}$ to the first excited states $n^2 P_{1/2,3/2}$. Since most alkali-metal atoms have a half-integer nuclear spin I, the diagrams are in the order $I = 1/2$, $3/2$, $5/2$ and $7/2$. The diagrams can be used for

I	Element				
$1/2$	^{1}H				
1	^{6}Li				
$3/2$	^{7}Li	^{23}Na	^{39}K	^{41}K	^{87}Rb
$5/2$	^{85}Rb				
$7/2$	^{133}Cs				

For each value of I, diagrams are shown for the D_1-line (left page) and D_2-line (right page). The diagram at the top of each page is for π-polarization, whereas the diagram on the bottom is for σ^+-polarization. The transition strength for σ^--polarization can be found by using the diagram for σ^+-polarization and replacing all M's by $-M$. The transition strength is normalized for each line so that the strength of the weakest allowed transition becomes an integer. The strength is calculated using the square of μ_{eg} in Eq. 4.33. In order to compare the strength of the D_1-line with the D_2-line, the numbers for the D_1-line have to be multiplied by a factor $2(I = 1/2)$, $10(I = 1)$, $5(I = 3/2)$, $140(I = 5/2)$, or $105(I = 7/2)$. The diagram for He* can be found on pg. 54.

D_1-line: $^2S_{1/2}$-$^2P_{1/2}$

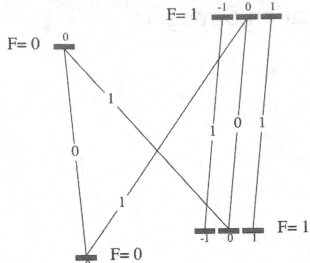
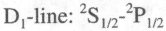

I=1/2

π-pol

D_1-line: $^2S_{1/2}$-$^2P_{1/2}$

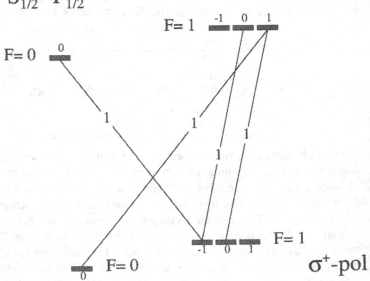

I=1/2

σ^+-pol

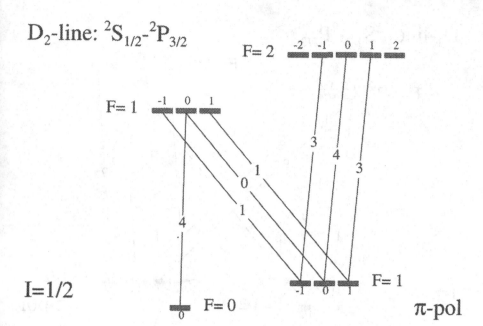

D_2-line: $^2S_{1/2}$-$^2P_{3/2}$

$I=1/2$

π-pol

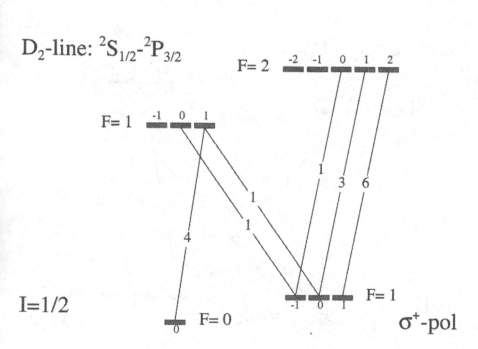

D_2-line: $^2S_{1/2}$-$^2P_{3/2}$

$I=1/2$

σ^+-pol

D_1-line: $^2S_{1/2}$-$^2P_{1/2}$

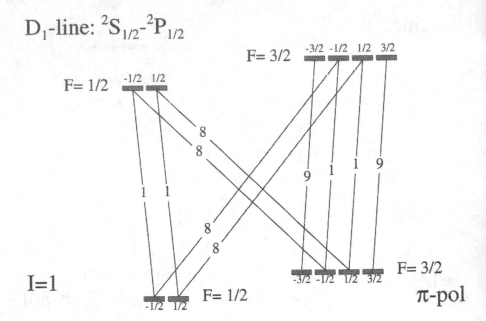

I=1

π-pol

D_1-line: $^2S_{1/2}$-$^2P_{1/2}$

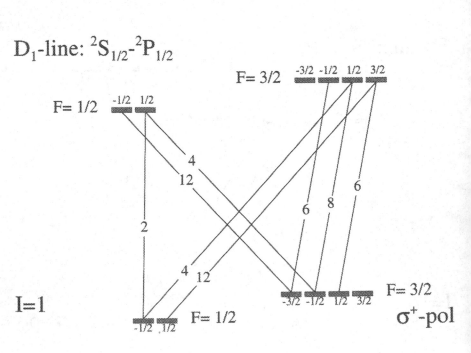

I=1

σ^+-pol

D$_2$-line: ^{2}S$_{1/2}$-^{2}P$_{3/2}$

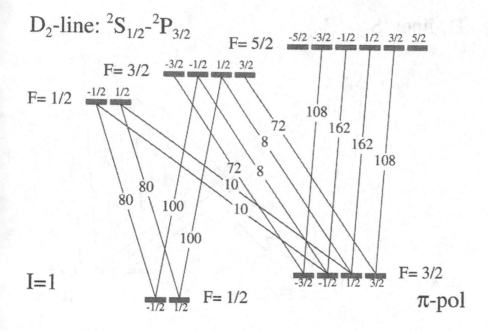

I=1

π-pol

D$_2$-line: ^{2}S$_{1/2}$-^{2}P$_{3/2}$

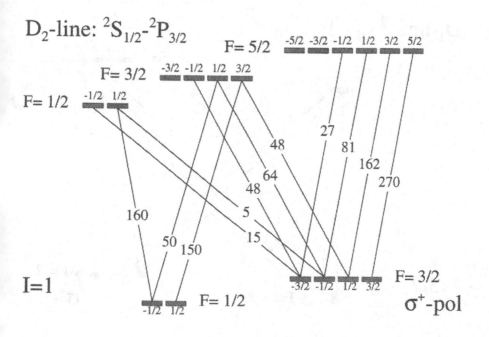

I=1

σ$^+$-pol

D_1-line: $^2S_{1/2}$-$^2P_{1/2}$

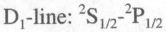

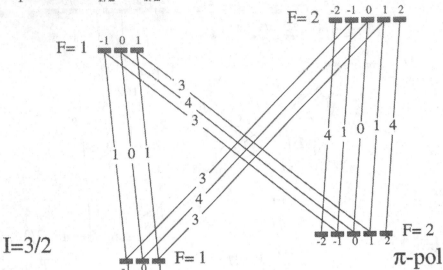

$I=3/2$

π-pol

D_1-line: $^2S_{1/2}$-$^2P_{1/2}$

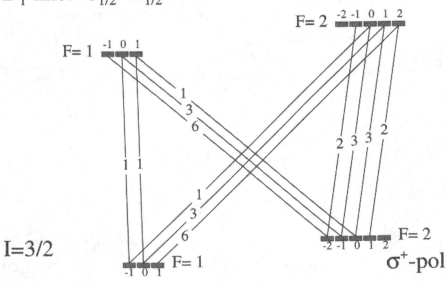

$I=3/2$

σ^+-pol

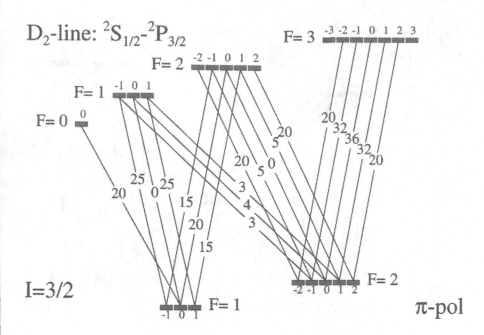

D_2-line: $^2S_{1/2}$-$^2P_{3/2}$

π-pol

$I=3/2$

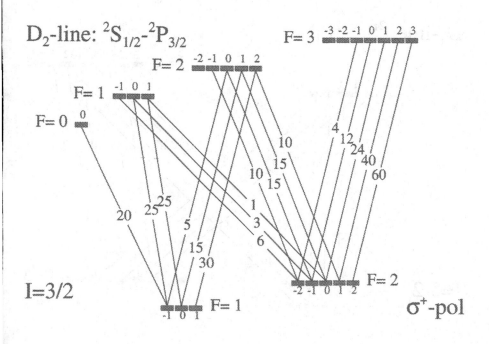

D_2-line: $^2S_{1/2}$-$^2P_{3/2}$

σ^+-pol

$I=3/2$

D_1-line: $^2S_{1/2}$-$^2P_{1/2}$

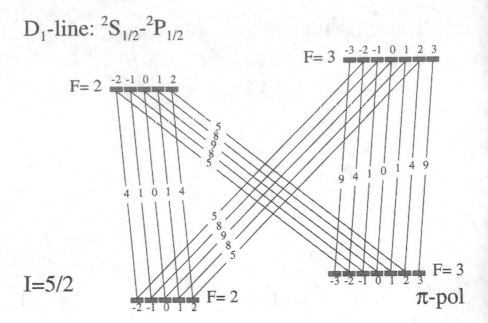

I=5/2

π-pol

D_1-line: $^2S_{1/2}$-$^2P_{1/2}$

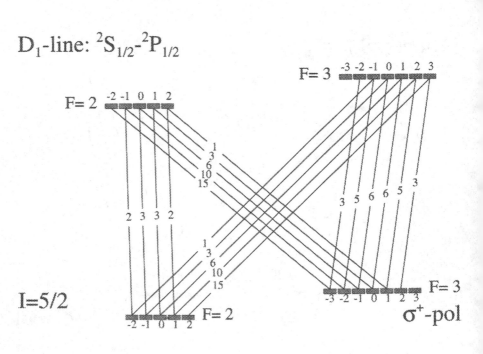

I=5/2

σ⁺-pol

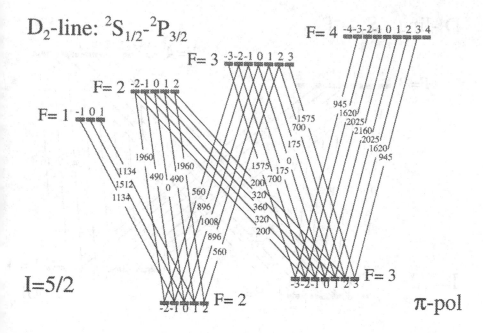

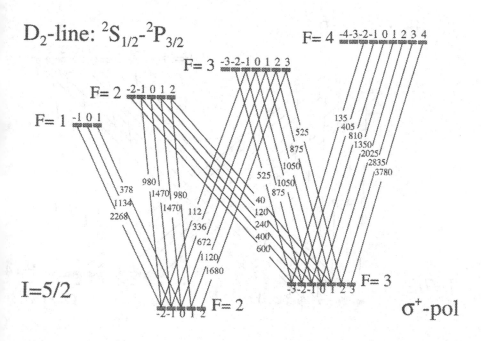

D_1-line: $^2S_{1/2}$-$^2P_{1/2}$

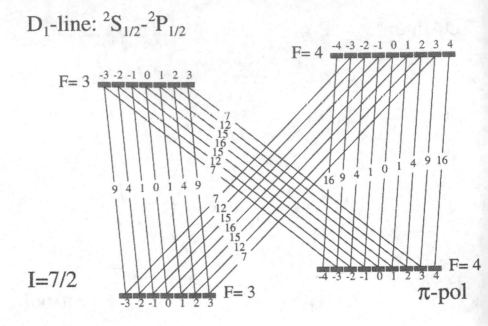

I=7/2

π-pol

D_1-line: $^2S_{1/2}$-$^2P_{1/2}$

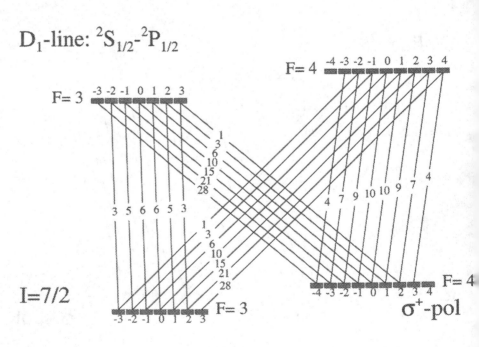

I=7/2

σ⁺-pol

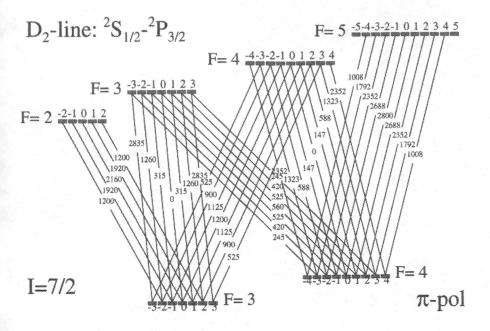

D_2-line: $^2S_{1/2}$-$^2P_{3/2}$

I=7/2

π-pol

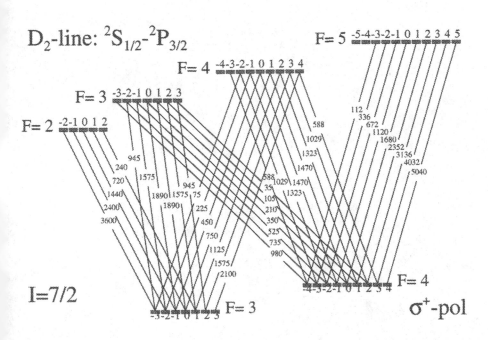

D_2-line: $^2S_{1/2}$-$^2P_{3/2}$

I=7/2

σ^+-pol

References

[1] L.D. Landau and E.M Lifshitz. *Quantum Mechanics (Non-Relativistic Theory)*. Pergamon Press, Oxford (1958).

[2] E. Merzbacher. *Quantum Mechanics*. Wiley & Sons, New York (1961).

[3] L.I. Schiff. *Quantum Mechanics*. McGraw-Hill, New York (1968).

[4] M. Sargent III, M.O. Scully, and Jr. W.E. Lamb. *Laser Physics*. Addison-Wesley, Reading (1974).

[5] C. Cohen-Tannoudji, B. Diu, and F. Laloë. *Quantum Mechanics*. Wiley, New York (1977).

[6] B.H. Bransden and C.J. Joachain. *Physics of Atoms and Molecules*. Wiley & Sons, New York (1983).

[7] S. Stenholm. *Foundations of Laser Spectroscopy*. J. Wiley & Sons, New York (1984).

[8] B.H. Bransden and C.J. Joachain. *Introduction to Quantum Mechanics*. Longman, New York (1989).

[9] D.J. Griffiths. *Introduction to Quantum Mechanics*. Prentice Hall, New Jersey (1995).

[10] I.I. Rabi. Space Quantization in a Gyrating Magnetic Field. *Phys. Rep.* **51**, 652 (1937).

[11] N.F. Ramsey. *Molecular Beams*. Clarendon Press, Oxford (1956).

[12] M.H. Mittleman. *Introduction to the Theory of Laser-Atom Interactions*. Plenum Press, New York (1982).

[13] D. Suter. *The Physics of Laser-Atom Interactions*. Cambridge University Press, Cambridge (1997).

[14] L. Allen and J.H. Eberly. *Optical Resonance and Two-Level Atoms*. Dover, New York (1975).

[15] R. Feynman, F. Vernon, and R. Hellwarth. Geometrical Representation of the Schrödinger Equation for Solving Maser Problems. *J. App. Phys.* **28**, 49 (1957).

[16] G.E. Pake. *Paramagnetic Resonance*. Benjamin, New York (1962).

[17] C.R. Ekstrom, C. Kurtsiefer, D. Voigt, O. Dross, T. Pfau, and J. Mlynek. Coherent Excitation of a He* Beam Observed in Atomic Momentum Distributions. *Opt. Commun.* **123**, 505 (1996).

[18] A. Einstein. On the Quantum Theory of Radiation. *Physik. Zeit.* **18**, 121 (1917).

[19] K. Molmer, Y. Castin, and J. Dalibard. Monte-Carlo Wave-Function Method in Quantum Optics. *J. Opt. Soc. Am. B* **10**, 524-538 (1993).

[20] K. Blum. *Density Matrix: Theory and Applications*. Plenum, New York (1981).

[21] C. Cohen-Tannoudji, J. Dupont-Roc, and G. Grynberg. *Photons and Atoms, Introduction to Quantum Electrodynamics*. John Wiley & Sons, New York (1989).

[22] R. Loudon. *The Quantum Theory of Light*. Clarendon Press, Oxford (1973).

[23] W. Heitler. *The Quantum Theory of Radiation*. Oxford Univ. Press, London (1954). Also published by Dover, New York, 1984 (paperback).

[24] V. Weisskopf and E. Wigner. Berechnung der natürlichen Linienbreite auf Grund der Diracschen Lichttheorie. *Zeit. f. Phys.* **63**, 54 (1930).

[25] C. Cohen-Tannoudji. Atoms in Strong Resonant Fields. In Balian *et al.*, editor, *Proceedings of Les Houches XXVII*, page 3, Amsterdam (1977). North Holland.

[26] P. Ehrenfest. Bemerkung über die angenäherte Gültigkeit der klassischen Mechanik innerhalb der Quantummechanik. *Zeit. f. Phys.* **45**, 455 (1927).

[27] J. Gordon and A. Ashkin. Motion of Atoms in a Radiation Trap. *Phys. Rev. A* **21**, 1606 (1980).

[28] E. Arimondo, M. Inguscio, and P. Violino. Experimental Determinations of the Hyperfine Structure in the Alkali Atoms. *Rev. Mod. Phys.* **49**, 31 (1977).

[29] A.R. Edmonds. *Angular Momentum in Quantum Mechanics.* Princeton University Press, Princeton (1957).

[30] H.A. Bethe and E.E. Salpeter. *Quantum Mechanics of One- and Two-Electron Atoms.* Springer-Verlag, Berlin (1957). Also published by Plenum, New York, 1977 (paperback).

[31] M. Abramowitz and I.A. Stegun. *Handbook of Mathematical Functions.* United States Department of Commerce, Washington (1964).

[32] D.R. Bates and A. Damgaard. The Calculation of the Absolute Strenghts of Spectral Lines. *Phil. Tr. R. Soc. A* **242**, 101 (1949).

[33] M. Rotenberg, N. Metropolis, R. Birins, and J. Wooten Jr. *The 3j and 6j Symbols.* Technology Press, Cambridge (1959).

[34] S.J. van Enk and G. Nienhuis. Entropy Production and Kinetic Effects of Light. *Phys. Rev. A* **46**, 1438-1448 (1992).

[35] I.S. Gradshteyn and I.M. Ryzhik. *Table of Integrals, Series and Products.* Academic Press, New York (1965).

[36] H. Risken. *The Fokker-Planck Equation.* Springer, Berlin (1984).

[37] D. Wineland and H. Dehmelt. Proposed $10^{14}\delta v/v$ Laser Fluorescence Spectroscopy on Tl^+ Mono-Ion Oscillator. *Bull. Am. Phys. Soc.* **20**, 637 (1975).

[38] T. Hansch and A. Schawlow. Cooling of Gases by Laser Radiation. *Opt. Commun.* **13**, 68-71 (1975).

[39] D. Wineland and W. Itano. Laser Cooling of Atoms. *Phys. Rev. A* **20**, 1521-1540 (1979).

[40] P. Meystre and S. Stenholm (Eds.). The Mechanical Effects of Light. *J. Opt. Soc. Am. B* **2**, 1705-1872 (1985).

[41] S. Chu and C. Wieman (Eds.). Laser Cooling and Trapping of Atoms. *J. Opt. Soc. Am. B* **6**, 1961-2288 (1989).

[42] W. Phillips and H. Metcalf. Laser Deceleration of an Atomic Beam. *Phys. Rev. Lett.* **48**, 596 (1982).

[43] J. Prodan, W. Phillips, and H. Metcalf. Laser Production of a Very Slow Monoenergetic Atomic Beam. *Phys. Rev. Lett.* **49**, 1149 (1982).

[44] T.E. Barrett, S.W. Dapore-Schwartz, M.D. Ray, and G.P. Lafyatis. Slowing Atoms with (σ^-)-Polarized Light. *Phys. Rev. Lett.* **67**, 3483-3487 (1991).

[45] M. Prentiss and A. Cable. Slowing and Cooling an Atomic-Beam Using an Intense Optical Standing Wave. *Phys. Rev. Lett.* **62**, 1354-1357 (1989).

[46] V.S. Bagnato, A. Aspect, and S.C. Zilio. Study of Laser Deceleration of an Atomic-Beam by Monitoring the Fluorescence Along the Deceleration Path. *Opt. Commun.* **72**, 76-81 (1989).

[47] J. Prodan and W. Phillips. Chirping the Light Fantastic — Recent NBS Atom Cooling Experiemnts. *Prog. Quant. Elect.* **8**, 231 (1984).

[48] W. Ertmer, R. Blatt, J.L.Hall, and M. Zhu. Laser Manipulation of Atomic Beam Velocities: Demonstration of Stopped Atoms and Velocity Reversal. *Phys. Rev. Lett.* **54**, 996 (1985).

[49] R. Watts and C. Wieman. Manipulating Atomic Velocities Using Diode Lasers. *Opt. Lett.* **11**, 291 (1986).

[50] V. Bagnato, G. Lafyatis, A. Martin, E. Raab, R. Ahmad-Bitar, and D. Pritchard. Continuous Stopping and Trapping of Neutral Atoms. *Phys. Rev. Lett.* **58**, 2194 (1987).

[51] R.J. Napolitano, S.C. Zilio, and V.S. Bagnato. Adiabatic Following Conditions for the Deceleration of Atoms with the Zeeman Tuning Technique. *Opt. Commun.* **80**, 110-114 (1990).

[52] R. Gaggl, L. Windholz, C. Umfer, and C. Neureiter. Laser Cooling of a Sodium Atomic-Beam Using the Stark-Effect. *Phys. Rev. A* **49**, 1119-1121 (1994).

[53] J.R. Yeh, B. Hoeling, and R.J. Knize. Longitudinal and Transverse Cooling of a Cesium Atomic-Beam Using the D1 Transition with Stark-Effect Frequency Compensation. *Phys. Rev. A* **52**, 1388-1393 (1995).

[54] W. Ketterle, A. Martin, M.A. Joffe, and D.E. Pritchard. Slowing and Cooling Atoms in Isotropic Laser-Light. *Phys. Rev. Lett.* **69**, 2483-2486 (1992).

[55] H. Batelaan, S. Padua, D.H. Yang, C. Xie, R. Gupta, and H. Metcalf. Slowing of ^{85}Rb Atoms with Isotropic Light. *Phys. Rev. A* **49**, 2780-2784 (1994).

[56] T.G. Aardema, R.M.S. Knops, S.P.L. Nijsten, K.A.H. van Leeuwen, J.P.J. Driessen, and H.C.W. Beijerinck. Transverse Diffusion in Isotropic Light Slowing. *Phys. Rev. Lett.* **76**, 748-751 (1996).

[57] L. Moi. Application of a Very Long Cavity Laser to Atom Slowing Down and Optical Pumping. *Opt. Commun.* **50**, 349 (1984).

[58] J. Hoffnagle. Proposal for Continuous White-Light Cooling of an Atomic Beam. *Opt. Lett.* **13**, 102 (1988).

[59] M. Zhu, C.W. Oates, and J.L. Hall. Continuous High-Flux Monovelocity Atomic-Beam Based on a Broad-Band Laser-Cooling Technique. *Phys. Rev. Lett.* **67**, 46-49 (1991).

[60] I.C.M. Littler, H.M. Keller, U. Gaubatz, and K. Bergmann. Velocity Control and Cooling of an Atomic-Beam Using a Modeless Laser. *Z. Phys. D* **18**, 307-308 (1991).

[61] R. Calabrese, V. Guidi, P. Lenisa, E. Mariotti, and L. Moi. Transverse Laser Cooling of Ions in a Storage Ring. *Opt. Commun.* **123**, 530-534 (1996).

[62] S.N. Atutov, R. Calabrese, R. Grimm, V. Guidi, I. Lauer, P. Lenisa, V. Luger, E. Mariotti, and L. Moi. "White-light" Laser Cooling of a Fast Stored Ion Beam. *Phys. Rev. Lett.* **80**, 2129 (1998).

[63] T. Breeden and H. Metcalf. Stark Acceleration of Rydberg Atoms in Inhomogeneous Electric Fields. *Phys. Rev. Lett.* **47**, 1726 (1981).

[64] G. Stevens, M. Widmer, F. Tudorica, C-H. Iu, and H. Metcalf. Coherent Excitation of Lithium to Rydberg States and Application to Atom Optics. *Bull. Am. Phys. Soc.* **41**, 1121 (1996).

[65] P.A. Molenaar, P. van der Straten, H.G.M. Heideman, and H. Metcalf. Diagnostic-Technique for Zeeman-Compensated Atomic-Beam Slowing — Technique and Results. *Phys. Rev. A* **55**, 605-614 (1997).

[66] R.J. Napolitano and V.S. Bagnato. The Effect of a Fluctuating Laser Field on the Process of Decelerating Atoms in the Zeeman Tuning Technique. *J. Mod. Opt.* **40**, 329-335 (1993).

[67] J. Dalibard and W. Phillips. Stability and Damping of Radiation Pressure Traps. *Bull. Am. Phys. Soc.* **30**, 748 (1985).

[68] S. Chu, L. Hollberg, J. Bjorkholm, A. Cable, and A. Ashkin. Three-Dimensional Viscous Confinement and Cooling of Atoms by Resonance Radiation Pressure. *Phys. Rev. Lett.* **55**, 48 (1985).

[69] P.D. Lett, R.N. Watts, C.E. Tanner, S.L. Rolston, W.D. Phillips, and C.I. Westbrook. Optical Molasses. *J. Opt. Soc. Am. B* **6**, 2084-2107 (1989).

[70] D. Sesko, C. Fan, and C. Wieman. Production of a Cold Atomic Vapor Using Diode-Laser Cooling. *J. Opt. Soc. Am. B* **5**, 1225 (1988).

[71] A. Aspect, N. Vansteenkiste, R. Kaiser, H. Haberland, and M. Karrais. Preparation of a Pure Intense Beam of Metastable Helium by Laser Cooling. *Chem. Phys.* **145**, 307-315 (1990).

[72] E. Riis, D.S. Weiss, K.A. Moler, and S. Chu. Atom Funnel for the Production of a Slow, High-Density Atomic-Beam. *Phys. Rev. Lett.* **64**, 1658-1661 (1990).

[73] M.R. Williams, M.J. Bellanca, L. Liu, C. Xie, W.F. Buell, T.H. Bergeman, and H.J. Metcalf. Atom Cooling in One Dimension with High-Intensity Laser Light. *Phys. Rev. A* **57**, 401 (1998).

[74] J. Umezu and F. Shimizu. Laser Cooling of an Atomic Beam by Spatial Doppler Tuning of a Resonance Transition. *Jpn. J. App. Phys.* **24**, 1655 (1985).

[75] M. Hoogerland. *Laser Manipulation of Metastable Neon Atoms*. Ph.D. thesis, Eindhoven University (1993).

[76] B. Sheehy, S.Q. Shang, P. van der Straten, and H. Metcalf. Collimation of a Rubidium Beam Below the Doppler Limit. *Chem. Phys.* **145**, 317-325 (1990).

[77] B. Sheehy, S.Q. Shang, R. Watts, H. Metcalf, and S. Hatamian. Diode-Laser Deceleration and Collimation of a Rubidium Beam. *J. Opt. Soc. Am. B* **6**, 2165-2170 (1989).

[78] T. Esslinger, A. Hemmerich, and T.W. Hansch. Imaging an Atomic-Beam in 2 Dimensions. *Opt. Commun.* **93**, 49-53 (1992).

[79] P. Gould, P. Lett, and W.D. Phillips. New Measurement with Optical Molasses. In W. Persson and S. Svanberg, editors, *Laser Spectroscopy VIII*, page 64, Berlin (1987). Springer.

[80] T. Hodapp, C. Gerz, C. Westbrook, C. Furtlehner, and W. Phillips. Diffusion in Optical Molasses. *Bull. Am. Phys. Soc.* **37**, 1139 (1992).

[81] C. Cohen-Tannoudji and W.D. Phillips. New Mechanisms for Laser Cooling. *Phys. Today* **43**, October, 33-40 (1990).

[82] P. Lett, R. Watts, C. Westbrook, W. Phillips, P. Gould, and H. Metcalf. Observation of Atoms Laser Cooled below the Doppler Limit. *Phys. Rev. Lett.* **61**, 169 (1988).

[83] J. Dalibard and C. Cohen-Tannoudji. Laser Cooling Below the Doppler Limit by Polarization Gradients — Simple Theoretical-Models. *J. Opt. Soc. Am. B* **6**, 2023-2045 (1989).

[84] P.J. Ungar, D.S. Weiss, S. Chu, and E. Riis. Optical Molasses and Multilevel Atoms — Theory. *J. Opt. Soc. Am. B* **6**, 2058-2071 (1989).

[85] B. Sheehy, S.Q. Shang, P. van der Straten, S. Hatamian, and H. Metcalf. Magnetic-Field-Induced Laser Cooling Below the Doppler Limit. *Phys. Rev. Lett.* **64**, 858-861 (1990).

[86] D.S. Weiss, E. Riis, S. Chu, P.J. Ungar, and Y. Shevy. Optical Molasses and Multilevel Atoms — Experiment. *J. Opt. Soc. Am. B* **6**, 2072-2083 (1989).

[87] J. Dalibard and C. Cohen-Tannoudji. Dressed Atom Approach to Atomic Motion in Laser Light: The Dipole Force Revisited. *J. Opt. Soc. Am. B* **2**, 1707 (1985).

[88] A. Aspect, J. Dalibard, A. Heidman, C. Salomon, and C. Cohen-Tannoudji. Cooling Atoms with Stimulated Emission. *Phys. Rev. Lett.* **57**, 1688 (1986).

[89] J. Dalibard. New Schemes in Laser Cooling. In S. Haroche, J-C. Gay, and G. Grynberg, editors, *Atomic Physics XI*, page 199, Singapore (1989). World Scientific.

[90] S.Q. Shang, B. Sheehy, P. van der Straten, and H. Metcalf. Sub-Doppler Laser Cooling in a Magnetic Field. In R. Lewis and J. Zorn, editors, *Atomic Physics XII*, pages 105–115, Singapore (1991). World Scientific.

[91] A. Kastler. Champ Lumineux a Structure Helicoidale dans un Cavite Laser. Possibilite d'Imprimer cette Structure Helicoidale a un Milieu Materiel Transparente Isotrope. *Compt. Rend. Acad. Scien., Paris* **271**, 999 (1971).

[92] G. Nienhuis, P. van der Straten, and S.Q. Shang. Operator Description of Laser Cooling Below the Doppler Limit. *Phys. Rev. A* **44**, 462-474 (1991).

[93] S.Q. Shang, B. Sheehy, P. van der Straten, and H. Metcalf. Velocity-Selective Magnetic-Resonance Laser Cooling. *Phys. Rev. Lett.* **65**, 317-320 (1990).

[94] P. van der Straten, S.Q. Shang, B. Sheehy, H. Metcalf, and G. Nienhuis. Laser Cooling at Low-Intensity in a Strong Magnetic-Field. *Phys. Rev. A* **47**, 4160-4175 (1993).

[95] C. Salomon, J. Dalibard, W.D. Phillips, A. Clairon, and S. Guellati. Laser Cooling of Cesium Atoms Below 3 μK. *Europhys. Lett.* **12**, 683-688 (1990).

[96] C. Gerz, T.W. Hodapp, P. Jessen, K.M. Jones, W.D. Phillips, C.I. Westbrook, and K. Molmer. The Temperature of Optical Molasses for 2 Different Atomic Angular Momenta. *Europhys. Lett.* **21**, 661-666 (1993).

[97] K. Molmer. Friction and Diffusion-Coefficients for Cooling of Atoms in Laser Fields with Multidimensional Periodicity. *Phys. Rev. A* **44**, 5820-5832 (1991).

[98] J. Javanainen. Numerical Experiments in Semiclassical Laser-Cooling Theory of Multistate Atoms. *Phys. Rev. A* **46**, 5819-5835 (1992).

[99] Y. Castin, J. Dalibard, and C. Cohen-Tannoudji. The Limits of Sisyphus Cooling. In Moi et al. [358], page 5.

[100] M. Kasevich and S. Chu. Laser Cooling Below a Photon Recoil with 3-Level Atoms. *Phys. Rev. Lett.* **69**, 1741-1744 (1992).

[101] V. Minogin and O. Serimaa. Resonant Light Pressure Forces in a Strong Standing Laser Wave. *Opt. Commun.* **30**, 373 (1979).

[102] E. Kyrola and S. Stenholm. Velocity Tuned Resonances as Multi-Doppleron Processes. *Opt. Commun.* **22**, 123 (1977).

[103] S.Q. Shang, B. Sheehy, H. Metcalf, P. van der Straten, and G. Nienhuis. Velocity-Selective Resonances and Sub-Doppler Laser Cooling. *Phys. Rev. Lett.* **67**, 1094-1097 (1991).

[104] R. J. Cook and R.K. Hill. An Electromagnetic Mirror for Neutral Atoms. *Opt. Commun.* **43**, 258 (1982).

[105] J.P. Dowling and J. Gea-Banacloche. Evanescent Light-Wave Atom Mirrors, Resonators Wave-Guides, and Traps. *Adv. Atom. Mol. Opt. Phys.* **37**, 1-94 (1996).

[106] V.I. Balykin, V.S. Lethokov, Yu.B. Ovchinnikov, and A.I. Sidorov. Reflection of an Atomic Beam from a Gradient of an Optical Field. *JETP Lett.* **45**, 353-356 (1987).

[107] V. Balykin, V. Letokhov, Yu.B. Ovchinnikov, and A. Siderov. Quantum-State-Selective Mirror Reflection of Atoms by Laser Light. *Phys. Rev. Lett.* **60**, 2137 (1988).

[108] M.A. Kasevich, D.S. Weiss, and S. Chu. Normal-Incidence Reflection of Slow Atoms from an Optical Evanescent Wave. *Opt. Lett.* **15**, 607-609 (1990).

[109] Yu.B. Ovchinnikov, I. Manek, and R. Grimm. Surface Trap for Cs Atoms Based on Evanescent-Wave Cooling. *Phys. Rev. Lett.* **79**, 2225 (1997).

[110] D. Pritchard. Cooling Neutral Atoms in a Magnetic Trap for Precision Spectroscopy. *Phys. Rev. Lett.* **51**, 1336 (1983).

[111] Yu.B. Ovchinnikov, J. Söding, and R. Grimm. Cooling Atoms in Dark Gravitational Laser Traps. *JETP Lett.* **61**, 21-26 (1995).

[112] P. Desbiolles, M. Arndt, P. Szriftgiser, and J. Dalibard. Elementary Sisyphus Process Close to a Dielectric Surface. *Phys. Rev. A* **54**, 4292-4298 (1996).

[113] Yu.B. Ovchinnikov, D.V. Laryushin, V.I. Balykin, and V.S. Letokhov. Cooling of Atoms on Reflection from a Surface Light-Wave. *JETP Lett.* **62**, 113-118 (1995).

[114] C. Tanner, B. Masterson, and C. Wieman. Atomic Beam Collimation Using a Laser Diode with a Self-Locking Power-Buildup Cavity. *Opt. Lett.* **13**, 357 (1988).

[115] J.J. Tollett, J. Chen, J.G. Story, N.W.M. Ritchie, C.C. Bradley, and R.G. Hulet. Observation of Velocity-Tuned Multiphoton Doppleron Resonances in Laser-Cooled Atoms. *Phys. Rev. Lett.* **65**, 559-562 (1990).

[116] A. P. Kazantzev and I.V.Krasnov. Rectification of the Gradient Force of Resonant Radiation Pressure. *JETP Lett.* **46**, 420 (1987).

[117] R. Grimm, Yu.B. Ovchinnikov, A.I. Sidorov, and V.S. Letokhov. Observation of a Strong Rectified Dipole Force in a Bichromatic Standing Light-Wave. *Phys. Rev. Lett.* **65**, 1415-1418 (1990).

[118] J. Söding, R. Grimm, Yu.B. Ovchinnikov, P. Bouyer, and C. Salomon. Short-Distance Atomic-Beam Deceleration with a Stimulated Light Force. *Phys. Rev. Lett.* **78**, 1420-1423 (1997).

[119] R. Grimm, G. Wasik, J. Söding, and Yu.B. Ovchinnikov. Laser Cooling and Trapping With Rectified Optical Dipole Forces. In A. Aspect, W. Barletta, and R. Bonifacio, editors, *Proceedings of the Fermi School CXXXI*, page 481, Amsterdam (1996). IOS Press.

[120] A. Goepfert, I. Bloch, D. Haubrich, F. Lison, R. Schütze, R. Wynands, and D. Meschede. Stimulated Focussing and Deflection of an Atomic Beam using Picosend Laser Pulses. *Phys. Rev. A* **56**, R3354 (1997).

[121] R. Grimm, J. Söding, and Yu.B. Ovchinnikov. Trapping Atoms by Rectified Forces in Bichromatic Optical Superlattices. *JETP Lett.* **61**, 367-372 (1995).

[122] J. Kawanaka, M. Hagiuda, K. Shimizu, F. Shimizu, and H. Takuma. Generation of an Intense Low-Velocity Metastable-Neon Atomic-Beam. *App. Phys. B* **56**, 21-24 (1993).

[123] H. Mastwijk. *Cold Collisions of Metastable Helium Atoms.* Ph.D. thesis, Utrecht University (1997).

[124] S. Chernikov, J. Taylor, N. Platonov, V. Gapontsev, P-J. Nacher, G. Tastevin, M. Leduc, and M. Barlow. 1083 nm Ytterbium-Doped Fiber Amplifier for Optical Pumping of Helium. *Electr. Lett.* **33**, 787-788 (1997).

[125] D. Wineland, W. Itano, J. Bergquist, and J. Bollinger. Trapped Ions and Laser Cooling. Technical Report 1086, N.I.S.T. (1985).

[126] A. Migdall, J. Prodan, W. Phillips, T. Bergeman, and H. Metcalf. First Observation of Magnetically Trapped Neutral Atoms. *Phys. Rev. Lett.* **54**, 2596 (1985).

[127] W. Wing. On Neutral Particle Trapping in Quasistatic Electromagnetic Fields. *Prog. Quant. Elect.* **8**, 181 (1984).

[128] T.H. Bergeman, N.L. Balazs, H. Metcalf, P. Mcnicholl, and J. Kycia. Quantized Motion of Atoms in a Quadrupole Magnetostatic Trap. *J. Opt. Soc. Am. B* **6**, 2249-2256 (1989).

[129] H. Friedburg and W. Paul. Optical Presentation with Neutral Atoms. *Naturwissenschaft* **38**, 159 (1951).

[130] H. Friedburg. Optische Abbildung mit Neutralen Atomen. *Zeit. f. Phys.* **130**, 493 (1951).

[131] T. Bergeman, G. Erez, and H. Metcalf. Magnetostatic Trapping Fields for Neutral Atoms. *Phys. Rev. A* **35**, 1535 (1987).

[132] A. Martin, K. Helmerson, V. Bagnato, G. Lafyatis, and D. Pritchard. rf Spectroscopy of Trapped Neutral Atoms. *Phys. Rev. Lett.* **61**, 2431 (1988).

[133] K. Helmerson, A. Martin, and D.E. Pritchard. Laser and RF Spectroscopy of Magnetically Trapped Neutral Atoms. *J. Opt. Soc. Am. B* **9**, 483-492 (1992).

[134] T. Bergeman. Classical Calculations of Atomic Trajectories in a Magnetostatic Trap. *Bull. Am. Phys. Soc.* **31**, 939 (1986).

[135] K. Helmerson, A. Martin, and D.E. Pritchard. Laser Cooling of Magnetically Trapped Neutral Atoms. *J. Opt. Soc. Am. B* **9**, 1988-1996 (1992).

[136] I.D. Setija, H.G.C. Werij, O.J. Luiten, M.W. Reynolds, T.W. Hijmans, and J.T.M. Walraven. Optical Cooling of Atomic-Hydrogen in a Magnetic Trap. *Phys. Rev. Lett.* **70**, 2257-2260 (1993).

[137] W. Petrich, M.H. Anderson, J.R. Ensher, and E.A. Cornell. Stable, Tightly Confining Magnetic Trap for Evaporative Cooling of Neutral Atoms. *Phys. Rev. Lett.* **74**, 3352-3355 (1995).

[138] M.H. Anderson, J.R. Ensher, M.R. Matthews, C.E. Wieman, and E.A. Cornell. Observation of Bose-Einstein Condensation in a Dilute Atomic Vapor. *Science* **269**, 198-201 (1995).

[139] J.J. Tollett, C.C. Bradley, C.A. Sackett, and R.G. Hulet. Permanent-Magnet Trap for Cold Atoms. *Phys. Rev. A* **51**, R22-R25 (1995).

[140] M.O. Mewes, M.R. Andrews, N.J. Vandruten, D.M. Kurn, D.S. Durfee, and W. Ketterle. Bose-Einstein Condensation in a Tightly Confining DC Magnetic Trap. *Phys. Rev. Lett.* **77**, 416-419 (1996).

[141] A. Ashkin. Acceleration and Trapping of Particles by Radiation Pressure. *Phys. Rev. Lett.* **24**, 156 (1970).

[142] S. Chu, J. Bjorkholm, A. Ashkin, and A. Cable. Experimental Observation of Optically Trapped Atoms. *Phys. Rev. Lett.* **57**, 314 (1986).

[143] A. Ashkin. Application of Laser Radiation Pressure. *Science* **210**, 1081-1088 (1980).

[144] A. Ashkin and J.M. Dziedzic. Observation of Radiation-Pressure Trapping of Particles by Alternating Light Beams. *Phys. Rev. Lett.* **54**, 1245 (1985).

[145] A. Ashkin and J.M. Dziedzic. Optical Trapping and Manipulation of Viruses and Bacteria. *Science* **235**, 1517 (1987).

[146] J.D. Miller, R.A. Cline, and D.J. Heinzen. Far-Off-Resonance Optical Trapping of Atoms. *Phys. Rev. A* **47**, R4567-R4570 (1993).

[147] C.S. Adams, H.J. Lee, N. Davidson, M. Kasevich, and S. Chu. Evaporative Cooling in a Crossed Dipole Trap. *Phys. Rev. Lett.* **74**, 3577-3580 (1995).

[148] T. Takekoshi and R.J. Knize. CO_2-Laser Trap for Cesium Atoms. *Opt. Lett.* **21**, 77-79 (1996).

[149] A. Ashkin. Trapping of Atoms by Resonance Radiation Pressure. *Phys. Rev. Lett.* **40**, 729 (1978).

[150] A. Ashkin and J. Gordon. Cooling and Trapping of Atoms by Resonance Radiation Pressure. *Opt. Lett.* **4**, 161 (1979).

[151] H. Metcalf and W. Phillips. Electromagnetic Trapping of Neutral Atoms. *Metrologia* **22**, 271 (1986).

[152] N. Davidson, H.J. Lee, C.S. Adams, M. Kasevich, and S. Chu. Long Atomic Coherence Times in an Optical Dipole Trap. *Phys. Rev. Lett.* **74**, 1311-1314 (1995).

[153] A. Siegman. *Lasers*. University Sciences, Mill Valley (1986).

[154] L. Allen, M. Beijersbergen, R. Spreeuw, and J. Woerdman. Orbital Angular Momentum of Light and the Transformation of Laguerre-Gaussian Laser Modes. *Phys. Rev. A* **45**, 8185 (1992).

[155] M. Beijersbergen, L. Allen, H. van der Veen, and J. Woerdman. Astigmatic Laser Mode Converters and Transfer of Orbital Angular Momentum. *Opt. Commun.* **96**, 123 (1993).

[156] G. Turnbull, D. Robertson, G. Smith, L. Allen, and M. Padgett. The Generation of Free-Space Laguerre-Gaussian Modes at mm-Wave Frequencies by Use of a Spiral Phaseplate. *Opt. Commun.* **127**, 183 (1996).

[157] N. Simpson, K. Dholakia, L. Allen, and M. Padgett. The Mechanical Equivalence of Spin and Orbital Angular Momentum of Light: an Optical Spanner. *Opt. Lett.* **22**, 52 (1997).

[158] D. McGloin, N. Simpson, and M. Padgett. Transfer of Orbital Angular Momentum from a Stressed Fiber Optic Waveguide to a Light Beam. *App. Opt.* **37**, 469 (1998).

[159] M. Beijersbergen. *Phase Singularities in Optical Beams*. Ph.D. thesis, University Leiden (1996).

[160] C.G. Aminoff, A.M. Steane, P. Bouyer, P. Desbiolles, J. Dalibard, and C. Cohen-Tannoudji. Cesium Atoms Bouncing in a Stable Gravitational Cavity. *Phys. Rev. Lett.* **71**, 3083-3086 (1993).

[161] Y. Castin and J. Dalibard. Quantization of Atomic Motion in Optical Molasses. *Europhys. Lett.* **14**, 761-766 (1991).

[162] P. Verkerk, B. Lounis, C. Salomon, C. Cohen-Tannoudji, J.Y. Courtois, and G. Grynberg. Dynamics and Spatial Order of Cold Cesium Atoms in a Periodic Optical-Potential. *Phys. Rev. Lett.* **68**, 3861-3864 (1992).

[163] P.S. Jessen, C. Gerz, P.D. Lett, W.D. Phillips, S.L. Rolston, R.J.C. Spreeuw, and C.I. Westbrook. Observation of Quantized Motion of Rb Atoms in an Optical-Field. *Phys. Rev. Lett.* **69**, 49-52 (1992).

[164] B. Lounis, P. Verkerk, J.Y. Courtois, C. Salomon, and G. Grynberg. Quantized Atomic Motion in 1D Cesium Molasses with Magnetic-Field. *Europhys. Lett.* **21**, 13-17 (1993).

[165] R. Gupta, S. Padua, C. Xie, H. Batelaan, T. Bergeman, and H.Metcalf. Motional Quantization of Laser Cooled Atoms. *Bull. Am. Phys. Soc.* **37**, 1139 (1992).

[166] R. Gupta, S. Padua, T. Bergeman, and H. Metcalf. Search for Motional Quantization of Laser-Cooled Atoms. In E. Arimondo, W. Phillips, and F. Strumia, editors, *Laser Manipulation of Atoms and Ions, Proceedings of Fermi School CXVIII, Varenna*, Amsterdam (1993). North Holland.

[167] C.C. Agosta, H.T.C. Stoof, I.F. Silvera, and B.J. Verhaar. Trapping of Neutral Atoms with Resonant Microwave-Radiation. *Phys. Rev. Lett.* **62**, 2361-2364 (1989).

[168] R.J.C. Spreeuw, C. Gerz, L.S. Goldner, W.D. Phillips, S.L. Rolston, C.I. Westbrook, M.W. Reynolds, and I.F. Silvera. Demonstration of Neutral Atom Trapping with Microwaves. *Phys. Rev. Lett.* **72**, 3162-3165 (1994).

[169] A. Ashkin and J. Gordon. Stability of Radiation-Pressure Particle Traps: an Optical Earnshaw Theorem. *Opt. Lett.* **8**, 511 (1983).

[170] D. Pritchard, E. Raab, V. Bagnato, C. Wieman, and R. Watts. Light Traps Using Spontaneous Forces. *Phys. Rev. Lett.* **57**, 310 (1986).

[171] P. Bouyer, P. Lemonde, M. Bendahan, A. Michaud, C. Salomon, and J. Dalibard. An Atom Trap Relying on Optical-Pumping. *Europhys. Lett.* **27**, 569-574 (1994).

[172] E. Raab, M. Prentiss, A. Cable, S. Chu, and D. Pritchard. Trapping of Neutral-Sodium Atoms with Radiation Pressure. *Phys. Rev. Lett.* **59**, 2631 (1987).

[173] H. Metcalf. Magneto-Optical Trapping and Its Application to Helium Metastables. *J. Opt. Soc. Am. B* **6**, 2206-2210 (1989).

[174] P. Kohns, P. Buch, W. Suptitz, C. Csambal, and W. Ertmer. Online Measurement of Sub-Doppler Temperatures in a Rb Magnetooptical Trap-by-Trap Center Oscillations. *Europhys. Lett.* **22**, 517-522 (1993).

[175] C.D. Wallace, T.P. Dinneen, K.Y.N. Tan, A. Kumarakrishnan, P.L. Gould, and J. Javanainen. Measurements of Temperature and Spring Constant in a Magnetooptical Trap. *J. Opt. Soc. Am. B* **11**, 703-711 (1994).

[176] C.G. Townsend, N.H. Edwards, C.J. Cooper, K.P. Zetie, C.J. Foot, A.M. Steane, P. Szriftgiser, H. Perrin, and J. Dalibard. Phase-Space Density in the Magnetooptical Trap. *Phys. Rev. A* **52**, 1423-1440 (1995).

[177] P. Molenaar. *Photoassociative Reactions of Laser-Cooled Sodium.* Ph.D. thesis, Utrecht University (1995).

[178] C. Monroe, W. Swann, H. Robinson, and C. Wieman. Very Cold Trapped Atoms in a Vapor Cell. *Phys. Rev. Lett.* **65**, 1571-1574 (1990).

[179] E.A. Cornell, C. Monroe, and C.E. Wieman. Multiply Loaded, AC Magnetic Trap for Neutral Atoms. *Phys. Rev. Lett.* **67**, 2439-2442 (1991).

[180] A.M. Steane and C.J. Foot. Laser Cooling Below the Doppler Limit in a Magnetooptical Trap. *Europhys. Lett.* **14**, 231-236 (1991).

[181] A.M. Steane, M. Chowdhury, and C.J. Foot. Radiation Force in the Magnetooptical Trap. *J. Opt. Soc. Am. B* **9**, 2142-2158 (1992).

[182] P. Gould. private communication.

[183] T. Walker, D. Sesko, and C. Wieman. Collective Behavior of Optically Trapped Neutral Atoms. *Phys. Rev. Lett.* **64**, 408-411 (1990).

[184] D.W. Sesko, T.G. Walker, and C.E. Wieman. Behavior of Neutral Atoms in a Spontaneous Force Trap. *J. Opt. Soc. Am. B* **8**, 946-958 (1991).

[185] K.E. Gibble, S. Kasapi, and S. Chu. Improved Magnetooptic Trapping in a Vapor Cell. *Opt. Lett.* **17**, 526-528 (1992).

[186] K. Lindquist, M. Stephens, and C. Wieman. Experimental and Theoretical Study of the Vapor-Cell Zeeman Optical Trap. *Phys. Rev. A* **46**, 4082-4090 (1992).

[187] W. Ketterle, K.B. Davis, M.A. Joffe, A. Martin, and D.E. Pritchard. High-Densities of Cold Atoms in a Dark Spontaneous-Force Optical Trap. *Phys. Rev. Lett.* **70**, 2253-2256 (1993).

[188] T. Walker, D. Hoffmann, P. Feng, and R.S. Williamson. A Vortex-Force Atom Trap. *Phys. Lett. A* **163**, 309-312 (1992).

[189] T. Walker, P. Feng, D. Hoffmann, and R.S. Williamson. Spin-Polarized Spontaneous-Force Atom Trap. *Phys. Rev. Lett.* **69**, 2168-2171 (1992).

[190] F. Shimizu, K. Shimizu, and H. Takuma. A High-Intensity Metastable Neon Trap. *Chem. Phys.* **145**, 327-331 (1990).

[191] F. Shimizu, K. Shimizu, and H. Takuma. 4-Beam Laser Trap of Neutral Atoms. *Opt. Lett.* **16**, 339-341 (1991).

[192] O. Emile, F. Bardou, C. Salomon, P. Laurent, A. Nadir, and A. Clairon. Observation of a New Magnetooptical Trap. *Europhys. Lett.* **20**, 687-691 (1992).

[193] R. Grimm, Yu.B. Ovchinnikov, A.I. Sidorov, and V.S. Letokhov. Dipole Force Rectification in a Monochromatic Laser Field. *Opt. Commun.* **84**, 18-22 (1991).

[194] N. Masuhara, J.M. Doyle, J.C. Sandberg, D. Kleppner, T.J. Greytak, H.F. Hess, and G.P. Kochanski. Evaporative Cooling of Spin-Polarized Atomic Hydrogen. *Phys. Rev. Lett.* **61**, 935 (1988).

[195] H.F. Hess. Evaporative Cooling of Magnetically Trapped and Compressed Spin-Polarized Hydrogen. *Phys. Rev. B* **34**, 3476 (1986).

[196] J.M. Doyle, J.C. Sandberg, I.A. Yu, C.L. Cesar, D. Kleppner, and T.J. Greytak. Hydrogen in the Sub-milliKelvin Regime: Sticking Probability on Superfluid ^{4}He. *Phys. Rev. Lett.* **67**, 603 (1991).

[197] O.J. Luiten, H.G.C. Werij, I.D. Setija, M.W. Reynolds, T.W. Hijmans, and J.T.M. Walraven. Lyman-Alpha Spectroscopy of Magnetically Trapped Atomic-Hydrogen. *Phys. Rev. Lett.* **70**, 544-547 (1993).

[198] K.B. Davis, M.O. Mewes, and W. Ketterle. An Analytical Model for Evaporative Cooling of Atoms. *App. Phys. B* **60**, 155-159 (1995).

[199] W. Ketterle and N.J. van Druten. Evaporative Cooling of Trapped Atoms. *Adv. Atom. Mol. Opt. Phys.* **37**, 181 (1996).

[200] V. Bagnato, D.E. Pritchard, and D. Kleppner. Bose-Einstein Condensation in an External Potential. *Phys. Rev. A* **35**, 4354 (1987).

[201] K. Huang. *Statistical Mechanics*. Wiley, New York (1963).

[202] C. Monroe, E. Cornell, C. Sackett, C. Myatt, and C. Wieman. Measurement of Cs-Cs Elastic Scattering at $T = 30 \ \mu K$. *Phys. Rev. Lett.* **70**, 414 (1993).

[203] O.J. Luiten, M.W. Reynolds, and J.T.M. Walraven. Kinetic Theory of the Evaporative Cooling of a Trapped Gas. *Phys. Rev. A* **53**, 381 (1996).

[204] V.I. Balykin, V.S. Letokhov, Yu.B. Ovchinnikov, and A.I. Sidorov. Quantum-State-Selective Mirror Reflection of Atoms by Laser Light. *Phys. Rev. Lett.* **60**, 2137 (1988).

[205] V.I. Balykin and V.S. Letokhov. Laser Optics of Neutral Atomic-Beams. *Phys. Today* **42**, 23-28 (1989).

[206] K. Helmerson, S.L. Rolston, L. Goldner, and W.D. Phillips. In *Optics and Interferometry with Atoms*, Konstanz (1992). WE-Heraeus-Seminar.

[207] F. Tudorica, O. Kritsun, G. Stevens, and H. Metcalf. Deflection Schemes Using Rydberg Atom Mirror. *Bull. Am. Phys. Soc.* **43**, 1365 (1998).

[208] H. Metcalf. Magnetic Trapping of Decelerated Neutral Atoms. *Prog. Quant. Elect.* **8**, 169 (1984).

[209] E. Hinds, M. Boshier, and I. Hughes. Magnetic Waveguide for Trapping Cold Atom Gases in Two Dimensions. *Phys. Rev. Lett.* **80**, 645 (1998).

[210] A. Lemonick, F. Pipkin, and D.R. Hamilton. Focusing Atomic Beam Apparatus. *Rev. Sci. Instrum.* **26**, 1112 (1955).

[211] R.L. Christensen and D.R. Hamilton. Permanent Magnet for Atomic Beam Focusing. *Rev. Sci. Instrum.* **30**, 356 (1959).

[212] H. Metcalf, W. Phillips, and J. Prodan. Focussing of Slow Atomic Beams. *Bull. Am. Phys. Soc.* **29**, 795 (1984).

[213] W.G. Kaenders, F. Lison, I. Muller, A. Richter, R. Wynands, and D. Meschede. Refractive Components for Magnetic Atom Optics. *Phys. Rev. A* **54**, 5067-5075 (1996).

[214] J. Bjorkholm, R. Freeman, A. Ashkin, and D. Pearson. Observation of Focussing of Neutral Atoms by the Dipole Forces of Resonance-Radiation Pressure. *Phys. Rev. Lett.* **41**, 1361 (1978).

[215] J.R. Zacharias. Precision Measurements with Molecular Beams. *Phys. Rev.* **94**, 751 (1954).

[216] R. Weiss. Contribution to "Festschrift for Jerrold R. Zacharias". private communication.

[217] O. Kritsun, J. Marburger, and H. Metcalf. Interference From a Source of Atoms Released in a Gravitational Field. *Bull. Am. Phys. Soc.* **43**, 1366 (1998).

[218] A. Scholz, M. Christ, D. Doll, J. Ludwig, and W. Ertmer. Magneto-optical Preparation of a Slow, Cold and Bright Ne* Atomic-Beam. *Opt. Commun.* **111**, 155-162 (1994).

[219] M.D. Hoogerland, J.P.J. Driessen, E.J.D. Vredenbregt, H.J.L. Megens, M.P. Schuwer, H.C.W. Beijerinck, and K.A.H. van Leeuwen. Bright Thermal Atomic-Beams by Laser Cooling — A 1400-Fold Gain in-Beam Flux. *App. Phys. B* **62**, 323-327 (1996).

[220] Z.T. Lu, K.L. Corwin, M.J. Renn, M.H. Anderson, E.A. Cornell, and C.E. Wieman. Low-Velocity Intense Source of Atoms from a Magnetooptical Trap. *Phys. Rev. Lett.* **77**, 3331-3334 (1996).

[221] K.G.H. Baldwin. private communication.

[222] M. Schiffer, M. Christ, G. Wokurka, and W. Ertmer. Temperatures Near the Recoil Limit in an Atomic Funnel. *Opt. Commun.* **134**, 423-430 (1997).

[223] F. Lison, P. Schuh, D. Haubrich, and D. Meschede. High Brilliance Zeeman Slowed Cesium Atomic Beam. *Phys. Rev. A* **61**, 013405 (2000).

[224] K. Dieckmann, R.J.C. Spreeuw, M. Weidemüller, and J.T.M. Walraven. Two-Dimensional Magneto-Optical Trap as a Source of Slow Atoms. *Phys. Rev. A* **58**, 3891 (1998).

[225] J. Nellessen, J.H. Muller, K. Sengstock, and W. Ertmer. Laser Preparation of a Monoenergetic Sodium Beam. *Europhys. Lett.* **9**, 133-138 (1989).

[226] J. Nellessen, J.H. Muller, K. Sengstock, and W. Ertmer. Large-Angle Beam Deflection of a Laser-Cooled Sodium Beam. *J. Opt. Soc. Am. B* **6**, 2149-2154 (1989).

[227] J. Nellessen, J. Werner, and W. Ertmer. Magnetooptical Compression of a Monoenergetic Sodium Atomic-Beam. *Opt. Commun.* **78**, 300-308 (1990).

[228] G. Timp, R.E. Behringer, D.M. Tennant, J.E. Cunningham, M. Prentiss, and K.K. Berggren. Using Light as a Lens for Submicron, Neutral-Atom Lithography. *Phys. Rev. Lett.* **69**, 1636-1639 (1992).

[229] K.K. Berggren, M. Prentiss, G. Timp, and R.E. Behringer. Neutral Atom Nanolithography. *Inst. Phys. Conf. Ser.* **127**, 79-84 (1992).

[230] J.J. Mcclelland, R.E. Scholten, E.C. Palm, and R.J. Celotta. Laser-Focused Atomic Deposition. *Science* **262**, 877-880 (1993).

[231] S. Nowak, T. Pfau, and J. Mlynek. Nanolithography with Metastabile Helium. *App. Phys. B* **63**, 203-205 (1996).

[232] K.K. Berggren, A. Bard, J.L. Wilbur, J.D. Gillaspy, A.G. Helg, J.J. Mcclelland, S.L. Rolston, W.D. Phillips, M. Prentiss, and G.M. Whitesides. Microlithography by Using Neutral Metastable Atoms and Self-Assembled Monolayers. *Science* **269**, 1255-1257 (1995).

[233] F. Lison, H-J. Adams, D. Haubrich, M. Kreis, S. Nowak, and D. Meschede. Nanoscale Atomic Lithography with a Cesium Atomic Beam. *App. Phys. B* **65**, 419-421 (1997).

[234] M.A. Kasevich, E. Riis, S. Chu, and R.G. Devoe. RF Spectroscopy in an Atomic Fountain. *Phys. Rev. Lett.* **63**, 612-616 (1989).

[235] A. Clairon, C. Salomon, S. Guellati, and W.D. Phillips. Ramsey Resonance in a Zacharias Fountain. *Europhys. Lett.* **16**, 165-170 (1991).

[236] K. Gibble and S. Chu. Future Slow-Atom Frequency Standards. *Metrologia* **29**, 201-212 (1992).

[237] K. Gibble and S. Chu. Laser-Cooled Cs Frequency Standard and a Measurement of the Frequency-Shift Due to Ultracold Collisions. *Phys. Rev. Lett.* **70**, 1771-1774 (1993).

[238] K. Gibble and B. Verhaar. Eliminating Cold-Collision Frequency Shifts. *Phys. Rev. A* **52**, 3370 (1995).

[239] R. Legere and K. Gibble. Quantum Scattering in a Juggling Atomic Fountain. *Phys. Rev. Lett.* **81**, 5780 (1998).

[240] Ph. Laurent, P. Lemonde, E. Simon, G. Santorelli, A. Clairon, N. Dimarcq, P. Petit, C. Audoin, and C. Salomon. A Cold Atom Clock in the Absence of Gravity. *Eur. Phys. J. D* **3**, 201 (1998).

[241] S.L. Rolston and W.D. Phillips. Laser-Cooled Neutral Atom Frequency Standards. *Proc. IEEE* **79**, 943-951 (1991).

[242] D. Wineland, R. Drullinger, and F. Walls. Radiation-Pressure Cooling of Bound resonant Absorbers. *Phys. Rev. Lett.* **40**, 1639-1642 (1978).

[243] W. Neuhauser, M. Hohenstatt, P. Toschek, and H. Dehmelt. Optical Sideband Cooling of Visible Atom Cloud Confined in Parabolic Well. *Phys. Rev. Lett.* **41**, 233-236 (1978).

[244] W. Neuhauser, M. Hohenstatt, P. Toschek, and H. Dehmelt. Localized Visible Ba$^+$ Mono-Ion Oscillator. *Phys. Rev. A* **22**, 1137-1140 (1980).

[245] D. Grison, B. Lounis, C. Salomon, J.Y. Courtois, and G. Grynberg. Raman-Spectroscopy of Cesium Atoms in a Laser Trap. *Europhys. Lett.* **15**, 149-154 (1991).

[246] J.W.R. Tabosa, G. Chen, Z. Hu, R.B. Lee, and H.J. Kimble. Nonlinear Spectroscopy of Cold Atoms in a Spontaneous-Force Optical Trap. *Phys. Rev. Lett.* **66**, 3245-3248 (1991).

[247] L. Hilico, C. Fabre, and E. Giacobino. Operation of a Cold-Atom Laser in a Magnetooptical Trap. *Europhys. Lett.* **18**, 685-688 (1992).

[248] E. Giacobino, J. Courty, C. Fabre, L. Hilico, and A. Lambrecht. In F. Ehlotsky, editor, *Fundamentals of Quantum Optics III*, page 133, New York (1993). Springer.

[249] A. Hemmerich and T.W. Hansch. 2-Dimensional Atomic Crystal Bound by Light. *Phys. Rev. Lett.* **70**, 410-413 (1993).

[250] G. Grynberg, B. Lounis, P. Verkerk, J.Y. Courtois, and C. Salomon. Quantized Motion of Cold Cesium Atoms in 2-Dimensional and 3-Dimensional Optical Potentials. *Phys. Rev. Lett.* **70**, 2249-2252 (1993).

[251] W.D. Phillips. private communication.

[252] G. Grynberg. private communication.

[253] J. Guo, P.R. Berman, B. Dubetsky, and G. Grynberg. Recoil-Induced Resonances in Nonlinear Spectroscopy. *Phys. Rev. A* **46**, 1426-1437 (1992).

[254] J. Guo and P.R. Berman. Recoil-Induced Resonances in Pump-Probe Spectroscopy Including Effects of Level Degeneracy. *Phys. Rev. A* **47**, 4128-4142 (1993).

[255] J.Y. Courtois, G. Grynberg, B. Lounis, and P. Verkerk. Recoil-Induced Resonances in Cesium — An Atomic Analog to the Free-Electron Laser. *Phys. Rev. Lett.* **72**, 3017-3020 (1994).

[256] D.R. Meacher, D. Boiron, H. Metcalf, C. Salomon, and G. Grynberg. Method for Velocimetry of Cold Atoms. *Phys. Rev. A* **50**, R1992-R1994 (1994).

[257] L. Hilico, P. Verkerk, and G. Grynberg. Optical-Phase Conjugation with Ultracold Cesium Atoms in a Magnetooptical Trap. *Compt. Rend. Acad. Scien. II* **315**, 285-291 (1992).

[258] J.R. Taylor. *Scattering Theory*. J. Wiley & Sons, New York (1972).

[259] C.J. Joachain. *Quantum Collision Theory*. North-Holland, Amsterdam (1975).

[260] B.H. Bransden. *Atomic Collision Theory*. Benjamin/Cummings, Reading (1983).

[261] T. Walker and P. Feng. Measurements of Collisions between Laser-Cooled Atoms. *Adv. Atom. Mol. Opt. Phys.* **34**, 125 (1993).

[262] P.S. Julienne, A.M. Smith, and K. Burnett. Theory of Collisions Between Laser Cooled Atoms. *Adv. Atom. Mol. Opt. Phys.* **30**, 141-198 (1992).

[263] J. Weiner. Advances in Ultracold Collisions: Experimentation and Theory. *Adv. Atom. Mol. Opt. Phys.* **35**, 45 (1995).

[264] K.A. Suominen. Theories for Cold Atomic-Collisions in Light Fields. *J. Phys. B* **29**, 5981-6007 (1996).

[265] P. Langevin. Une Formule Fondamentale de Théorie Cinétique. *Ann. Chim. Phys.* **5**, 245 (1905).

[266] P.S. Julienne and F.H. Mies. Collisions of Ultracold Trapped Atoms. *J. Opt. Soc. Am. B* **6**, 2257-2269 (1989).

[267] G.F. Gribakin and V.V. Flambaum. Calculation of the Scattering Length in Atomic-Collisions Using the Semiclassical Approximation. *Phys. Rev. A* **48**, 546-553 (1993).

[268] C.D. Wallace, T.P. Dinneen, K.Y.N. Tan, T.T. Grove, and P.L. Gould. Isotopic Difference in Trap Loss Collisions of Laser Cooled Rubidium Atoms. *Phys. Rev. Lett.* **69**, 897-900 (1992).

[269] A. Gallagher and D.E. Pritchard. Exoergic Collisions of Cold Na*-Na. *Phys. Rev. Lett.* **63**, 957-960 (1989).

[270] P.S. Julienne and J. Vigue. Cold Collisions of Ground-State and Excited-State Alkali-Metal Atoms. *Phys. Rev. A* **44**, 4464-4485 (1991).

[271] H. Mastwijk, J. Thomsen, P. van der Straten, and A. Niehaus. Optical Collisions of Cold, Metastable Helium Atoms. *Phys. Rev. Lett.* **80**, 5516-5519 (1998).

[272] R.N. Zare. *Angular Momentum*. Wiley, New York (1988).

[273] L.P. Ratliff, M.E. Wagshul, P.D. Lett, S.L. Rolston, and W.D. Phillips. Photoassociative Spectroscopy of 1_g-State, 0_u^+-State and 0_g^--State of Na_2. *J. Chem. Phys.* **101**, 2638-2641 (1994).

[274] M. Movre and G. Pichler. Resonance Interaction and Self-broadening of Alkali Resonance Lines I. Adiabatic Potential Curves. *J. Phys. B* **10**, 2631 (1977).

[275] P.A. Molenaar, P. Van der Straten, and H.G.M. Heideman. Long-Range Predissociation in Two-Color Photoassociation of Ultracold Atoms. *Phys. Rev. Lett.* **77**, 1460 (1996).

[276] K.M. Jones, S. Maleki, L.P. Ratliff, and P.D. Lett. Two-colour Photoassociation Spectroscopy of Ultracold Sodium. *J. Phys. B* **30**, 289-308 (1997).

[277] K.M. Jones, P.S. Julienne, P.D. Lett, W.D. Phillips, E. Tiesinga, and C.J. Williams. Measurement of the Atomic Na(3P) Lifetime and of Retardation in the Interaction Between 2 Atoms Bound in a Molecule. *Europhys. Lett.* **35**, 85-90 (1996).

[278] A. Fioretti, D. Comparat, A. Crubellier, O. Dulieu, F. Masnou-Seeuws, and P. Pillet. Formation of Cold Cs_2 Molecules through Photoassociation. *Phys. Rev. Lett.* **80**, 4402-4405 (1998).

[279] H.J. Metcalf. Highly excited atoms. *Nature* **284**, 127-130 (1980).

[280] W.R. Anderson, J.R. Veale, and T.F. Gallagher. Resonant Dipole-Dipole Energy Transfer in a Nearly Frozen Rydberg Gas. *Phys. Rev. Lett.* **80**, 249 (1998).

[281] I. Mourachko, D. Comparat, F. De Tomasi, A. Fioretti, P. Nosbaum, V.M. Akulin, and P. Pillet. Many-Body Effects in a Frozen Rydberg Gas. *Phys. Rev. Lett.* **80**, 253 (1998).

[282] J. Mlynek, V. Balykin, and P. Meystre (Eds.). Optics and Interferometry with Atoms. *App. Phys. B* **54**, 319-485 (1992).

[283] D. Keith, M. Schattenburg, H. Smith, and D. Pritchard. Diffraction of Atoms by a Transmission Grating. *Phys. Rev. Lett.* **61**, 1580 (1988).

[284] M. Chapman, T. Hammond, A. Lenef, J. Schmiedmayer, R. Rubenstein, E. Smith, and D. Pritchard. Photon Scattering from Atoms in an Atom Interferometer: Coherence Lost and Regained. *Phys. Rev. Lett.* **75**, 3783 (1995).

[285] P.E. Moskowitz, P.L. Gould, S.R. Atlas, and D.E. Pritchard. Diffraction of an Atomic Beam by Standing-Wave Radiation. *Phys. Rev. Lett.* **51**, 370 (1983).

[286] P. Gould, G. Ruff, and D. Pritchard. Diffraction of Atoms by Light: The Near-Resonant Kapitza-Dirac Effect. *Phys. Rev. Lett.* **56**, 827 (1986).

[287] P. Martin, P. Gould, B. Oldaker, A. Miklich, and D. Pritchard. Diffraction of Atoms Moving through a Standing Light Wave. *Phys. Rev. A* **36**, 2495 (1987).

[288] T. Sleator, T. Pfau, V. Balykin, and J. Mlynek. Imaging and Focusing of an Atomic-Beam with a Large Period Standing Light-Wave. *App. Phys. B* **54**, 375-379 (1992).

[289] T. Sleator, T. Pfau, V. Balykin, O. Carnal, and J. Mlynek. Experimental Demonstration of the Optical Stern-Gerlach Effect. *Phys. Rev. Lett.* **68**, 1996-1999 (1992).

[290] T. Pfau, C.S. Adams, and J. Mlynek. Proposal for a Magnetooptical Beam Splitter for Atoms. *Europhys. Lett.* **21**, 439-444 (1993).

[291] J. Söding and R. Grimm. Stimulated Magnetooptical Force in the Dressed-Atom Picture. *Phys. Rev. A* **50**, 2517-2527 (1994).

[292] T. Pfau, C. Kurtsiefer, C.S. Adams, M. Sigel, and J. Mlynek. Magnetooptical Beam Splitter for Atoms. *Phys. Rev. Lett.* **71**, 3427-3430 (1993).

[293] R. Grimm, J. Söding, and Yu.B. Ovchinnikov. Coherent Beam Splitter for Atoms Based on a Bichromatic Standing Light-Wave. *Opt. Lett.* **19**, 658-660 (1994).

[294] J. Lawall, S. Kulin, B. Saubamea, N. Bigelow, M. Leduc, and C. Cohen-Tannoudji. Three-Dimensional Laser Cooling of Helium Beyond the Single-Photon Recoil Limit. *Phys. Rev. Lett.* **75**, 4194 (1995).

[295] F. Riehle, T. Kisters, A. Witte, J. Helmcke, and C.J. Borde. Optical Ramsey Spectroscopy in a Rotating Frame — Sagnac Effect in a Matter-Wave Interferometer. *Phys. Rev. Lett.* **67**, 177-180 (1991).

[296] M. Kasevich and S. Chu. Measurement of the Gravitational Acceleration of an Atom with a Light-Pulse Atom Interferometer. *App. Phys. B* **54**, 321-332 (1992).

[297] D.S. Weiss, B.C. Young, and S. Chu. Precision-Measurement of the Photon Recoil of an Atom Using Atomic Interferometry. *Phys. Rev. Lett.* **70**, 2706-2709 (1993).

[298] O. Carnal and J. Mlynek. Youngs Double-Slit Experiment with Atoms — A Simple Atom Interferometer. *Phys. Rev. Lett.* **66**, 2689-2692 (1991).

[299] D.W. Keith, C.R. Ekstrom, Q.A. Turchette, and D.E. Pritchard. An Interferometer for Atoms. *Phys. Rev. Lett.* **66**, 2693-2696 (1991).

[300] O. Carnal, M. Sigel, T. Sleator, H. Takuma, and J. Mlynek. Imaging and Focusing of Atoms by a Fresnel Zone Plate. *Phys. Rev. Lett.* **67**, 3231-3234 (1991).

[301] E. Rasel, M. Oberthaler, H. Batelaan, J. Schmiedmayer, and A. Zeilinger. Atom Wave Interferometry with Diffraction Gratings of Light. *Phys. Rev. Lett.* **75**, 2633 (1995).

[302] D.M. Giltner, R.W. Mcgowan, and S.A. Lee. Atom Interferometer Based on Bragg Scattering from Standing Light Waves. *Phys. Rev. Lett.* **75**, 2638-2641 (1995).

[303] S. Kunze, S. Dürr, and G. Rempe. Bragg Scattering of Slow Atoms from a Standing Light Wave. *Europhys. Lett.* **34**, 343 (1996).

[304] M. Sagnac. L'éther lumineux démontré par léffet du vent relatif d'éther dans un intefèromètre en rotation uniforme. *Compt. Rend. Acad. Scien., Paris* **157**, 708 (1913).

[305] T. Gustavson, P. Bouyer, and M. Kasevich. Measurements with an Atom Interferometer Gyroscope. *Phys. Rev. Lett.* **78**, 2046 (1997).

[306] M. Kasevich and S. Chu. Atomic Interferometry Using Stimulated Raman Transitions. *Phys. Rev. Lett.* **67**, 181-184 (1991).

[307] M. Snadden, J. McGuirk, P. Bouyer, K. Haritos, and M. Kasevich. Measurement of the Earth's Gravity Gradient with an Atom Interferometer-Based Gravity Gradiometer. *Phys. Rev. Lett.* **81**, 971 (1998).

[308] J. Schmiedmayer, M.S. Chapman, C.R. Ekstrom, T.D. Hammond, S. Wehinger, and D.E. Pritchard. Index of Refraction of Various Gases for Sodium Matter Waves. *Phys. Rev. Lett.* **74**, 1043-1047 (1995).

[309] C.R. Ekstrom, J. Schmiedmayer, M.S. Chapman, T.D. Hammond, and D.E. Pritchard. Measurement of the Electric Polarizability of Sodium with an Atom Interferometer. *Phys. Rep.* **51**, 3888 (1995).

[310] Ch. Miniatura, O. Gorceix J. Robert, V. Lorent, S. Le Boiteux, J. Reinhardt, and J. Baudon. Atomic Interferences and the Topological Phase. *Phys. Rev. Lett.* **69**, 261 (1992).

[311] K. Zeiske, G. Zinner, F. Riehle, and J. Helmcke. Atom Interferometry in a Static Electric Field: Measurement of the Aharonov-Casher Effect. *App. Phys. B* **60**, 205-209 (1995).

[312] V.S. Lethokov. Narrowing of the Doppler Width in a Standing Light Wave. *JETP Lett.* **7**, 272 (1968).

[313] C. Salomon, J. Dalibard, A. Aspect, H. Metcalf, and C. Cohen-Tannoudji. Channeling Atoms in a Laser Standing Wave. *Phys. Rev. Lett.* **59**, 1659 (1987).

[314] K.I. Petsas, A.B. Coates, and G. Grynberg. Crystallography of Optical Lattices. *Phys. Rev. A* **50**, 5173-5189 (1994).

[315] P.S. Jessen and I.H. Deutsch. Optical Lattices. *Adv. Atom. Mol. Opt. Phys.* **37**, 95-138 (1996).

[316] A. Rauschenbeutel, H. Schadwinkel, V. Gomer, and D. Meschede. Standing Light Fields for Cold Atoms with Intrinsically Stable and Variable Time Phases. *Opt. Commun.* **148**, 48 (1998).

[317] P. Verkerk, D.R. Meacher, A.B. Coates, J.Y. Courtois, S. Guibal, B. Lounis, C. Salomon, and G. Grynberg. Designing Optical Lattices — An Investigation with Cesium Atoms. *Europhys. Lett.* **26**, 171-176 (1994).

[318] G. Birkl, M. Gatzke, I.H. Deutsch, S.L. Rolston, and W.D. Phillips. Bragg Scattering from Atoms in Optical Lattices. *Phys. Rev. Lett.* **75**, 2823-2826 (1995).

[319] C.I. Westbrook, R.N. Watts, C.E. Tanner, S.L. Rolston, W.D. Phillips, P.D. Lett, and P.L. Gould. Localization of Atoms in a 3-Dimensional Standing Wave. *Phys. Rev. Lett.* **65**, 33-36 (1990).

[320] R. Dicke. The Effect of Collisions upon the Doppler Width of Spectral Lines. *Phys. Rev.* **89**, 472 (1953).

[321] M. Wilkens, E. Schumacher, and P. Meystre. Band Theory of a Common Model of Atom Optics. *Phys. Rev. A* **44**, 3130 (1991).

[322] M. Doery, M. Widmer, J. Bellanca, E. Vredenbregt, T. Bergeman, and H. Metcalf. Energy-Bands and Bloch States in 1D Laser Cooling and Their Effects on the Velocity Distribution. *Phys. Rev. Lett.* **72**, 2546-2549 (1994).

[323] M. Dahan, E. Peik, J. Reichel, Y. Castin, and C. Salomon. Bloch Oscillations of Atoms in an Optical Potential. *Phys. Rev. Lett.* **76**, 4508 (1996).

[324] H. Metcalf. Laser Cooling as a Form of Optical-Pumping in the Quantum Domain of Atomic Motion. *Phys. Scripta* **T70**, 57-63 (1997).

[325] T. Bergeman. Quantum Calculations for One-Dimensional Laser Cooling — Temporal Evolution. *Phys. Rev. A* **48**, R3425-R3428 (1993).

[326] H.T.C. Stoof. Atomic Bose-Gas with a Negative Scattering Length. *Phys. Rev. A* **49**, 3824-3830 (1994).

[327] T. Bergeman. Hartree-Fock Calculations of Bose-Einstein Condenstation of ^{7}Li Atoms in a Harmonic Trap for $T > 0$. *Phys. Rev. A* **55**, 3658 (1997).

[328] T. Bergeman. Erratum: Hartree-Fock Calculations of Bose-Einstein Condenstation of ^{7}Li Atoms in a Harmonic Trap for $T > 0$. *Phys. Rev. A* **56**, 3310 (1997).

[329] D. Fried, T. Killian, L. Willmann, D. Landhuis, S. Moss, D. Kleppner, and T. Greytak. Bose-Einstein Condensation of Atomic Hydrogen. *Phys. Rev. Lett.* **81**, 3811 (1998).

[330] K. Davis, M-O. Mewes, M. Andrews, M. van Druten, D. Durfee, D. Kurn, and W. Ketterle. Bose-Einstein Condensation in a Gas of Sodium Atoms. *Phys. Rev. Lett.* **75**, 3969 (1995).

[331] C.C. Bradley, C.A. Sackett, J.J. Tollett, and R.G. Hulet. Evidence of Bose-Einstein Condensation in an Atomic Gas with Attractive Interactions. *Phys. Rev. Lett.* **75**, 1687-1690 (1995).

[332] D. Kleppner. The Fuss About Bose-Einstein Condensation. *Phys. Today* **49**, August, 13 (1996).

[333] C. Cesar, D. Fried, T. Killian, A. Polcyn, J. Sandberg, I. Yu, T. Greytak, D. Kleppner, and J. Doyle. Two-Photon Spectroscopy of Trapped Atomic Hydrogen. *Phys. Rev. Lett.* **77**, 255 (1996).

[334] M.R. Andrews, D.M. Kurn, H.-J. Miesner, D.S. Durfee, C.G. Townsend, S. Inouye, and W. Ketterle. Propagation of Sound in a Bose-Einstein Condensate. *Phys. Rev. Lett.* **79**, 553 (1997).

[335] M.R. Andrews, C.G. Townsend, H.-J. Miesner, D.S. Durfee, D.M. Kurn, and W. Ketterle. Observation of Interference Between Two Bose Condensates. *Science* **275**, 637 (1997).

[336] M.-O. Mewes, M.R. Andrews, D.M. Kurn, D.S. Durfee, C.G. Townsend, and W. Ketterle. Output Coupler for Bose-Einstein Condensed Atoms. *Phys. Rev. Lett.* **78**, 582 (1997).

[337] M. Kozuma, L. Deng, E.W. Hagley, J. Wen, R. Lutwak, K. Helmerson, S.L. Rolston, and W.D. Phillips. Coherent Splitting of Bose-Einstein Condensed Atoms with Optically Induced Bragg Diffraction. *Phys. Rev. Lett.* **82**, 871 (1999).

[338] E.A. Burt, R.W. Ghrist, C.J. Myatt, M.J. Holland, E.A. Cornell, and C.E. Wieman. Coherence, Correlations, and Collisions — What One Learns About Bose-Einstein Condensates from Their Decay. *Phys. Rev. Lett.* **79**, 337-340 (1997).

[339] H.-J. Miesner, D.M. Stamper-Kurn, M.R. Andrews, D.S. Durfee, S. Inouye, and W. Ketterle. Bosonic Stimulation in the Formation of a Bose-Einstein Condensate. *Science* **279**, 1005 (1998).

[340] D. Jin, J. Ensher, M. Matthew, C. Wieman, and E. Cornell. Collective Excitations of a Bose-Einstein Condensate in a Dilute Gas. *Phys. Rev. Lett.* **77**, 420 (1996).

[341] M. Mewes, M. Andrews, N. van Druten, D. Kurn, D. Durfee, C. Townsend, and W. Ketterle. Collective Excitations of a Bose-Einstein Condenstate in a Magnetic Trap. *Phys. Rev. Lett.* **77**, 988 (1996).

[342] D.S. Jin, M.R. Matthews, J.R. Ensher, C.E. Wieman, and E.A. Cornell. Temperature-Dependent Damping and Frequency-Shifts in Collective Excitations of a Dilute Bose-Einstein Condensate. *Phys. Rev. Lett.* **78**, 764-767 (1997).

[343] D.M. Stamper-Kurn, M.R. Andrews, A. Chikkatur, S. Inouye, H.-J. Miesner, J. Stenger, and W. Ketterle. Optical Confinement of a Bose-Einstein Condensate. *Phys. Rev. Lett.* **80**, 2072 (1998).

[344] C.J. Myatt, E.A. Burt, R.W. Ghrist, E.A. Cornell, and C.E. Wieman. Production of 2 Overlapping Bose-Einstein Condensates by Sympathetic Cooling. *Phys. Rev. Lett.* **78**, 586-589 (1997).

[345] A. Aspect, E. Arimondo, R. Kaiser, N. Vansteenkiste, and C. Cohen-Tannoudji. Laser Cooling below the One-Photon Recoil Energy by Velocity-Selective Coherent Population Trapping. *Phys. Rev. Lett.* **61**, 826 (1988).

[346] F. Bardou, J.P. Bouchaud, O. Emile, A. Aspect, and C. Cohen-Tannoudji. Subrecoil Laser Cooling and Levy Flights. *Phys. Rev. Lett.* **72**, 203-206 (1994).

[347] A. Aspect, C. Cohen-Tannoudji, E. Arimondo, N. Vansteenkiste, and R. Kaiser. Laser Cooling Below the One-Photon Recoil Energy by Velocity-Selective Coherent Population Trapping — Theoretical-Analysis. *J. Opt. Soc. Am. B* **6**, 2112-2124 (1989).

[348] M.S. Shahriar, P.R. Hemmer, M.G. Prentiss, P. Marte, J. Mervis, D.P. Katz, N.P. Bigelow, and T. Cai. Continuous Polarization-Gradient Precooling-Assisted Velocity-Selective Coherent Population Trapping. *Phys. Rev. A* **48**, R4035-R4038 (1993).

[349] M. Widmer, M.J. Bellanca, W. Buell, H. Metcalf, M. Doery, and E. Vredenbregt. Measurement of Force-Assisted Population Accumulation in Dark States. *Opt. Lett.* **21**, 606-608 (1996).

[350] M. Olshanii and V. Minogin. Three Dimensional Velocity-Selective Coherent Population Trapping. *Opt. Commun.* **89**, 393 (1992).

[351] M.T. Widmer, M.R. Doery, M.J. Bellanca, W.F. Buell, T.H. Bergeman, and H.J. Metcalf. High-Velocity Dark States in Velocity-Selective Coherent Population Trapping. *Phys. Rev. A* **53**, 946-949 (1996).

[352] H. Batelaan, E. Rasel, M. Oberthaler, J. Schmeidmayer, and A. Zeilinger. Classical and Quantum Atom Fringes. In *Atom Interferometry* [362], pages 85–120.

[353] W.D. Phillips (Ed.). Laser Cooled and Trapped Atoms. *Prog. Quant. Elect.* **8**, 115-259 (1984).

[354] H.C.W. Beijerinck and B.J. Verhaar (Eds.). Dynamics of Inelastic Collisions of Electronically Excited Atoms. *Chem. Phys.* **145**, 153-331 (1990).

[355] E. Arimondo and H-A. Bachor (Eds.). Special Issue on Atom Optics. *J. Eur. Opt. Soc.* **8**, 495-753 (1996).

[356] V. Minogin and V. Letokhov. *Laser Light Pressure on Atoms.* Gordon and Breach, New York (1987).

[357] A. Kazantzev, G. Surdutovich, and V. Yakolev. *Mechanical Action of Light on Atoms.* World Scientific, Singapore (1990).

[358] L. Moi, S. Gozzini, C. Gabbanini, E. Arimondo, and F. Strumia, editors. *Light Induced Kinetic Effects on Atoms, Ions and Molecules*, Pisa (1991). ETS Editrice.

[359] A. Arimondo, W. Phillips, and F. Strumia, editors. *Laser Manipulation of Atoms and Ions, Proceedings of the International School of Physics "Enrico Fermi", Course CXVII*, Amsterdam (1993). North Holland.

[360] V.I. Balykin and V.S. Lethokov. *Atom Optics with Laser Light.* Harvard Academic Publishers, Chur (1995).

[361] A. Aspect, W. Barletta, and R. Bonifacio, editors. *Proceedings of the Fermi School CXXXI*, Amsterdam (1996). IOS Press.

[362] P. Berman (Ed.). *Atom Interferometry.* Academic Press, New York (1997).

[363] C. Cohen-Tannoudji. Laser Cooling and Trapping of Neutral Atoms — Theory. *Phys. Rep.* **219**, 153-164 (1992).

[364] H. Metcalf and P. van der Straten. Cooling and Trapping of Neutral Atoms. *Phys. Rep.* **244**, 204-286 (1994).

[365] C.S. Adams, M. Sigel, and J. Mlynek. Atom Optics. *Phys. Rep.* **240**, 143-210 (1994).

[366] C.S. Adams and E. Riis. Laser Cooling and Trapping of Neutral Atoms. *Prog. Quant. Elect.* **21**, 1-79 (1997).

[367] G. Nienhuis. Impressed by Light: Mechanical Action of Radiation on Atomic Motion. *Phys. Rep.* **138**, 151-192 (1986).

[368] A. Aspect. Manipulation of Neutral Atoms — Experiments. *Phys. Rep.* **219**, 141-152 (1992).

[369] J. Thomas and L. Wang. Precision Position Measurement of Moving Atoms. *Phys. Rep.* **262**, 311-368 (1995).

[370] H. Wallis. Quantum-Theory of Atomic Motion in Laser-Light. *Phys. Rep.* **255**, 204-287 (1995).

Index

absorption cross section, 27, 275
adiabatic motion, 140, 141
adiabatic rapid passage, **12–14**
adiabaticity, 146
Aharonov-Casher effect, 229
alkali-metal atom, **39–41**, 49, 52, 80,
        86, 201, 214, 273, 277, 279
alkaline-earth atom, 273
amplitude gradient, 31
amplitude modulator, 221
angular momentum, 139
        orbital, 40, 43, 45
        spin, 40, 43, 45
        total, 40
asymmetric anharmonic potential, 142
atom interferometry, 219, 223, **227–**
        229
atom laser, 248
atom optics, 57, 70, **179–198**
        coherent, 247
atomic beam brightening, **186–190**
atomic beam collimation, **90–94**, 105,
        135–136
atomic beam deceleration, **73–86**, 135–
        136

atomic beam splitter, **223–224**
atomic cavity, 180
atomic clock, **192–194**
atomic density, 68, 162, 232
atomic density correlation, 248
atomic fountain, **185–186**, **193–194**,
        227
atomic funnel, 188
atomic lens, **181–185**, 224, 227
atomic lithography, 190
atomic mirror, 80, 124, **180–181, 225–**
        226
atomic nanofabrication, 187, **190–192**
atomic orbit, 140–143
atomic polarizers, **226–227**
atomic sources, **224–225**
atomic time, 192
atomic trampoline, 124, 180

ballistic technique, 237, 245
band structure, **238–239**
BEC, *see* Bose-Einstein condensa-
        tion
Berry's phase, 229
bichromatic force, **131–135**

Bloch equation, 24
Bloch oscillation, 239
Bloch sphere, 12, 13
Bloch state, 238
Bloch vector, **11–12**, 13
Bloch's theorem, 231
Boltzmann equation, **171**
Bose-Einstein condensation, 138, 140,
     147, 165, 171, 175, 220,
     225, **241–250**
  first-order coherence, **246–248**
  formation, 249
  higher order coherence, **248–249**
  observation of, **244–246**
  oscillation, 250
  phase of a, 246
  second sound, 250
  sound waves, 250
bosonic stimulation, 249
Bragg diffraction, 228, 235
Bragg reflection, 225, 228, 238, **259–261**
Bragg regime, 222, 223
brightness, 91, 92, 94, 186
brightness theorem, 70
brilliance, 186
Brillouin zone, 225, 238, 239, 261
broadband slowing, **79**
butterfly trap, 148

capture range, 58, 161
centrifugal barrier, 201, 202, 211
cesium, 76, 107, 112
chaotic, 142
chemical potential, 169
chirp slowing, **76**
classical motion, 140, **143–145**
Clebsch-Gordan coefficient, 53, 101,
     106
clock, 192
closed family, 254, 258
cloverleaf trap, 148
coherence, 18, 25, 107, 127
cold molecule, 217
collision, 165, 193, 248

  bad, 204
  excited state, 204, **207–218**
  fine-structure changing, 208
  good, 204
  ground state, **204–206**
  inelastic, 206
  optical, **209–213**
  reactive, 203
  ultra-cold, **199–218**
collision rate, 165
  elastic, 173
Condon point, 207, 211
controlled NOT gate, 261
cooling limit, **66–67**, **113–114**
  capture, 58, 276
  Doppler, 58, 65, 82, 89, 90, 96,
     111, 113, 121, 276
  evaporative, **174–175**
  recoil, 59, 111, 114, 276
correlation function, 248
cross section, 203, 206
  differential, 201
  total, 201
crystallography, 233
Cs, *see* cesium
cycling transition, 49
cylindrical microlens, 190

damping coefficient, 35, 36, 65, 90,
     109
damping force, **102–103**, 106, 254
damping rate, 35
dark states, **251–262**
deBroglie wave optics, **219–229**, 254
  with gratings, **220–223**
Debye-Waller factor, 235
density limit, 162
density matrix, 15, **17–27**, 107, 109,
     110, 118
density of states, 168, 170
density operator, 17, 18
detailed balance, 173
Dicke narrowing, 237
diffraction, 220, 222
diffuse light, 78, 79, 85

diffuse slowing, **78–79**
diffusion coefficient, 65, 89, 109
dipolar relaxation, 174
dipole force, 8, 33, **123–136**, 151,
    152, 231
    fluctuations, 152
dipole force lens, 190
dipole force operator, 108
dipole force rectification, **129–131**
dipole force trap, *see* optical trap, dipole
    force
dipole matrix element
    angular part, **52–53**
    geometrical part, 51
    physical part, 51
    radial part, **51–52**
dipole moment, 6, 21, 50, 149
dipole-dipole interaction, 202, 203,
    217
dispersion coefficient, 201
dispersive measurement, 245
dissipative force, *see* scattering force
dissipative process, 117, 179
Doppler laser cooling, 100, 103, 109,
    113
Doppler shift, 13, 73–80, 83, 87, 90,
    100, 127, 133, 135, 161,
    237
Doppler temperature, *see* cooling limit,
    Doppler
Doppler width, 195
Doppleron, 128, 135
dressed state, **9–10**, 126, 127, 129,
    133, 223

Earnshaw's theorem, 137, 156
egg-crate potential, 231
Ehrenfest theorem, 29, 30
Einstein coefficient, 6
Einstein-Podolsky-Rosen, 261
elastic collision, 166, 171, 174
elastic collision rate, 168, 170
electric dipole approximation, 5
energy band structure, 155, 236, 238
energy conservation, 222, 253

entangled state, 255, **261–262**
entropy, 60
equilibrium state, 67
ergodic, 167
evanescent wave, **124**, 180, 226
evaporation
    speed of, **171–173**
evaporative cooling, 146, **165–175**,
    244
excited state decay, **14–16**
exciton, 243

family momentum, 254, 256, 257
far-off-resonance trap, *see* optical trap,
    far-off-resonance
far-off-resonant lattice, 232
Fermi Golden Rule, 4
field quantization, 20, 21
fine structure, 40, **53–56**
fine-structure changing collision, 209
Fock states, 261
Fokker-Planck equation, 57, 64, **66–
    67**, 110, 111, 113
force on two-level atoms, **29–37**
    at rest, **31–34**
    in motion, **34–37**
force operator, 30, 107, 108
FORT, *see* optical trap, far-off-resonance
fountain, *see* atomic fountain
four-photon Raman transition, 259
Fresnel zone plate, 227
frozen Rydberg gas, 218

gain, 197
gamma function, 169
Gaussian laser beam, 150
geometrical optics, 70, 179
Global Positioning System, 192
gradient force, *see* dipole force
gravitational acceleration, 227, 228
gravitational field, 220
gravito-optical trap, 126, 154, 180
gravito-optics, 186
Gross-Pitaevski equation, 242, 243
gyroscope, 228

H, *see* hydrogen, 56
He, *see* helium
helium, 44, 45, 50, 76, 80, 135
HeNe laser, 225
hfs, *see* hyperfine structure
high intensity optical molasses, *see* optical molasses, at high intensity
high-resolution spectroscopy, 138
hydrogen, 51, 76, 166
    spin-polarized, 243
hyperfine state, 82, 86
hyperfine structure, 40, 43, 49, **53–56**

ideal gas, 168
impact parameter, 201, 203
index of refraction, 229
inelastic collision, 173, 174
interaction potential, 200, 203
interference fringe, 246
Ioffe trap, 139, 140, 144, 147, 246
ion trap, **194–195**
island of stability, 142
isotope, 112, 119

K, *see* potassium
Kronig-Penney model, 220, 231

Λ transition, 243
Landé *g*-factor, 42, 77
Landau-Zener model, 134, 212
Langevin model, 201, 202
large period standing wave, 223
laser cooling, **57–70**
launch velocity, 185
length standard, 191
Li, *see* lithium
lifetime, 22, 217, 274
light
    circularly polarized, 46, 48, 53, 106
    linearly polarized, 46–48, 53
light pressure force, *see* scattering force
light shift, **7–8**, 9, 33, **101**, 102, 104, 109, 111, 112, 114, 117, 123, 124, 129, 130, 151, 155, 196, 223, 231
limit, *see* cooling limit
lin-angle-lin, 258
lin-perp-lin, *see* polarization gradient, lin ⊥ lin
linewidth, 26, 274
    power-broadened, 26, 75, 89
Liouville's theorem, **68–70**
lithium, 76
lithography, *see* atomic lithography
long-range molecular state, 217
low saturation limit, 8

magnetic dipole transition, 155
magnetic hexapole lens, 138
magnetic mirror, 181
magnetic trap, **138–140**, 244
    quadrupole, 138, 139, 141, 146, 147
magnetic trapping, **137–148**
magnetically induced laser cooling, 100, **104–105**, 118, 121
magneto-optical force, 164
magneto-optical lens, 189
magneto-optical trap, **156–164**, 244
    atom capture, **159–162**
    atom cooling, **158–159**
Maxwell-Boltzmann distribution, **61–63**, 65, 67, 68
metastable beam, **189–190**
metastable noble gas atom, **43–45**, 49, 201, 273, 278
microlenses, 190
microwave cavity, 155, 193
microwave clock, 194
MILC, *see* magnetically induced laser cooling
mixing angle, 10
momentum conservation, 222, 253
momentum space, 63
momentum space compression, 90, 91
Monte Carlo, 115

Monte Carlo wavefunction method, 15, 24
MOT, *see* magneto-optical trap
multilevel atom, **39–56**, 97

Na, *see* sodium
nanofabrication, *see* atomic nanofabrication
navigation, 192
Ne, *see* neon
neon, 44, 45, 49
neutrality of matter, 187
non-adiabatic, 99, 100, 105, 106, 117
non-linear optical effect, 259
non-linear optics, **195–198**
nuclear spin, 40
number of modes, 21

OBE, *see* optical Bloch equation
OM, *see* optical molasses
optical Bloch equation, **23–24**, 115, 118
optical brightening, 91
optical crystal, 232
optical Earnshaw theorem, 156
optical lattice, 225, **231–239**
optical molasses, 36, **87–97**, 100, 109, **111–113**, 126, 127, 188, 237, 244
    lin ⊥ lin, 237
    at high intensity, 123, **126–128**
optical potential, 155
optical pumping, **86**, 99–106, 116, 117, 162, 163, 255
optical transition, **50–52**
optical trap, **149–164**
    blue detuned, **153–155**
    dipole force, **150–155**
    far-off-resonance, 151
    hybrid dipole radiative, **152–153**
    microscopic, **155**
    radiation pressure, **156**
    single beam, **150–152**
orbit
    atomic, 147

orbital frequency, 141

paraxial domain, 185
partial wave, 167, 201, 203, 211
partial wave analysis, 200
PAS, *see* photo-associative spectroscopy
pendulösung, 228
periodic potential, 231, 233, 236
permanent magnet, 148, 188
perturbation theory, **3–4**
phase conjugate reflection, 198
phase fluctuation, 233
phase gradient, 31
phase shift, 201, 204
phase space density, **68–70**, 165, 168, 170, 175
phase space volume, 132
phase transition, 243
photo-associative spectroscopy, **213–218**
photon recoil, 59, 227
Planck spectrum, 241
Poincaré plot, 142, 143
polarization, **45–47**, 229
polarization gradient, 46, 47, 100, 102, 104, 111, 113, 117, **120–122**
    lin ⊥ lin, 46, 47, 102
    σ^+-σ^-, 47, 110, 120, 122
polarization gradient cooling, 108, 114
    lin ⊥ lin, **100–103**, 109, 111, 122
    σ^+-σ^-, **106–107**, 111
potassium, 76
potential scattering, **200–204**
power broadening, **24–27**
pure state, **18–20**

quadrupole field, 156
quadrupole-quadrupole interaction, 202
quantization axis, 255, 257
quantum beat, 227
quantum behavior, 203
quantum computing, 261

quantum state of motion, 111, 113, 138, **145–148**, 155, 219, 220, **235–238**, 252
quantum threshold, 203
quantum view of laser cooling, **239–240**

Rabi frequency, 5, 50, 107, 108, 132
    vacuum, 21
Rabi two-level problem, **4–6**, 23
radiance, 186
radiation pressure force, *see* scattering force
radiative escape, 208, 209
Raman cooling, 114, 225
Raman transition, 114, 118–121, 196, 238, 253
Raman-Nath regime, 222
Ramsey fringes, 194
Ramsey oscillation, 227
random walk, **63–65**, 66, 89, 95, 113, 256, 259
rate coefficient, 210
Rb, *see* rubidium
reactive force, *see* dipole force
recoil energy, 89
recoil frequency, 35, 59, 228, 275
recoil temperature, *see* cooling limit, recoil, 112
recoil velocity, 73
recoil-induced resonances, 197
redistribution force, *see* dipole force
reduced mass, 200
resonances, 118
rf spectroscopy, 144, 145, 193
rf transition, 175
rotating frame transformation, 7
rotating wave approximation, 5–7, 152
rovibrational state, 215
rubidium, 76, 107, 112, 119
Russell-Saunders, 40, 43
RWA, *see* rotating wave approximation
Rydberg atom, **79–80**, 117, 149, 180, 218, 226

s-wave, 167
s-wave scattering, 202
Sagnac effect, 228
saturation, **24–27**
saturation intensity, 25, 275
saturation parameter, 25, 26
scattering force, 32, 151, 152
scattering length, 167, 171, 194, 204, 205, 242
    negative, 242
scattering rate, 25, 26
Schrödinger's cat, 261
second
    definition, 192
second-order Doppler effect, 193
selection rule, **47–50**
self-assembled monolayer, 190
sheet of light, 94
single mode, 224
Sisyphus laser cooling, 102, 105, 106, **116–118**, 128, 129
sodium, 40, 41, 76, 80, 82, 95, 96, 107
spatial compression, 91
spectral brightness, 186
spherical unit vector, 50
spin alignment, 163
spin relaxation, 174
spontaneous decay rate, 22
spontaneous emission, 14, 15, 17, 19, **20–22**, 32, 49, 63, 74, 85, 101, 102, 113, 114, 117, 123, 124, 126, 127, 129, 132, 212, 214, 221, 223
standing wave, 45, 155
Stark compensation, 85
Stark shift, 77–79, 181
Stark slowing, **77–78**
    Rydberg, **79–80**
stationary state, 256, 258
statistical mixture, 18–20
steady-state, 58
Stern-Gerlach, 138
stimulated emission, 14, 74, 88, 117, 123, 126, 127, 131, 197

stopping length, 75
sub-Doppler laser cooling, **99–122**
    theory, **107–111**
surface scattering, 187
survival rate, 210
sympathetic cooling, 250

temperature, **58–59**, 67, 96, 97
tetrahedral configuration, 234
tetrahedral symmetry, 163
thermal equilibrium, 58, 61, 67
thermodynamics, 166, 168, 172
three-body collision, 248
time orbiting potential, 147
time-of-flight, 80–82, 86
time-orbiting potential, 147
TOF, *see* time-of-flight
TOP, *see* time-orbiting potential
totally internally reflected, 180
transient molecule, 214
transition amplitude, 4
transition rate, 4
translational partition function, 211
transverse motion, **85**
trap depth, 166
trap loss, **207–208**
trap stability, 140
trapped ion, 262
trapping potential, 167, 168
trapping time, 138
two-color spectroscopy, 217
two-frequency light field, 128, 129,
    131, 132

uncertainty principle, 114

vacuum, 138, 141, 143
vacuum field, 14, 15
vacuum fluctuation, 20
van der Waals interaction, 201
vapor cell MOT, 161
velocity
    average, 62
    most probable, 63
    root mean square, 63
    thermal, 76
velocity compression, 90
velocity selective coherent popula-
    tion trapping, 225, 251, 256,
    259–261
    in two-level atom, **252–254**
velocity selective resonance, 110, 118,
    119, **120–122**
vibrational level, 155, 206, 237, 238
VSCPT, *see* velocity selective coher-
    ent population trapping
VSR, *see* velocity selective resonance

Wigner threshold, 206
Wigner-Eckart theorem, 50, 53
Wigner-Weisskopf theory, 20

X-ray Bragg reflection, 260
X-ray diffraction, 235

Zeeman shift, 42, 77, 78, 84, 118,
    161
Zeeman slowing, **77**
Zeeman-tuning, 80

Graduate Texts in Contemporary Physics

S.R.A. Salinas: **Introduction to Statistical Physics**

B.M. Smirnov: **Clusters and Small Particles: In Gases and Plasmas**

B.M. Smirnov: **Physics of Atoms and Ions**

M. Stone: **The Physics of Quantum Fields**

F.T. Vasko and A.V. Kuznetsov: **Electronic States and Optical Transitions in Semiconductor Heterostructures**

A.M. Zagoskin: **Quantum Theory of Many-Body Systems: Techniques and Applications**